T0138399

Red Revolution,
Green Revolution

Published with support of the Susan E. Abrams Fund

Red Revolution, Green Revolution

Scientific Farming in Socialist China

SIGRID SCHMALZER

The University of Chicago Press

Chicago and London

Sigrid Schmalzer is professor of history at the University of Massachusetts, Amherst. She is the author of *The People's Peking Man*, also published by the University of Chicago Press, and coeditor of *Visualizing Modern China*.

The University of Chicago Press, Chicago 60637
The University of Chicago Press, Ltd., London
© 2016 by The University of Chicago
All rights reserved. Published 2016.
Printed in the United States of America

25 24   5

ISBN-13: 978-0-226-33015-0 (cloth)
ISBN-13: 978-0-226-33029-7 (e-book)
DOI: 10.7208/chicago/9780226330297.001.0001

Library of Congress Cataloging-in-Publication Data

Schmalzer, Sigrid, author.
    Red revolution, green revolution : scientific farming in socialist China / Sigrid Schmalzer.
        pages cm
    Includes bibliographical references and index.
    ISBN 978-0-226-33015-0 (cloth : alk. paper) — ISBN 978-0-226-33029-7 (e-book)
 1. Communism and agriculture.   2. Agriculture and state—China.   I. Title.
    HX550.A37S36 2016
    338.1'851—dc23
                                                    2015017858

♾ This paper meets the requirements of ANSI/NISO Z39.48–1992 (Permanence of Paper).

In memory of my father,
Victor Schmalzer (1941–2015)

Contents

Illustrations

INTRODUCTION

People who follow the news today are very familiar with the fierce, global debates among farmers, corporations, scientists, activists, and governments over the use of genetically modified organisms in agriculture. These are but the most recent in a long series of unresolved arguments about farming and food: we are not close to consensus on the agricultural technologies introduced in the previous century—the "high-yield varieties" of seeds and the agrochemicals they require to fulfill their promise. Against those who have applauded this "green revolution" for solving hunger around the world, environmentalists have decried the consequences for wildlife, ecological sustainability, and human health; social activists have blamed the new technologies for economic hardship and social ruptures constituting violence against poor farmers and Third World peoples; and academics have exposed the corporate interests driving the development of the new technologies and the political interests behind US promotion of them overseas.[1]

Beyond these concerns lies a more general and less well-recognized problem produced by, and implicit in the very concept of, the green revolution. It is the problem of how to understand the relationship between science and technology on one hand and sociopolitical transformation on the other. Does science offer an alternative to political solutions for the world's problems, as the US architects of green revolution intended? Or is science inseparable from its political context, as Marxists have long argued and scholars in the field of science and technology studies continue to maintain?[2]

The history of the green revolution as it unfolded in socialist-era China represents a critical piece of the puzzle we must assemble to address these and other pressing questions about our common future. New agricultural technologies have been of central importance in the transformation of China's economy, from the increase in the quantity and variety of food on people's tables to the diversification of industry and thus the far greater availability of consumer goods. At the same time, agricultural issues—from water resources to pollution to food safety—play a significant role in the widely recognized environmental and health crises that now confront not only China itself but the larger world connected to China through global economic and ecological networks. For those reasons alone, we need to know more about the environmental and social consequences of China's green revolution. But this history also speaks more generally to the relationship between China's revolutionary politics and its insatiable appetite for science and technology.[3] Moreover, it facilitates a needed critique of the fundamental assumptions about science and society that undergirded the green revolution, and that continue to undergird the dominant political ideology of the world today.

The goal of this book is to bring into view China's unique intersection of red and green revolutions. Socialist Chinese agricultural science will not serve as a model: it is always important to exercise caution when looking to history for models, and socialist China offers perhaps even more than the usually large set of complications found in any real society. But neither will it be merely a cautionary example: the Mao era was not the simple picture of totalitarian oppression and ecological disaster that is presented in many accounts available in English. Rather, in this book socialist Chinese agricultural science will be called upon to challenge dominant assumptions about what constitutes science, how science relates to politics, who counts as a scientific authority, and how agriculture should be organized or transformed. In the process of stretching our minds to grasp the different answers to these questions offered by socialist Chinese history, along with the limits and consequences of those answers, we will be better able to think critically and inspirationally about the prospects for agriculture and science in our own times and places.

Revolutions, Red and Green

In 1968 the director of the US Agency for International Development (USAID), William Gaud, coined the term "green revolution." He said:

> Record yields, harvests of unprecedented size and crops now in the ground demonstrate that throughout much of the developing world—and particularly in Asia—we are on the verge of an agricultural revolution. . . . It is not a violent Red Revolution like that of the Soviets, nor is it a White Revolution like that of the Shah of Iran. I call it the Green Revolution."[4]

As William Gaud's geopolitical color-coding made clear, the green revolution was about far more than saving lives and improving standards of living. Born of the Cold War, it was a strategy for preventing the spread of ideologies opposed by the United States. If farmers around the world could be raised from poverty through technological improvements to agriculture, they might be less likely to seek political solutions.[5] This was a quintessentially technocratic vision: it relied on technical experts to provide technological fixes for social and political problems. And it was the very premise behind USAID, which John F. Kennedy founded in 1961 to encourage economic development in impoverished countries lest communist rivals exploit the potential for revolution and "ride the crest of its wave—to capture it for themselves."[6]

The significance was not lost on observers in China, where Mao Zedong had brought a "red revolution" to victory in 1949, and where, still under his leadership, an even more tumultuous transformation had begun under the banner of "Cultural Revolution" (1966–1976). In 1969, *People's Daily* bemoaned the pursuit of "green revolution" in India, defining it as "the so-called 'agricultural revolution' that the reactionary Indian government is using to hoodwink the people." The article made clear just why the green revolution represented a "reactionary" choice: the Indian Minister of Food and Agriculture had reportedly "cried out in alarm that if the 'green revolution' . . . does not succeed, a red revolution will follow."[7]

Does this mean that socialist China opposed the new technologies of the green revolution or agricultural modernization more generally? No. Contrary to common perception, even the most radical leaders in socialist China embraced the causes of science and modernization, and so in some important ways, the green revolution in red China looked strikingly similar to the green revolution as Gaud imagined it. The goal there as elsewhere was to transform the material conditions of agriculture through mechanization, the introduction of new seeds, and the application of modern chemicals in order to increase production and raise standards of living. The organization of these efforts in China was shaped in part by Soviet experience: Soviet advice influenced much of socialist China's 1950s work in science and economic development.[8] However, perhaps surprisingly, the Chinese approach exhibited far more striking parallels to the research and extension system embraced in the early twentieth-century United States and promoted abroad (including in pre-1949 China) by the Rockefeller Foundation and other US organizations: in each case, research centers focused on meeting the perceived needs of farmers and turned to local experiment stations for testing and disseminating (i.e., "extending") the new technologies they developed.[9]

Still, China's agricultural transformation was embedded in a philosophy of science profoundly different from that driving the green revolution promoted by the United States. In contrast with the technocratic position so clearly articulated by William Gaud, the dominant position in socialist China was that science could not be divorced from politics, and modernization could not be separated from revolution. What was anathema to Chinese radicals about proponents of the green revolution was not their support for "modernization" or "development," but rather their assumption that science and technology were inherently apolitical forces, and worse yet, their attempts to use these forces to circumvent social and political revolution. Thus, the term "green revolution" was never adopted in socialist China; the same set of agricultural technologies was instead called "scientific farming" (科学种田).

The dominant perspective on science and politics in socialist China was epitomized by an oft-quoted 1963 statement by Chairman Mao:

> Class struggle, the struggle for production, and scientific experiment are the three great revolutionary movements for building a mighty socialist country. These movements are a sure guarantee that Communists will be free from bureaucracy and immune against revisionism and dogmatism, and will forever remain invincible.[10]

For Mao and other radicals in socialist China, science was a "revolutionary movement" alongside the more familiar political commitments to class struggle (that is, the effort to combat the reemergence of power inequities favoring the formerly elite classes) and the struggle for production (that is, the effort to increase the material base of the economy through a socialist organization of labor). And so for Mao and others in China, the introduction of green revolution technologies could be politically legitimate only if it proceeded through red revolutionary means.

Hence the launching in the mid-1960s of the "rural scientific experiment movement" (农村科学实验运动), with grassroots "scientific experiment groups" organized throughout the countryside on a "three-in-one" basis: "old peasants" with practical experience, "educated youth" with revolutionary zeal, and local cadres with correct political understanding would work together to identify needs and develop solutions.[11] They would overturn "technocratic" approaches promoted by scientific elites and "capitalist roaders"; instead, they would place "politics in command." The significance the state accorded to their work demonstrated the inseparability of society and politics on one hand and science and technology on the other. For example, when a team of teenage girls, named the March 8 Agricultural Science Group in honor of International Women's Day, used pig manure as fertilizer to increase production in a lackluster field, they were understood to have struck a blow for scientific farming, not because their technology was new (it was old as the hills) and not, sad to say, because it was ecologically sustainable (this was not a political value at the time), but because it helped them overturn unscientific, old, sexist ideas about women's farming abilities.[12] Far from being viewed as an apolitical force capable of solving problems without revolution, "scientific farming" was embraced in socialist China as a means for the radical transformation of society.

The radical approach to agricultural transformation extended beyond national borders as China sought to lead the Third World in its struggles with the legacies of colonialism and ongoing imperialist forces. African countries were not among the recipients of the original US green revolution (Barack Obama proposed to rectify this shortly before his first official visit to the continent).[13] This left the field open for China, supposedly isolated by Cold War geopolitics but actually active in many corners of the globe less trodden by the superpowers. In the West African nations of Liberia, Sierra Leone, and The Gambia, Chinese experts on the ground supervised the production of locally made rice threshers, demonstrated composting and the use of animal manure for fertilizer, and raised chickens and pigs to feed themselves, all the while calling attention to these activities as examples of the Maoist principle of anti-imperialist self-reliance.[14] Maoist approaches to science had clear influence in the East African country of Mozambique as well, where the revolutionary leader Samora Machel celebrated the wisdom of peasants and mechanics, and decried the "arrogance" of experts who kept themselves apart from the masses, making themselves into a "privileged class," whose intelligence thus became "sterile, like those seeds locked in the drawer."[15]

India's experience with the green revolution offers an interesting counterpoint. The Indian government was a willing partner in promoting green revolution, but this did not mean that its interests were identical to those of the United States. "If the major motives of the U.S. agencies were to forestall communist insurrections and promote a free-market economy, the Indian government was more concerned to avoid a 'crisis of sovereignty' and retain the moral legitimacy to rule, distinguishing Indian from British rule through better food security and thus avoidance of famine-related suffering."[16] Nor did Indian leaders limit themselves to the US model. In the mid-1950s, India sent delegations to China specifically to learn about its approach to agricultural modernization, which was considered more advanced in China than in India. According to Akhil Gupta, "China functioned [for India] as model, competitor, and alternative, a country with 'essentially similar' problems, resources, and goals but (and this was crucial to someone like Nehru) a different—nondemocratic—political system."[17] Nonetheless, the Indian state followed the technocratic approach to agricultural transformation articulated in Gaud's "green revolution," with efforts to address social and economic change largely serving to shore up rather than challenge the larger technocratic structure in which they were incorporated. Moreover, while Chinese policies favored the development of a nationwide network of peasant technicians supporting socialist agriculture, Indian leaders sought to "transform what they considered to be traditional and backward farmers into 'risk-taking,' profit-making individuals."[18]

The embeddedness of the green revolution in US foreign policy and in global capitalism has made it vulnerable to criticism by leftist academics and activists in South Asia and Latin America—groups with greater numbers, more latitude for expression, and a more active diaspora than leftists in China enjoy. A mark of the advanced (so to speak) state of Indian critiques of green revolutionary development can be seen in the existence already in 1998 of a sophisticated critique of the critique: in *Postcolonial Developments*, Gupta, while himself critical of capitalism and colonialism, has offered a thoughtful analysis of how proponents of "indigenous knowledge" have been guilty of essentialism in their misrepresentation of the actual agricultural practices and priorities of Indian people.[19] In contrast, the history of China's green revolution has barely registered on the radars of China scholars, and the relatively small diaspora of leftist Chinese academics has only just begun to explore its political significance.[20] That said, in recent years some organizations in China have begun questioning the relationship between market capitalism and agricultural science, advocating for recognition of the value of "indigenous knowledge," and rallying around the cause of "food sovereignty." The epilogue will offer a discussion of these movements; the task between now and then is to understand their thus-far largely unarticulated Mao-era roots.

The Green Revolution and the Transformation of Chinese Agriculture

The green revolution in China has been an elusive historical object in part because of the diversity of technologies—chemical and organic, modern and traditional—that enjoyed state promotion during the 1960s and 1970s. And so one observer might be impressed by the speed with which chemical fertilizer plants were being built, while another might just as reasonably celebrate the ecological sensibility of the ubiquitous collection of night soil and use of pig manure (figures 1 and 2).[21] A propaganda poster might foreground the (very rare) use of airplanes to dust crops, while a poster designed for

FIGURE 1. The Great Leap Forward was the high point of official state encouragement, which continued throughout the Mao era, to raise swine for fertilizer production. Reproduced from "Zhu shi 'huafeichang' you shi 'jubaopen'" (Pigs are "fertilizer factories" as well as "treasure bowls") (Shanghai renmin meishu chubanshe, December 1959). Stefan R. Landsberger Collection, International Institute of Social History, Netherlands, http://chineseposters.net.

an elementary school classroom might highlight the role of frogs, birds, and ladybugs in controlling insect pests (figures 3 and 4). Other visual representations evoking this diversity of approaches can be seen in figures 6 and 7 in chapter 1.

China's patchwork of agricultural practices emerged for reasons both practical and political. The enthusiasm with which localities used chemical fertilizers, insecticides, and tractors when they were available suggests that the oft-highlighted emphasis on composting, biological control of insect pests, and other "sustainable" methods gained much strength from sheer economic necessity. At the same time, both new and old technologies spoke to socialist-era political values. Modernization of agricultural inputs resonated with the ideal of building a new, prosperous countryside, while promotion of long-standing, labor-intensive practices helped valorize peasant wisdom and the immensity of what could be accomplished through mass collective effort.

FIGURE 2. As more chemical fertilizer plants came on line, propaganda encouraging its use increased. The people carrying baskets are scattering chemical fertilizer, and the people with hoes are distributing it properly in the soil. This painting appeared in a book produced for foreign audiences of paintings from Huxian, Shanxi. Thanks largely to Jiang Qing's patronage, the "peasant painters of Huxian" rose to fame and served as a model for others to follow. Their paintings were collected and exhibited in Beijing in 1973 and subsequently in France, England, and other foreign countries. Zhang Fangxia, "Fertilizing the Cotton Fields," in Fine Arts Collection Section of the Cultural Group under the State Council of the People's Republic of China, *Peasant Paintings from Huhsien County* (Peking: Foreign Languages Press, 1974), 34.

Figures 1 and 2 add to the sense of "patchwork" created by the juxtaposition of "old" and "new" technology. Note that though manure is an "old" fertilizer technology, in figure 1 it takes on an industrial quality, while the "new" technology of chemical fertilizer appears almost old-fashioned in this pastoral scene where a large number of people work a small area without mechanization.

Some of the most dramatic and remarked-upon elements of China's agricultural transformation involved changes to the physical landscape. Terracing and other forms of radical land reconstruction brought new areas under cultivation, and dams and irrigation systems made water available where it was needed. Such restructuring of physical resources in turn allowed for farming on a larger scale and greater use of mechanization. Of all efforts to modernize agriculture, mechanization stood in first place for Mao. Mechanization provided the material basis for revolutionary social reorganization: adoption of tractors would enable larger field sizes, and so the transition from family farming to communal agriculture.[22] However, throughout the Mao era, tractors remained in short supply, and securing machinery for a production team required some ingenuity or even entrepreneurship on the part of local leaders. And so it should not be surprising that alongside promotion of

为农业生产服务

FIGURE 3. Peasants hail and applaud an airplane spraying their crops from the air. The use of airplanes in agriculture was extremely rare; here it offers a vision for the future that stands in contrast with the support for biological control of insect pests seen in figure 4. It is also worth comparing and contrasting the vision displayed here with that found in figure 24. Reproduced from Xinhua tongxun she, ed., *Wei nongye shengchan fuwu* (Serve agricultural production) (Beijing: Xinhua tongxun she, 1964), cover.

mechanization as the "way out" of China's agricultural dilemmas, the state simultaneously promoted the approach known as "intensive cultivation" (精耕细作), through which prodigious amounts of labor made small areas of land capable of generating large yields of varied products. In the early 1960s the state began specifically endorsing the practice of intercropping—that is planting more than one crop in a single plot of land. While this made good use of limited acreage and took advantage of the ways different organisms could benefit one another, it was not conducive to mechanization. Instead, intercropping and other elements of traditional, intensive Chinese styles of farming have inspired the work of people around the world pursuing sustainable forms of agriculture (including, for example, agroecological farming, permaculture, and rice intensification). Thus, modern mechanization and traditional intensification coexisted as strategies for increasing production in Mao-era China and both approaches enjoyed Mao's stamp of approval.[23]

In China as elsewhere, the key agricultural innovation of the green revolution lay in the development of new varieties of cereal crops—especially the dwarf or semi-dwarf varieties that made effective use of soil nutrients to produce large quantities of grain on strong, short stalks that did not collapse under their weight. These are typically called high-yield varieties (HYV) in

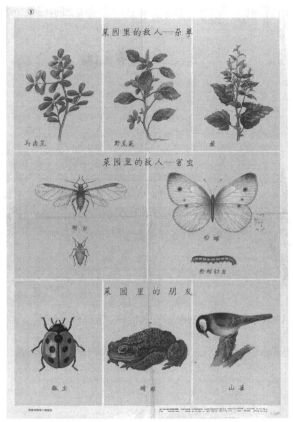

FIGURE 4. A poster for use in elementary schools as part of the language curriculum. This is a typical merging of instruction in core subjects with material related to economic production and other state priorities. The top section identifies three examples of "enemy" weeds: purslane, amaranth, and lamb's quarters. The middle section identifies two examples of "enemy" insects: aphids and butterflies of the Pieridae family. The bottom section identifies three examples of "friends" (i.e., biological control agents): ladybugs, toads, and titmice. Renmin jiaoyu chubanshe, ed., "Caiyuanli de diren he pengyou" (Enemies and friends in the vegetable garden) (Beijing: Jiaoyu tupian chubanshe, 1956).

English, though some critics of the green revolution have suggested changing this name to "high-responsive varieties" to make clear that they have been bred specifically to respond to the application of chemical fertilizer and would not otherwise necessarily outperform traditional varieties.[24] In Chinese, they have typically been referred to more generally as "improved varieties" (良种). Important Chinese breakthroughs in this area were Hong Qunying and Hong Chunli's Aijiao nante (矮脚南特) variety in 1956 and Huang Yaoxiang's more successful Guangchangai (广场矮) in 1959. Breeding work continued to be a key emphasis throughout the Mao era at research institutes and in production teams across the country, especially in grains but also in vegetables and livestock. And despite China's relatively isolated position during the Cold War, it was by no means entirely cut off from the stream of new varieties being produced at key research centers abroad. For example, in 1966, the Ford Foundation–funded, Philippines-based International Rice Research Institute developed an important dwarf variety of rice, IR8, with parent stock from Taiwan; by 1967 it had made its way to China.[25] As will be fully explored in chapter 4, the development of hybrid rice varieties in the early 1970s represented the next important transformation of the genetic landscape of Chinese agriculture.

The green revolution changed what people ate: the new varieties tasted different, and when food was abundant enough to permit a certain level of choosiness, people often perceived them to be lacking in flavor or nutrition. The diversity of grains in particular decreased in China (as in the rest of the world) because of the proliferation of the relatively small number of "improved" varieties promoted by the green revolution. At the same time, the new technologies allowed more people to eat foods traditionally considered to be of higher quality. Han Chinese culture has long valued rice (and secondarily, wheat) above other grains, but ecological and economic necessity had for millennia compelled most people to eat millet or sorghum—and in more recent centuries, corn and sweet potatoes.[26] As Jeremy Brown notes of the northern region surrounding the city of Tianjin, during the Mao era "the difference between village and city came down to coarse grain [粗粮] versus fine grain [细粮], meaning cheap cornmeal versus expensive processed wheat flour."[27] Mao-era introductions of improved varieties raised per-acre yields, while investments in terracing and irrigation increasingly made possible the cultivation of rice in places where the natural environment was better suited to other crops. When "father" of the green revolution Norman Borlaug visited China in 1977 and saw rice growing in the arid North, he wrote in his journal, "Rice—why the hell they grow it here—except in low-lying flood areas—I can't believe. Perhaps it's because Taichai [sic, Dazhai] said 'do it.' "[28] Whether or not Dazhai was the direct inspiration, the more important point is that rice was in high demand—and it still is. If a community grows other, less valuable grains instead, it is often assumed that the reason is necessity: if only flatter, more fertile land or better irrigation were available, they would certainly plant rice or wheat instead.[29]

One of the crucial inspirations historians have noted for the development of high-yield varieties was the new availability of ammonia in the United States after World War II, as factories that had supplied the vast quantities used for military explosives now produced previously undreamt-of amounts of nitrogen-rich ammonia fertilizer.[30] In China, however, chemical fertilizer remained in short supply for decades as the state promoted "improved varieties." Beyond this economic pressure, organic fertilizer had an important political ally in Mao, who in 1959 declared pigs in particular to be "small scale, organic fertilizer factories" that—in addition, of course, to their other valued attributes—produced fertilizer "ten times better" than chemical types.[31] This did not, however, stop localities from using all the chemical fertilizer they could get, or from building chemical fertilizer plants whenever possible. And in 1973—just five years after *People's Daily* lambasted India for following the US model of "green revolution"—China signed contracts with the US Kellogg Corporation along with two companies in Holland and Japan to build ten large ammonia factories for the production of nitrogenous fertilizer.[32] Still, throughout the 1950s, 1960s, and 1970s, efforts to build chemical fertilizer plants coincided with the promotion of expanded use of animal manure, human waste, and "green manure" cover crops; composting; dredging rivers and ponds; and in some places mining bat guano from caves.

The other major type of agrochemical that transformed twentieth-century agriculture was chemical pesticides—especially herbicides and insecticides. Herbicides target weeds and save the labor of hand weeding. Insect pests have literally plagued farmers since ancient times, but modern agricultural practices—including monoculture (growing a single species over a wide area), breeding for responsiveness to fertilizer rather than pest resistance, and ironically the use of chemical insecticides themselves—often exacerbate the situation, creating larger outbreaks that are harder to control. The result is a "pesticide treadmill," in which ever-larger amounts of chemical insecticide are required, but themselves create resistance in pests, larger outbreaks, and thus still more insecticide.[33] And of course, the pesticide- and fertilizer-laced runoff from fields had the predictable effect on water quality, killing fish and crustaceans that in the 1970s still provided many rural people with much-valued supplements to their diet.

As with chemical fertilizers, the timing of China's experience with chemical pesticides differed somewhat from that in other countries. In the 1950s and 1960s, and even into the 1970s, the "problem" associated with pesticides was often described as a lack of supply: whenever they became available, people flocked to take advantage of their efficiency in dispatching insect pests. However, people in places where insecticides were available quickly began observing the downside to their overuse. Moreover, Chinese insect scientists had deep connections to international science and were alert to the problems of pest resistance and chemical toxicity that their colleagues abroad were documenting. And so, lack of access to chemical insecticides and *simultaneously* concerns about resistance and toxicity combined to encourage scientists and agricultural officials to explore other pest-control solutions, including biological controls (cultivating natural predators, parasites, and diseases of insect pests) and cultural controls (adapting cultivation practices to interrupt pest life cycles) in a way that appeared "advanced" to environmentalists abroad.

If China's green revolution comprised a patchwork of methodologies, the patches themselves cannot easily be characterized as "modern" or "traditional." In many cases, "traditional" methods were in fact new to the localities where they were being disseminated, and they often differed at least to some extent from their earlier form. For example, the use of nitrogen-rich cover crops to nourish the soil between harvests had been known in China for centuries, but this did not mean that farmers all over China were uniformly growing them. Many experienced this as a new technology, or at least were introduced to new types of plants. For them, the term "scientific farming" might call to mind the colorful cover crops that flowered in early spring on the terraced hills—a level of red on the bottom, purple in the middle, and green at the top, representing the different varieties especially suited to the different conditions.[34] Moreover, even the most familiar methods might be experienced as radically new when the state mobilized people to pursue them on unprecedented scales or in unfamiliar ways. Generations of farmers had practiced night soil collection and crushed insect eggs by hand, but digging out three years of accumulated manure in the latrines for a single application in a poorly performing field and training a corps of young secondary school graduates to monitor pests and manually eliminate them at strategic times in their life cycles represented changes in practice as dramatic as the introduction of modern chemicals.

Mao is famous for exhorting Chinese people to "never forget class struggle" and promoting the idea of "continuous revolution." Without constant vigilance, China's red revolution apparently could not maintain its victory. This was certainly no less true of the green revolution. Tractors and irrigation systems needed constant repair. Improved crop varieties degenerated without periodic fresh injection of new genetic material. Parasitic wasps bred in captivity as biological control agents suffered the same fate unless breeders frequently returned to the wild for more parent stock. And soil wore out unless its nutrients were constantly replenished. All this maintenance required considerable human labor. As Edward Melillo has urged us to remember, there are no "agricultural miracles" that represent purely technological fixes innocent of the sweat—and yes, the blood—of laboring people.[35] In the end, a catalog of technologies cannot help but return to social and political history.

Interpreting the Sources, Recapturing the Past

Today it may seem strange to think of Chinese socialism as a way forward. We live in what some have called a postsocialist world. That is to say, while the possibilities of socialism and the potential for socialist revolution remain, the dominant verdict on the Cold War is that the capitalist West won, and the socialist histories of places like China and the former Soviet Union are read through a prism of failure that makes it hard to imagine what seemed so exciting to so many people—including many Americans—at the time.[36] It is perhaps even stranger to think of Western scientists holding up China as a model of sensible and sustainable environmental policy, as chapter 2 will document, amid reports of infants dying from poisoned milk supplies and desperate attempts to move rivers from the South to the parched North. The historian's goal becomes to recapture a past time from what, though separated from the present by just a few decades, has already become a very murky history. This requires careful attention to the perspectives of diverse historical actors, presented in a variety of problematic but nonetheless essential sources.

The vast majority of sources available on the Mao era were to one extent or another produced by the state. Many of these fall under the generally accepted category of state "propaganda"—that is, material produced by the state specifically to promote the state's priorities. Even as they simultaneously served a variety of other purposes, government documents, technical handbooks, and even journal articles ostensibly authored by scientists also bear some of these marks of the state's policy craftsmen. The rhetoric they employed was so blatant and pervasive that terms like "the masses of poor and lower-middle peasants" risk becoming meaningless. Nonetheless, historians must resist becoming so jaded that we fail to take seriously the political ideology that in fact meant a great deal to, and had very real consequences for, people on the ground. In the chapters that follow, I will frequently work to re-create the pictures painted in propaganda and other state sources. This is not because I believe them to be "true," but because they articulated visions of science that challenged dominant perspectives in the capitalist West and served as important ideological influences and inspirations within China and around the world. As such, they offer invaluable evidence for understanding the ideology state actors sought to project to their audiences.[37]

"Ideology" is a term so loaded, and with such different implications to scholars in different fields (not to mention the general public), that its use requires some explanation. Ideology exists in every society. In socialist China, not only was ideology (意识形态, literally "pattern of consciousness") not perceived in negative terms, but the state actively and explicitly engaged in what was called "thought work" (思想工作) to produce "correct" ideological orientations and promote them among the population.[38] In liberal-capitalist countries, the state is not expected to be so unified, nor do state agents typically take on such explicit roles in "thought work." But whether state-socialist or liberal-capitalist, every historical context produces ideologies that reflect and reinforce the perspectives and interests of the social groups that generate them—much as Karl Marx originally explained in his analysis of capitalist ideology. With respect to the role of ideology in science, I adopt what William Lynch has called a "neutral" concept, in that "the identification of ideology alone does not imply critique," and that "there may be true and false, progressive and reactionary ideologies."[39] However, if identifying ideology does not imply a critique, neither does it prevent one. Instead of mounting the critique against the fact that ideology is present, we may mount it against the role it plays in social oppression, cultural imperialism, environmental destruction, or other negative forces.[40]

Numerous scholars have debated the question of how ideology emerges and what people do with it. While some have followed Marx in emphasizing the power of a governing ideology to shape individuals' minds without their consciously knowing it, others—like Michel de Certeau—have emphasized instead the agency that people possess as they resist dominant ideologies or alter them to suit their own desires.[41] This critical debate will especially inform chapters 6 and 7, which use diaries, memoirs, and interviews to explore the meanings that science held for "educated youth" involved in the scientific experiment movement. These chapters contain evidence of active resistance, and also reason to suspect more unconscious acceptance of ideological messages conveyed in propaganda. At the same time, I heed Judith Farquhar and Qicheng Zheng's warning not to "romanticize resistance by presuming that a desire to fight back against instituted power is natural, inevitable, or even the most interesting aspect of social life."[42] Indeed, youth often actively embraced elements of state propaganda in ways that were meaningful and empowering to them. And Gail Hershatter has suggested something similar with respect to peasant women: state agents literally put words in the mouths of women selected to be model laborers, but Hershatter warns us not to imagine that beneath the "layer of state manipulation . . . [lay] an inchoate 'real' China where long-suffering peasants resided under socialism, by turns pliant and resistant, but always distinct from something called 'the state.'" Rather, "village women themselves made considerable efforts to learn new skills and overcome personal terrors" to become the models the state asked them to be.[43]

Although the primary usefulness of propaganda is what it tells us about state ideology, read carefully propaganda sources may also provide clues as to the experiences of actual people, the functioning of their communities, and their culture. Despite their best efforts, even the most careful propagandists could not help but reveal some of the tensions of the larger political and social context. Reading against the grain permits glimpses of what concerned them. For example, hearing about "class enemies" denigrating efforts to modernize agriculture, historians may rightly be skeptical about the specifics (it is highly unlikely that people resisted new technologies because they were descendants of landlords seeking to sabotage the revolution), but we may safely infer that the state worried about people not being quick enough to accept change, and that suggests some kind of resistance on the ground. Determining what in fact motivated that resistance usually requires engagement with other kinds of sources. And so we must simultaneously read at the "surface" level of rhetoric to understand official values (which were variably already shared, newly accepted, or resisted and rejected by people on the ground) and probe below that level to explore the society it purports to describe.

Take as an example figure 5, from a series of posters on the "four-level agricultural scientific experiment network" of Huarong County, Hunan Province (discussed below). The posters were distributed to communes across the country to promote the approach to agricultural science that Huarong perfected. This particular poster—titled "Self-Reliance; Practice Scientific Research with Diligence and Frugality"—celebrates educated youth for making do with locally available materials in their agricultural experiments. Historians can analyze such sources in multiple layers. Since these are photographs, they provide some insight into who participated in the experiments, how they dressed, and the materials with which they had to work. However, these were not candid shots—they were undoubtedly carefully prepared for maximum effectiveness as propaganda.[44] It is always possible that the clothing was chosen especially for the purpose, that the notebook was provided to fit the propaganda worker's understanding of scientific practice, or even that these particular people were chosen for their gender or other aspects of their appearance and in fact did not participate in the experiments. On the level of propaganda, analysis of this poster is less problematic, and therefore it becomes a valuable source for understanding state priorities and values. The streaked windows, clay pots, rolled-up pants, and muddy feet all speak to the state's interest in presenting rural scientific experiment as humble, earthy, and self-reliant—something that ordinary peasants could and should be doing. On the other hand, there is a lower limit to humbleness: by no means would the authorities want to suggest that people—especially educated youth, for whom the state bore special responsibility—were badly clothed or undernourished. The ways people displayed and viewed such posters also

FIGURE 5. The sixth of a set of posters from 1975 on the Huarong County four-level agricultural science network. They were designed to be displayed in common areas to inspire scientific experiment in communes around China.

The poster's title is "Self-Reliance; Practice Scientific Research with Diligence and Frugality." The caption celebrates the policy of "self-reliance and arduous struggle," and praises Huarong for "persistently drawing on local resources, using local methods, and improvising equipment, such that they met the needs of agricultural scientific research and drove forward mass-based scientific farming activities." The explanation for the left picture reads, "In spring 1971, in order to popularize cultivating seedlings in greenhouses, Huarong County established a 'model' greenhouse, but because it was too expensive to build, they could not popularize it. Xinjian Brigade in Xinhe Commune substituted mud bricks and wood for red bricks and reinforced concrete, membrane to replace glass, and reeds for seedling trays, thus spending little more than ten yuan. This kind of 'native [*tu*] greenhouse' was warmly welcomed by the masses and very quickly became popularized throughout the county." The explanation for the right picture reads: "At each level of the agricultural science organization, the masses are mobilized to select methods that are crude and simple, substituting the native for the foreign, and in this way resolve the equipment needs of scientific experiment. They use [old-fashioned] balance scales to replace [scientific] scales, clay bowls for seedling containers, and warming on the stove in place of incubators. These are educated youth from Jinggang Commune using clay bowls to conduct scientific experiment."

Xinhua tongxun she, ed. *Dagao kexue zhongtian, jiasu nongye fazhan* (Greatly undertake scientific farming, accelerate agricultural development) (Beijing: Renmin meishu chubanshe, 1975).

bears analysis. Hung in public areas, posters served obvious political and educational purposes; as a spot of color, they were also undoubtedly consumed for their aesthetic value.

Whether sources depict "real people" or "poster children" for a political agenda can sometimes be ambiguous. The characters appearing in posters, newspaper articles, schoolbooks, or even government documents were presented with the expected reactions of real people in mind; and real people often modeled themselves on the characters appearing in such materials. Propaganda imitates life even as it seeks to shape it; and life imitates propaganda at least as often as it resists. Hence the uncanny appearance of similar images and voices in propaganda posters, memoirs, interviews, official documents, scientific articles, and works of literature. So it should not be surprising that the values of self-reliance and struggle expressed in the Huarong poster can also be found in the writings of ordinary people from that time. For example, the diary of a young man "sent down" from the city to the countryside records that in 1971 he volunteered to participate in the scientific experiment movement. In a summary report of his work, he wrote that his research "brings into play the proletarian revolutionary spirit of using local methods, starting from scratch, self-reliance, hard work, not fearing failure, and overcoming hardships."[45] In the effort to move beyond the state's vision to capture the meaning of scientific experiment for real people, it is tempting to treat diaries and other "grassroots" sources as correctives to the unreliable pictures painted in propaganda materials. What this example shows is just how deeply intertwined propaganda and experience could be.

State-produced materials are not the only sources that require careful, critical analysis. Diaries, memoirs, biographies, interviews, and academic publications alike emerge from specific contexts of production and are transformed through specific contexts of circulation. Moreover, scholars cannot escape ideology by limiting research to sources produced in the postsocialist era: such sources may reflect more familiar political priorities and therefore be less obviously ideological to the reader's eyes, but they are ideological nonetheless.[46] Stories people tell today are filtered through today's politics and shaped too by the intervening time. A warbling echo of the Huarong poster's themes sounds in a 2007 biography of the "father of hybrid rice," Yuan Longping. A teacher of agricultural science in the Hunan hinterland, Yuan is said to have made do with whatever he could find to support his research, including discarded earthenware pots from a nearby kiln factory. But rather than emphasize "self-reliance" or other Mao-era mass-science values, the biography highlights Yuan's unwillingness to burden his family by using their savings to purchase equipment—a value resonant with the post-Mao state's encouragement of people to enrich their own households.[47]

Following the fall of the "Gang of Four" and the discrediting of Cultural Revolution radicalism, the colorful, inspiring image of Mao-era science visible in propaganda materials faded from view remarkably quickly. It was replaced by an equally vivid picture of violence and suffering, with stories of scientists beaten, humiliated, forced to labor, or driven to suicide; patients in hospitals failing to receive medical care because the doctors were in the fields and their places taken by janitors; rural people worked within an inch of their lives to reshape landscapes in futile, and ultimately destructive, attempts to emulate Dazhai. Postsocialist writings on Mao-era science frequently portray the history as a travesty of broken bodies and dreams, with slim victories emerging thanks only to the selfless dedication of noble and brilliant individuals like Yuan Longping.

However, in the 2010s many people in China are taking an interest—one I share—in identifying aspects of the collectivist era that deserve reconsideration. People who worked in agricultural extension (农业推广, the system of bringing new technologies to farmers) during the Mao era express a sharp sense that something went right under Mao that has since been lost because of the difficulty of organizing all the private households, the distance that has emerged between rich officials and poor peasants, and the erosion of collectivist values.[48] This specific contemporary political context provides opportunities for researchers like myself, since people are primed to tell just the kinds of stories that most interest us. However, for this very reason, interpreting the interview data requires caution and especially critical thinking about which stories are told and how they are framed. The problem is not simply that people present the history in overall positive or negative terms, but rather that the specific politics of today make room for specific narratives about the past. Interviews are not only likely to produce narratives about concerns plaguing officials today (irrigation systems and extension top the list); they are also typically framed in ways that highlight the dominant developmentalist paradigm.

Chapters 4 through 7 rely extensively on interviews with people who participated in the scientific experiment movement, most of which I conducted during a trip to multiple locations in Guangxi Province in 2012. The specific contexts of those encounters shaped the conversations and the knowledge they produced, and they deserve some discussion here. Most of the interviews I conducted with Mao-era agricultural technicians came about through the assistance of an agricultural historian in Beijing, Cao Xingsui. I first met Cao in 2010, when I interviewed him about his experiences in scientific farming as an urban youth "sent down" to northwestern Guangxi Province during the Cultural Revolution. Two years later, he introduced me to many friends and acquaintances in the agricultural research and extension system in Guangxi. Cao had from the beginning characterized agricultural extension as one of the few successes of the Mao era, and most of the others we met in Guangxi framed the history in the same way.

Although the stories they told were compelling and I have no reason to doubt their accuracy, their interpretation is clearly influenced by their pro-found sense of disappointment with what they see as a decline in commit-ment on the part of both state and society to agricultural extension and to maintaining the improvements in agriculture (especially irrigation systems) introduced in the 1970s.[49] Many of these interviews occurred in group set-tings. Cao, his assistant, and I visited three state agricultural units where we had the opportunity to speak with five to ten people in each place who worked in agricultural extension during the Mao era. Although I was free to ask whatever questions I wanted, and people appeared mostly uninhibited in their responses, the stories people told were clearly influenced by the group dynamic. In two of the three cases the overall tone for the sessions was very effectively established by the leader of the group, who was seen to have some authority and who introduced the topic with a frame that emphasized the successes of Mao-era agricultural extension that have been lost in more re-cent years. However, in the third case, although it began on the same positive note, the first interview subject to deliver his personal narrative decisively established a different frame for the history, and this set the tone for the rest of the discussion. The differences in the stories we heard at this site com-pared with the other sites were probably at least partly due to economic and cultural factors: the site is considerably poorer and the local people are more dominantly ethnic minorities and viewed as more "backward" by the mostly Han technicians. However, we should not discount the power of the first nar-rator to open the way for more negative memories to surface. Here again the impulse may be to dismiss the interviews as flawed by virtue of their settings, but it is more productive to recognize the way the interviews speak not only to Mao-era experiences but also to the layered interpretive frames used by people who have lived through the dynamic decades.

I conducted interviews with peasants at other sites in Guangxi, accompa-nied only by an assistant who comes from a peasant family. These interviews were far less obviously influenced by the ongoing conversation among of-ficials and technicians about the lagging state of affairs in agricultural ex-tension and the virtues of a bygone era. Nonetheless, the interviewees did sometimes express such thoughts on their own in the middle of interviews. Moreover, like the technicians, peasants offered a strongly developmentalist narrative in which some places (including their own) were "backward" and others "advanced."

In addition to formal interviews, I had more casual opportunities to chat with agricultural technicians, former educated youth, and peasants over meals, in cars, and in homes. Informal conversations encourage an appre-ciation for the subtleties of memory and the complex feelings people have about their pasts. When talking about nothing in particular, one person might mention an aversion to the taste of pumpkin—a consequence of having eaten too much of it during the 1970s when she was in Guangxi and it was almost the only vegetable in her diet. Another might speak nostalgically of the delicious pumpkin vines that people eat as greens in Guangxi but are almost impossible to find in Beijing where he now lives (he resorted to growing his own in the yard outside his office building).

Taken together the sources offer an almost blinding kaleidoscope of bright and dark slices. As the historian Zhu Xueqin said of the Cultural Revolution, "It was an age ruled by both the poet and the executioner. The poet scattered roses everywhere, while the executioner cast a long shadow of terror."[50] The bright and the dark slices of history do not blur together in some murky gray, nor would it make sense for historians to try to average them out, listing pros on one side, cons on the other, and coming up with some kind of balance sheet to assess the net effect of this complicated period of history. Rather, it was a time and place that presented extraordinary opportunities for reenvisioning the world, some successful and others not, with victories and failures alike earned through tremendous hard work, no small amount of violence, but often also pleasure and kindness.

Summary

One of the premises of this book is that scientific farming meant different things to different people, and so the experiences of Chinese scientists, peasants, local cadres, technicians, and "educated youth" will each receive individual attention. I have intentionally focused the narrative around a handful of different people and places to achieve a level of human detail otherwise unattainable, but the examples in this book come from many parts of China. Some were "models" (a variant of the "poster child") in one way or another, and some were "ordinary" (to the extent that any place in all its uniqueness can be ordinary). Some were places of international exchange, and some were hinterlands where visitors from the county seat were the height of cosmopolitanism. A number of very interesting paths—Chinese and foreign, intellectual and peasant, animal and human—came together at Big Sand (Dasha, 大沙) Commune in the southeastern province of Guangdong; this site, and especially its famed program of insect control, will receive extensive discussion in the chapters that follow. Some other sites appear prominently simply because materials about them are available—thanks to their national importance, the vagaries of the used book market, or the personal connections of generous colleagues and friends in China.[51] The method of source collection was not intended to produce a comprehensive catalog of scientific farming activities across the nation; nor will it permit the systematic comparison of a few exhaustively studied cases. However, the evidence is broad and, in places, deep enough both to explore the diversity of people's experiences and also to demonstrate certain strong patterns across widely divergent sites.

Chapter 1 introduces the elements of state policy and ideology that bore most directly on agricultural science. I urge a reconsideration of the common wisdom that credits political "moderates" with promoting science while associating "radicals" with an antiscience, antimodernization agenda. Instead, I argue that science and agricultural development were values widely shared across historical periods and political perspectives, with important contradictions lying inside that broad area of agreement. The ideal of a revolutionary bottom-up experiment process existed in tension with the impulse to impose national models on local communities. In a related way, the radical insistence on the primacy of mass science and the technocratic privileging of elite, professional science occupied the two poles of the commonly referenced binary *tu* and *yang*. These contradictions endured despite state efforts to resolve them through formulas like "raising *tu* and *yang* together" and through the "three-in-one" configuration that brought cadres, technicians, and peasants together in scientific experiment groups.

Chapters 2 and 3 consider the experiences of Chinese agricultural scientists and the historical narratives of their lives and work. The focus on scientists represents the approach most readers are likely to expect in a history of science; nonetheless, even here I seek to challenge the tendency to define scientific achievement solely in terms of professional circuits and research institutions. Instead, I explore the different ways two scientists navigated the *tu/yang* binary in science. Chapter 2 introduces Pu Zhelong, who received his PhD in entomology at the University of Minnesota in 1949, then dedicated the rest of his career to aiding Chinese agriculture through the biological control of insect pests. A research scientist at prestigious Sun Yat-sen University, he was without doubt an elite, *yang* scientist. However, his strong political commitment to socialism and his affinity for peasants led him to success in the realm of *tu* science also, and he fared comparatively well even amid the anti-intellectual politics of the Cultural Revolution. Chapter 3 turns to Yuan Longping, who epitomized a humbler face of Mao-era science, though in the end he became far more famous than Pu ever would. Educated in China and assigned to a backwater agricultural college, Yuan nonetheless conducted important research on hybrid rice technology during the Cultural Revolution and came to fame in the late 1970s, eventually gaining the moniker "father of hybrid rice." His story highlights the importance of politics in historical narratives: during Hua Guofeng's brief reign (1976–1978), his research was cast in the familiar terms of "mass science"; in postsocialist China, he has typically been portrayed as an intellectual beleaguered by the radical politics of the Cultural Revolution—though nostalgia for Mao-era get-your-hands-dirty humility is often also in evidence.

Chapters 4 and 5 move beyond scientists to explore the experiences of people in rural communities, the testing ground for both green and red revolutionary transformations. The state, for both political and practical reasons, could not do without the active cooperation of rural people. Scientific farming radically reorganized agricultural authority in rural communities, creating a vast corps of "peasant technicians" with knowledge of a wide range of new farming practices, including both chemical and organic, labor-intensive and labor-saving. In some cases, scientific farming helped codify and promote existing forms of knowledge possessed by peasants, but in other cases, it threatened to replace them. The rural scientific experiment movement helped the state push through certain desired changes while avoiding responsibility for material assistance, but it also provided local communities with tools for resistance. And when they did resist, rural people pushed the state even further to invest in education and engage in inclusive practices. Chapter 4 explores peasant participation in scientific farming, focusing especially on the dual view of peasants as "experienced" and "backward." Chapter 5 looks at many of these same issues, but specifically from the perspective of local political cadres and agricultural technicians caught between state mandates from above and the realities of the rural communities they served. At the same time, the two chapters offer insight into the surprising historical path of the rural scientific experiment movement, a program based on the party's own policy process of selecting successful innovations achieved at "experimental points" (试验点) and extending them across other regions—which was, in a striking twist, originally inspired by 1930s agricultural extension efforts modeled on US practices.

Chapters 6 and 7 take up the story of the "educated youth" who participated in vast numbers in socialist China's rural scientific experiment movement, and for whom agricultural science was at times an amusing diversion, and at other times a much graver undertaking. Of all the historical characters appearing in accounts of the Cultural Revolution, youth are undoubtedly the best represented. Yet even here the story is far from fully told. The bitterness infusing the dominant narrative has obscured the significance of the real, lasting issues that people of that time wrestled to resolve and has impeded a nuanced understanding of the complex and varied ways young people experienced the era. Still more importantly, our knowledge of youth is almost entirely that of urban educated youth "sent down" to the countryside; we know far less about the experiences of the far greater number of rural educated youth who "returned" to their villages after graduating from urban secondary schools.[52] These chapters examine the experiences of both urban and rural youth—and though their outcomes often differed, they shared an aspiration to accomplish something important and a sense that participation in scientific experiment offered a valuable opportunity. Science mattered to many youth because it was both revolutionary and intellectual, and so offered opportunities for both political glory and personal advancement. Chapter 6 examines what I call the Lei Feng paradox, in which youth faced conflicting calls to be revolutionary heroes and simultaneously mere "bolts" in the revolutionary machine. Chapter 7 explores the tension between "opportunity" and "failure" that participation in the scientific experiment movement represented for educated youth.

The epilogue takes up the legacies of both green and red revolutions under dramatically altered political, economic, social, and cultural conditions. The technological changes effected by scientific farming are an underappreciated source of China's current economic growth; at the same time, they are to blame for much environmental destruction. But despite the near-blanket condemnation of Mao-era radicalism, people in China today continue to find inspiration in past practices for solutions to today's problems. The apparent victory of technocratic, green revolutionary agricultural policies masks the continuing relevance of more radical, red revolutionary approaches to agricultural science inherited from the Mao era.

This history presents all the subtleties and complexities created as real people—and poster children—grappled with the enormous upheavals that accompanied the political and technological revolutions of 1960s–1970s China. And so contradictions abound. Agricultural science in socialist-era China was highly transnational with deep connections to Western, especially US, scientific knowledge and institutional networks; it was also self-consciously self-reliant, as China sought to create *tu* science as an alternative to that wielded by capitalist and imperialist nations.[53] *Tu* science represented a serious bottom-up challenge to technocracy, but it was always forced to compete with the tendency toward dogmatism and the insistence on imposing models from the top down.[54] Moreover, though the alternative vision of *tu* science was deeply inspiring to people around the world, it was also very easily co-optable by state bureaucracies and, especially after 1978, by global corporate powers.[55] Like their counterparts in other fields, agricultural scientists were at once victims of political persecution and active agents who often shared key values with the socialist state and found *tu* science a rich resource for pursuing their profession in the service of social needs.[56] In China as elsewhere, the introduction of green revolution technologies undermined more sustainable technologies that had been practiced for centuries and so resulted in "deskilling" of rural people; but the same agricultural science networks that promoted the quick fixes of pesticides and chemical fertilizers also introduced complex new skills involved in raising wasps for biological control and spread "traditional" knowledge of green manure technologies into new areas.[57] For youth, participation in agricultural science was simultaneously a genuine opportunity to exercise their intellectual talents in the service of noble goals, and a tragically limited endeavor that too frequently ended in failure for themselves and the communities they sought to help. Stepping back to assess the history at a broader level, we see that the green revolution in red China was striking in its epistemological and political expectations, but the overall similarities in outcome for China and other parts of the world are difficult to ignore.[58]

Despite such irresolvable ambiguities, an analysis of the experiences recounted in this book supports a number of strong conclusions about the history of China's red and green revolutions, and the relationship between science and politics more broadly. To begin with, I argue that the political fluctuations of Mao-era China cannot be characterized as struggles between proscience and antiscience factions. Technocrats and radicals had different perspectives on how science should work, but both groups embraced science as a core value. By the same token, excessive faith in the possibilities of science and modernization presented very similar dangers in the hands of radicals and technocrats. The radicals' insistence on putting "politics in command" of science and technology did not result in the kind of critique of green revolution technologies that was needed from the standpoint of environmental health, and it fell short also in the realm of labor and social justice. There is perhaps no more vivid example of this than the oft-mocked Dazhai-emulating efforts to transform landscapes through brute force: as one participant explains it, he and his comrades "wreaked unprecedented havoc on the grasslands, working like fucking beasts of burden, only to commit unpardonable crimes against the land."[59] Here and in many other cases, we see the environmental consequences of the development orientation shared by radicals and technocrats alike, and the costs to human beings of the coercive politics employed in its pursuit.

This does not, however, suggest that science and technology should ideally be separate from politics. As Susan Greenhalgh has shown, the one-child policy of the Deng era is a chilling example of the consequences of technocracy: rather than trust Chinese demographers (many of whom came from a Marxist humanist tradition) to develop population policy, state leaders turned to experts in the supposedly more objective field of ballistic missile science, who crunched the numbers in favor of the ruthless plan China has followed since 1980.[60] Of course, the one-child policy is no less political for its reliance on the hard calculations of ballistic missile science, but it is certainly less just, less democratic, and less humane. In agriculture as well, technocratic approaches mask politics and so inhibit positive political engagement. In James Ferguson's words, the technocratic concept of "development" is "an 'anti-politics machine,' depoliticizing everything it touches, everywhere whisking political realities out of sight, all the while performing, almost unnoticed, its own preeminently political operation of expanding bureaucratic state power."[61] In a moving, antitechnocratic conclusion to his political economy of the plant biotechnology industry, Jack Kloppenburg proposed the need for "agricultural science finally [to] generate a cohort of internal critics"—and he suggested that they start by contemplating a 1961 statement by former head of the US Department of Agriculture Henry Wallace, "Scientific understanding is our joy. Economic and political understanding is our duty."[62] Looking at the history as it played out in China, I argue that the Mao era did not have too much politics in science, but rather too little in the way of rigorous political critique of technological triumphalism—and too much in the way of violent factionalism and persecution.

The people of the world still need to figure out how to do agricultural science differently. We need to be able to trust that technologies ostensibly developed to increase our capacity to feed people are not in fact riding roughshod over social and environmental needs to profit private interests. We need to be able to insist that social and political relations matter in the equation. And we need to be able to understand these issues as they relate to the history of a place home to one-fifth of the world's population and poised to make or break our global future.

Socialist Chinese efforts to effect a politically engaged philosophy of science fell far short of the hopes they kindled. Nor does any historical example yet meet the task.[63] However, the history explored here contains much that may inspire a rethinking of dominant assumptions about science and society. Having engaged in such reconsideration, we will be better positioned to confront problems of hunger and sustainability in appropriately social and political ways, and avoid the pitfalls of imagining purely technological solutions to the problems we face together.

Agricultural Science and the Socialist State

Introduction

The dominant historical narrative of science in Mao-era China charts a pendulum-like alternation between "radical" periods (the Great Leap Forward and most of the Cultural Revolution) when political struggle stifled intellectual pursuits and economic development, making science virtually impossible, and "moderate" (or technocratic) periods when steadier minds—especially those of Zhou Enlai, Liu Shaoqi, and Deng Xiaoping—prevailed and more liberal policies rekindled the hopes of beleaguered scientists.[1] David Zweig depicts Maoist "radical policies" on agriculture to have been "fueled by an anti-modernization mentality that saw economic development as the antithesis of revolution."[2] In fact, however, the history of agricultural science in socialist China is marked by a great deal of continuity across radical and moderate periods, and modernization based on scientific development was a value embraced by leaders across the political spectrum. Indeed, the move to develop "scientific farming" began circa 1961 during the heyday of the moderate technocrats, but it built on important precedents set during the Great Leap Forward, came into its own amid the intensifying radical politics of 1965, flourished throughout the Cultural Revolution, and remains relevant even today.[3] The green revolution thus progressed along much the same timeline in China as elsewhere, and it did so in the very middle of China's continually unfolding red revolution.

The Cold War presented at least three competing development paradigms, including the one embraced by Mao and his followers.[4] The attractiveness to Third World nations of the Marxist-Leninist model of state-led economic development alarmed many academics and political leaders in the United States, inspiring Walt Rostow's tremendously influential "non-

communist manifesto," *The Stages of Economic Growth* (1959). The parallels
between Leninism and Rostow's "modernization theory" are clear.[5] Both
were committed to modernization through technological development,
and both depended on deterministic expectations that development would
proceed through specific "stages." Soviet agricultural policy embraced the
goal of progress through modernization and even adopted the US strategy of Taylorism to increase efficiency in farming practices.[6] Though Mao
considered himself a Leninist and never questioned the progressive value
of modernization, his economic and political program—and the philosophy of science that went with it—departed in dramatic ways from modernization as pursued in the Soviet Union. Frustrated with the bureaucratic
and technocratic structures of authority that formed in China during the
period of Soviet learning, and with the rigid expectation of "stages" that
slowed China's progress toward communism, Mao sought to abandon the
determinism of staged growth and instead embrace a voluntarist faith in the
power of the masses to channel their collective revolutionary will into rapid
achievement of a truly communist economy. His was an explicitly political
vision of development that promised to eliminate the "three great differences" that privileged mental over manual labor, cities over countryside,
and workers over peasants.

This chapter moves away from the pendulum narrative to focus on questions that promote a fuller understanding of the political significance of agricultural science for the socialist Chinese state, and the significance of the state
in agricultural science. Whereas in later chapters the chief protagonists are
people at the grassroots grappling with mandates descending from above, this
first chapter focuses on the policies and ideological priorities developed at the
upper levels of the state. Agricultural science, and specifically the philosophy
and practice of agricultural extension, had a deep historical relationship with
state policy and ideology. The central tensions found in agricultural science
policy resonated with the broader tensions faced by the socialist Chinese state
as its leaders strove to resolve dilemmas related both to internal political and
economic conditions and to the geopolitical contexts of colonialism and the
Cold War. In the "point-to-plane" system of policy experimentation and
implementation, in the *tu/yang* binary that informed Mao-era politics of science, in the emergence of the rural scientific experiment movement from the
priorities of both radical and technocratic state leaders, and in the "three-in-
one" epistemology that dominated state writings on agricultural science at
the grassroots, the threads of China's red and green revolutions were tightly
interwoven.

Agricultural Knowledge and the State

As Francesca Bray has demonstrated, imperial-era China "was from its inception an agrarian state in the strong sense of the term," and so "dissemination of technical agricultural knowledge was considered an essential technique of the state."[7] Embracing a similar mandate, the socialist-era state created an extensive knowledge network premised on the idea that science is relevant to agriculture, and thus that knowledge of how to farm in any particular village can and should benefit from outside institutions. Even the term *laonong* (literally "old farmer"), which held such political potency in the socialist era, was used similarly in eighteenth-century China, when, in William Rowe's words, "activist governors . . . nominat[ed] 'experienced farmers' (*laonong*) from the local population itself to serve as exemplars of technological proficiency."[8] And Peter Perdue's account of nineteenth-century officials in Hunan attempting to convince farmers to plant two crops of rice a year reads as strikingly similar to what we find in 1960s–1970s China with respect to both state ambitions and local resistance.[9]

Despite its explicit hostility toward what it called "feudalism" (a term meant to capture both the class oppression and the religious "superstitions" of imperial-era society), the socialist state was not above borrowing from the traditional symbolic universe. A telling example is the "Eight-Character Charter" for agriculture (八字宪法, figures 6 and 7). Sanctified by Mao during the mid-1950s and widely popularized beginning in the Great Leap Forward (1958–1960), this was an easy mnemonic that organized agricultural knowledge and practice under the headings of eight Chinese characters that stood for landscaping, fertilizer, water, seeds, close planting, crop protection, tools, and management (土、肥、水、种、密、保、工、管).[10] The formulation was new, but the Eight-Character Charter strongly evoked the "eight-character fortune-telling" (八字算命) popular in rural areas, which used characters derived from the date and time of a person's birth to make predictions about the person's fate. Mao knew this practice well: not only was he born and raised in a rural village, but he criticized eight-character fortune-telling in his famous 1927 essay "Report on an Investigation of the Peasant Movement in Hunan." The Chinese Communist Party frequently adopted this strategy—using popular customs to further state priorities, while simultaneously seeking to replace the "superstitious" or otherwise undesirable elements of the old practices with scientific or otherwise ideologically correct meanings. And for their part, as Steve Smith has shown, peasants "were perfectly capable of combining magico-religious elements with secular elements

FIGURE 6. In this depiction of the eight aspects of agriculture identified by Mao, fertilizer is represented by an old technology (dredging rivers and ponds) greatly intensified during the socialist era, while crop protection is represented by a new technology (spraying chemical insecticides). Note also the appeal to "tradition" in the use of the Double Happiness character around the border—a symbol of future happiness that adds to the resonance with rural culture already emphasized by the semantic connection between the "Eight-Character Charter" and "eight-character fortune-telling." Xu Jiping. "Nongye bazi xianfa" (Eight-Character Charter) (Shanghai: Shanghai renmin meishu chubanshe, December 1959). Stefan R. Landsberger Collection, International Institute of Social History, Netherlands, http://chineseposters.net.

FIGURE 7. The vision of agriculture depicted here involves a greater number of newer technologies than in the 1959 poster. The panel on fertilizer includes chemical fertilizers in addition to river dredging and pig manure; the panel on industry now includes motorized machinery instead of the wooden, human-powered machines shown in 1959. Also significant is the emphasis on cotton, a crop highly promoted during this period to support the textile industry. Ren Meijun, Li Zuowan, and Liu Yushan, "Nongye 'bazi xianfa' hao" (The Eight-Character Charter for agriculture is best) (Shanghai: Shanghai renmin chubanshe, October 1974). Stefan R. Landsberger Collection, International Institute of Social History, Netherlands, http://chineseposters.net.

from the Party's own discourse . . . so if their world-view was rooted in an essentially religious cosmology, it was nevertheless powerfully shaped by revolutionary policies and by official propaganda."[11]

Its roots in imperial-era Chinese precedents notwithstanding, the socialist state's mechanisms for agricultural extension derived more directly from an influence geographically more distant and politically even more suspect. Despite all their struggles and failures, the work of John Lossing Buck and other Americans who pursued agricultural reform in early twentieth-century China left a profound legacy.[12] In 1953, an American agricultural economist observed with alarm, "Refugees making their way out of China bring constant reports that experimental farms established and financed by the United Nations and the United States have been taken over by Communists, the fruits of their experiments accepted, and their teachings forced upon Chinese farmers."[13] Indeed, the extension system the Chinese state adopted in the 1950s bore clear resemblance to that of the United States.

An article penned by Buck in 1918 demonstrates how strongly his approach to agricultural transformation prefigured that of the Mao era. Buck wrote, "In order to carry on this work it seems to me necessary that it be divided into three parts: an experiment farm, demonstration work, and school work. . . . Scientific principles of agriculture can best be instilled in the school boys. They will be much more ready to accept new ideas as compared with the ignorant farmer, who can be best reached through farm demonstration work."[14] Buck's integration of experiment, demonstration, and "school work" would take new form in the Mao-era emphasis on integrating experiment, demonstration, and extension.[15] And, as will be explored fully in later chapters, Buck's faith in the younger generations found a loud and clear echo in Mao's conviction that youth were the "least conservative" of social actors and so the most valuable for spearheading change.

Chinese-American agricultural extension expert Hsin-Pao Yang's 1945 "Promoting Cooperative Agricultural Extension Service in China" reviewed the work of a number of Chinese extension projects based on the American system, among them the rural reconstruction project in Dingxian by James Yen (Yan Yangchu). Yang's analysis highlighted themes that were to emerge again strongly in Mao-era extension work.[16] He proclaimed extension work a "grass-root operation" and decried the situation in which Chinese agronomists trained in the United States could "relate vividly how cotton is raised in Mississippi, corn cultivated in Iowa, wheat harvested by combines in Kansas, but are unable to help the hard-struggling farmers in their potato patches or in their rice paddies." According to Yang, successful extension workers should not only understand local issues but should adopt a humble attitude to win farmers' respect: "A decade ago a sensation was created among the villagers when a college professor took off his shoes and got himself dirty in a rice seedling plot where he demonstrated the proper way of transplanting. He put his teaching across because he followed the most natural way of working with the people."[17]

On a few other points, however, Yang's prescriptions spoke to a very different set of political priorities than those later adopted by the socialist Chinese state. In contrast with Mao's emphasis on self-reliance, Yang argued that "no one can live exclusively unto himself" and so "China cannot attain these [agricultural] objectives entirely by her own efforts. Help from and cooperation with other countries are indispensable." He also strongly urged extension workers to recognize that Chinese villagers "live by well-established behavior patterns" and that "customs and habits are dynamic stabilizers of community life."[18] If the Chinese extension system failed to look utterly familiar to US agricultural scientists when they arrived on delegations in the 1970s, it was no doubt because of the infusion of revolutionary politics that insisted on self-reliance and social transformation, that placed political value on the mobilization not only of technical experts but also youth, party cadres, and old peasants, and that emphasized not only top-down "extension" of technologies developed by experts but also experiments and innovations pursued by peasants at the grassroots to meet local needs and suit local conditions.[19]

Testifying to just how tightly China's green and red revolutions intertwined, the US agricultural extension system influenced not just socialist Chinese agriculture but key political processes of the Chinese Communist Party itself. When he traveled to China in 1974 on the coattails of the US Plant Studies Delegation, China scholar Philip Kuhn observed that the term "experiment" was a highly potent element in "current Chinese ideology."[20] Much more recently, the political scientist Sebastian Heilmann has traced the historical roots of China's "distinctive policy process" that emphasizes local experimentation at "experimental points," from which the center can select the most promising for widespread application. One of the influences Heilmann identifies for this policy process was the work of agricultural reformers in the 1910s and 1920s, who advocated "experimental extension"— that is, trying out new technologies and, on the basis of those trials, extending the ones that worked. By the 1960s, the system of "using one place's experience to lead a whole area" (以点带面) or "moving from point to plane" (由点到面) was such an accepted part of the policy process that its roots in agricultural extension were no longer noticed, even as it was adopted as the guiding philosophy for agricultural technicians themselves.

Heilmann argues that during the Mao era, and particularly during the radical periods of the Great Leap Forward and Cultural Revolution, the experimental policy process shifted decisively. In place of genuine encouragement of local innovation, political pressures to enforce ideological correctness favored the heavy-handed imposition of national models across locales, whether those models suited the locales or not. However, Heilmann further notes, "Certain programs of the 1960s and 1970s allowed meaningful experimentation to find new policy instruments when the policy context was more relaxed and top-level backing was present."[21] I find considerable evidence that scientific experiment continued to play an important role in agricultural extension; moreover, the commitment to experiment and local self-reliance as revolutionary values offered an antidote to inappropriate models imposed from above. Interestingly, the political ideals most useful in combating excessive imposition of models were those most often trumpeted by radical leaders.

Tu and Yang

In the terms of Mao-era scientific discourse, radical political and scientific leaders emphasized *tu* (土) over *yang* (洋). *Tu* denoted a cluster of related meanings (native, Chinese, local, rustic, mass, crude) that contrasted with *yang* (foreign, Western, elite, professional, ivory-tower) to form a radical vision of science in Mao-era China, that is, a science produced by the broad masses for the fulfillment of socialist revolutionary goals. Official policy encouraged harnessing *tu* and *yang* together in productive partnership (土洋并举 or 土洋结合). However, radicals harbored suspicions of scientists with foreign connections and consistently pushed for *tu* to lead *yang*, while technocrats took every opportunity to secure the leadership of the professional scientists whose skills they trusted to modernize China. *Tu* and *yang* mapped well onto other, more famous binaries that structured Maoist approaches to science—for example, red versus expert (i.e., commitment to socialist revolutionary politics versus technical expertise) and theory versus practice—and also onto the binary at the heart of this study, the green and red revolutions.

That the history as it emerges from the sources falls so easily into binaries tells us something important about the time and place: it is highly characteristic of Maoist dialectical materialism, and also of course characteristic of the geopolitics of the Cold War. My intention in employing these binaries is not to fall back on dichotomies but rather to think critically about them in their historical contexts and also to consciously "try them on" to see what they reveal and what they obscure about science in socialist-era China. While historians of science are very familiar with the pair "science and technology" and often debate the meanings of these terms and their relationships to one another in different times and places, they did not evoke the same degree of provocative contradiction in China as they have in the West. In China, *tu* and *yang* were considerably more important, speaking simultaneously to transnational relationships (foreign versus native) and to cross-class relationships (intellectual versus peasant). As such, *tu*/*yang* offers insight into how people in Mao-era China understood the way scientific knowledge is made and how it travels from one group of people to another—what James Secord has called "knowledge in transit."[22]

The *tu*/*yang* binary has further value in expanding our understanding of science in the linked global contexts of colonialism and modernization. In his highly influential work on this subject in relation to Egypt, Timothy Mitchell uses the term "binarism" to shed light on the implicit, or even hidden, ways in which twentieth-century social science divided the world. He writes, "Overlooking the mixed way things happen, indeed producing the effect of neatly separate realms of reason and the real world, ideas and their objects, the human and the nonhuman, was how power was coming to work in Egypt, and in the twentieth century in general."[23] The history of *tu* and *yang* in socialist China offers something different: a chance to explore the explicit adoption of analytical categories that simultaneously grew out of, challenged, and subtly reinforced the binaries of colonialism.

The concept of *tu* and *yang*—especially the way it ties together native/peasant on one hand and foreign/elite on the other—was undeniably a product of colonialism. As a number of scholars have argued, a key conceptual transformation of the early twentieth century in China, one with deep and lasting consequences for Chinese society, was the invention of the Chinese "peasantry." What had been "farmers" and "villagers" were now a mass of "peasants," defined by their oppression and their backwardness. But not only did the identity of rural people undergo transformation, China's own identity became increasingly that of a "rural" or even "peasant" nation. This conflation of "China" and "rural" or "peasant" owed much to the theorizing of Li Dazhao, who solved the problem of how an economically backward country like China could be expected to produce a communist revolution by positing that colonialism had turned the Chinese nation as a whole into the world's proletariat.[24] The Chinese nation itself thus took on a class character in relation to the rest of the world, a concept that only grew in strength after the 1949 revolution. And so for Mao-era China, transnational science necessarily traveled across terrain marked simultaneously by nation and by class.

Tu science has an important place in the history of the Chinese revolution.

The values associated with *tu*—self-reliance, mass mobilization, practical application—constituted a set of dovetailing priorities that emerged during the 1940s as the Chinese Communist Party struggled to mobilize people in their base areas to fight two wars: the War of Resistance against Japan and the civil war against Chiang Kai-shek's Nationalist Party. With the emerging leaders of the Cold War either outright supporting Chiang Kai-shek (in the case of the United States) or at least committed to a policy of nonaggression with him (in the case of the Soviets), Chinese communists determined that the only sure course lay in the development of indigenous resources—material, methodological, and human—to meet pressing economic and military needs. In the revolutionary "cradle" of Yan'an, the commitment to self-reliance, applied science, native methods, and mass mobilization became linked in ways that were to last throughout the Mao era.[25]

In 1939, Chinese communists responded to economic blockade by launching a movement for self-reliance in industry and defense.[26] Scientific knowledge had an obvious and important role to play in developing the means to produce material necessities such as matches, soap, candles, and explosives. Despite the inevitable orientation toward practical applications that this situation implied, for several years the party maintained a commitment to basic (that is, fundamental or theoretical) scientific knowledge. This changed in mid-1942 with the major political upheaval of the Party Rectification Movement. As Mao was consolidating his power through criticism of "bourgeois" intellectuals and party officials associated with the Soviet Union, the scientific leadership also underwent a profound shift.

The transformation centered on two figures: Xu Teli and Le Tianyu. Xu was the head of the Yan'an Academy of Natural Sciences. His approach was rooted in a belief that teaching and research in basic science formed a necessary foundation for the development of revolutionary China's science and economy. The commitment to following the masses and learning from practical experience that came with Rectification doomed Xu's program. The chairman of the Biology Department at the Natural Science Institute, Le Tianyu, had embraced an approach far more consistent with what was newly in vogue. His success in establishing a factory for producing beet sugar entirely with local beets and handmade equipment had already made him something of a "local hero."[27] During the Rectification Campaign, Le took advantage of the political wind to argue for his own work as the model that the entire institute should follow. Le's criticisms focused on the institute's use of foreign textbooks, problematic in terms of both self-reliance and learning through practice. In contrast, under Le's direction, the Biology Department required students to go among the peasants, learning from them how to manufacture dyes and medicines from local plants. This was mass-based, applied science that made full use of local resources. Many faculty and students rallied to the defense of Xu and basic science as a whole, but by early 1943 Le's approach to science had won the day, and the Natural Science Institute became a part of People's University, which was fully under party control.[28]

From its origins in the Yan'an period, the *tu/yang* binary took off during the Great Leap Forward (1958–1960), when Mao began to pull away from the path laid out by the Soviet advisers who had guided China through the first stages of building a socialist economy.[29] At that point, the stakes were raised for distinguishing between "*tu* experts" (土专家, native experts, especially from the laboring classes) and "*yang* experts" (洋专家, which variably meant Soviet experts, Chinese experts trained in foreign countries, or Chinese experts otherwise associated with institutions or bodies of knowledge somehow markable as "foreign").[30] Though official policy prescribed "uniting" *tu* and *yang*, propaganda more often trumpeted the value of *tu* to the point of denigrating *yang*. There was officially never any shame or political danger in being *tu*, whereas, especially during the Cultural Revolution, people overly associated with *yang* frequently found themselves attacked for being bourgeois and associated with foreign imperialism. On the other hand, there is no denying that the class privilege of *yang* was in the end much stronger than that of *tu*— and even during the Cultural Revolution, the nation's need for their expertise meant that professional scientists enjoyed many special privileges along with their special punishments.

As Warwick Anderson has explained, dichotomies produced by colonial regimes—including "global/local, first-world/third-world, Western/ Indigenous, modern/traditional, developed/underdeveloped, big-science/ small-science, nuclear/non-nuclear, and even theory/practice"—have played a critical role in the structure of science as it has developed globally, and always involve an imbalance of power.[31] What Mao and other Chinese communist leaders did was to make such dichotomies—including theory/practice, red/expert, *tu/yang*, and others—explicit parts of the governing ideology. Moreover, the state staked its legitimacy on representing the disempowered *tu* side, what in postcolonial theory is termed the "subaltern."[32] Embracing China's allegedly "poor and blank" condition as a virtue, Mao simultaneously reinforced the dichotomies of colonial modernity and sought to turn them upside down, claiming the subaltern as power.[33]

These epistemological interventions had material consequences, especially for scientists. The conflation of native and peasant meant that scientists' status as intellectuals, and thus nonpeasants, threatened to brand them further as nonnative—and while *tu* was the disempowered side of colonial modernity, *yang* was the dangerous side of Maoist political culture. As will be explored fully in chapters 2 and 3, Chinese scientists thus experienced the hybrid identities typical of postcolonial contexts: they embodied both *tu* and *yang* simply in the process of becoming both "Chinese" and "scientist." Beyond this spontaneous occurrence, the state further actively sought to create such hybridity with policies aimed at turning intellectuals into peasants and peasants into intellectuals, or at harnessing native and foreign resources to one yoke.

Through the celebration of *tu* science, the Chinese socialist state created for the Chinese nation a subaltern voice. Although that voice was deeply inspiring to many people, it did not necessarily do justice to the people it claimed to represent. Nor did it ever touch the developmentalist modernization paradigm that presumed human mastery over nature: indeed, in this respect Maoism was thoroughly consistent with the fundamental assumptions of modern technoscience everywhere. And so the question becomes, Was the *tu*/*yang* configuration a significant and efficacious way of "decolonizing" science, or did it merely reproduce the terms of colonial epistemology?[34] The chapters that follow will provide opportunities to revisit this thorny issue.

Three in One and the Emergence of the Scientific Experiment Movement

The problem of how to balance the practical need for professional expertise with the ideological significance of peasant experience inspired not only the *tu*/*yang* binary but also two of the legs of the "three-in-one" (三结合) formulation that guided the constitution of agricultural science groups—the third was the political authority of the state itself. It was during the Great Leap Forward that *People's Daily* first reported on efforts to bring political leaders, agricultural technicians, and peasants together in three-in-one groups to establish experiment fields and popularize new technologies.[35] By 1960, one county in the northeastern province of Jilin alone reported the existence of 645 such groups.[36] That year China was in the midst of the largest famine in history: the Great Leap mandate to achieve fully developed communism, at a speed and geographic scale the world had never seen, drove state agents to violent measures in a futile effort to meet unrealistic production quotas.[37] In the aftermath of the famine, Mao temporarily took a backseat and left open the possibility that Liu Shaoqi and other so-called "moderates" might steer China down a different path. Scientists and other intellectuals advanced an agenda more in line with a professional, even technocratic model of research and education. At the same time, the agricultural economy accommodated more family-based farming and sideline industries. In many areas, families were allocated small plots where they could grow vegetables for their own consumption, and the pressure to collectivize livestock lessened, such that chickens and pigs could return to something of their former privileged position in the family subsistence economy.[38] Still, Mao's seat was not very far back and not for long. In 1962, the Tenth Plenum of the Central Committee produced a mixture of economic pronouncements incorporating both post–Great Leap moderation and political language that signaled a clear leftward turn.[39] The move to develop "scientific farming" spoke to both sets of priorities, setting the stage for the intersection of green and red revolutions at the center of this book.

Policy directives in late 1962 ordered increased investments in the system of agrotechnical extension stations (农业技术推广站) initiated in 1950. Each station was required to have three to ten cadres with degrees from agricultural schools and experience in production and extension.[40] From February to April 1963, the National Conference on Agricultural Science and Technology Work, mandated by the Tenth Plenum, undertook a "major planning session for agricultural science and technology as well as for overall agricultural development in the 1960s."[41] Among the conference's most influential decisions was the expansion of demonstration farms (样板田, literally "model fields"), where newly introduced seeds and agricultural methods could be tested for suitability to local conditions and their worth demonstrated to local people. A technocratic vision dominated the conference. Nie Rongzhen, head of the State Science and Technology Commission, gave a speech in which he called for the realization of the "Four Modernizations" in agriculture, industry, national defense, and science and technology.[42] In an April article entitled "This Is the Time for Scientists to Do Their All," *People's Daily* reporters trumpeted the research of distinguished scientists like marine biologist Zeng Chengkui (C. K. Tseng) and approvingly quoted Liu Shaoqi as saying that the Four Modernizations "will depend on the hard work of everyone in the nation, will depend on the hard work of the scientists, and especially will require the leadership of the old scientists." The "masses" were an afterthought at most, appearing only ceremonially in the last line.[43]

But soon the rhetoric shifted profoundly to reflect the new politics of the Socialist Education Movement, born of Mao's urgent need to reestablish faith in socialism in the wake of the Great Leap famine. Staples of the Socialist Education Movement were stories of corruption and ideological errors on the part of local officials and members of politically bad classes, followed by their redemption through political reeducation. Mao's 1963 grouping of scientific experiment together with class struggle and production made room for many stories of the failure of certain wrongheaded officials and members of unfavorable classes to grasp the importance of science and the new agricultural technologies that came with it.[44]

In May 1964, *People's Daily* reported the appearance of a "new thing": mass scientific experiment small groups (群众性科学实验小组).[45] The demonstration farms retained their centrality but were now framed as a means to bring together "expert research" and "mass science."[46] In February 1965, the National Conference on Agricultural Experiment rode the Socialist Education Movement tide to launch a new "agricultural scientific experiment movement" (农业科学实验运动).[47] A report on the conference in *People's Daily* tied these efforts to both the Tenth Plenum and Mao's call to pursue scientific experiment as a revolutionary movement, thus bringing technological solutions and radical politics together under one umbrella. The article highlighted the multiple forms of revolutionary "three-in-one integration" reportedly in progress: not only were cadres, science workers, and the rural masses coming together, but also "demonstration farms, laboratories, and experiment fields," and "experiment, demonstration, and extension."[48] All of this, "under party leadership," was said to be resulting in "a revolutionary movement with demonstration fields as the center, specialized science and technology teams as the backbone, and mass scientific experiment activities as the foundation."[49]

The precise structure and activities associated with the scientific experiment movement varied from place to place, but certain patterns were widespread. Available educational resources were cobbled together to provide the necessary training for peasants to become technicians. Sometimes this meant night courses at local schools; sometimes the party branch secretary selected a few peasants or educated youth to attend a short course at an agricultural school in a nearby city. Periodic conferences at the local, provincial, and national levels brought members of grassroots scientific experiment groups together to exchange experiences. They also helped party officials identify and popularize model individuals or teams that could serve as inspiration for others.

Another very common pattern was for experiment groups to establish a "three field" method for demonstration, experiment, and seed propagation.[50] (Again, this concept had roots in earlier years, but it became much more clearly codified and widespread after 1965. And everything seemed to come in threes in the scientific experiment movement—itself the third of the "three great revolutionary movements.") The 1965 Conference of Activists in Beijing Municipal Rural Scientific Experiment Groups explained the system. The seed fields were meant to speed up the process of developing new

varieties and ensure that, once developed, they did not deteriorate. By organizing the production of seed at the local level, individual farmers would not
need to select their own seed for planting, but neither would the community
be dependent on outside organizations for seeds. Demonstration fields allowed the "masses" to "see" and "touch" new technologies so that they would
more quickly recognize their benefits and accept their use. Only proven and
noncontroversial technologies were to be introduced in the demonstration
fields; newer technologies that had not yet gained acceptance locally were to
first be tried in the experiment fields.[51] Bringing the point-to-plane system
full circle back to agriculture, demonstration fields were expected to play a
key role in increasing production "from a single point to many points, and
from many points to the whole plane."[52]

No other aspect of the rural scientific experiment movement was more
pronounced than the "three-in-one" formulation of the experiment groups.
A 1969 report from Guangdong explained the value of three-in-one combi-
nations: "Cadres have confidence, youth have technology, and old poor peasants have experience." It went on to elaborate that cadres grasp all aspects of
the situation, such that they can determine what kinds of experiments will
best serve production needs; old peasants are down-to-earth and uncon-
cerned with profit and fame, understand the rhythms of production, and
have a wealth of practical knowledge; and youth have technological knowledge and are accepting of new technologies, such that they "dare to think and
dare to act."[53]

Sometimes the specific groups that constituted the "three-in-one" formulation differed. Most notably, where agricultural technicians were available,
such "science and technology personnel" often substituted for youth. On the
other hand, according to one former production team leader I interviewed,
in his team "three-in-one" referred to urban youth who had been sent down
to the village, local rural youth, and cadres like himself (who were deemed
sufficiently "old" to balance out the youthfulness of the other two groups).[54]
But whichever specific three groups were involved, the overall emphasis was
on mobilizing the "broad masses" to participate in agricultural science. This
commitment arose from both political and practical concerns. Especially
during the more radical periods, it was essential that new technologies not
appear to be handed down from elite scientists in ivory towers. However,
even had the central authorities wanted to rely on agricultural scientists to
transform Chinese agricultural practices, they would have found it impossible to gather enough professionals to do the job. The 1965 Beijing conference
reported the deployment of some 1,200 science and technology personnel to
the countryside to work directly with peasants. But their expertise was spread

very thin: in the same area, there were already more than 8,000 grassroots scientific experiment groups and 4,000 rural schools organizing night classes for peasants to study agricultural science.[55] The goal from the beginning was thus to provide basic training for rural people—especially rural youth— who would then become "peasant technicians" capable of testing and popularizing new seeds, chemicals, and other technologies.[56] Youth participation was of tremendous significance in the scientific experiment movement. The idea that science should be pitched to youth is common in modern societies, but the Chinese case stands out because of the degree to which science itself was characterized as youthful, and youth themselves understood as agents of revolutionary scientific transformation.[57] Moreover, the sheer number of youth involved in the scientific experiment movement testifies to their central importance: at the 1965 Beijing conference for activists in rural scientific experiment groups, participants included 55 cadres, 23 old peasants, and 371 educated youth.[58] A representative scientific experiment group from Henan Province described in a 1966 publication included 40 cadres, 10 old peasants, and 120 youth.[59]

The terms "educated youth," "science and technology personnel," "cadres," and "old peasants" worked in multiple ways. Real people existed behind these terms, and one of the goals in later chapters will be to piece together the diverse ways that they experienced the formation of scientific experiment groups and the spread of scientific farming. However, the terms were also somewhat slippery, and the categories often overlapped in various ways. Youth could sometimes be cadres, especially when they were promoted to production team leader. They could also sometimes be technicians, when they had received special training. And some youth were originally peasants—while others (the "sent-down youth") hailed from urban areas. Beyond referring to specific people, the terms were loaded with highly potent political symbolism: there were clear ideological reasons for associating education with youth, and old age with peasants.

The "three-in-one" system constituted a highly articulated, structured "standpoint epistemology"—that is, a notion that people contribute differently to the production of knowledge based on their social position. Peasants and lab scientists have different experiences with rice, and rice matters to them in different ways; thus, they know different things about rice and communicate that knowledge differently. Three-in-one scientific experiment groups were meant to bring three different kinds of people together to ensure that scientific knowledge would benefit from multiple forms of expertise and so be properly revolutionary. Along with the *tu/yang* binary, the three-in-one system thus represented the socialist Chinese state's efforts to grapple with enduring contradictions of colonialism and class in the production of scientific knowledge.

Models, Networks, and Knowledge

To give the abstract political values of the mass scientific experiment movement a convincing material form, the state cultivated specific people, programs, and communities to serve as national models. The emphasis on models was to some extent a logical outcome of the point-to-plane system: once a local experiment had been proven effective, it could be tried in other areas and, if suitable, extended there. In fact, however, models often bolstered a top-down approach to agricultural transformation that undermined the commitment to local innovation of solutions suited to local conditions. Here again the political tensions of the socialist state—especially the contradiction between reliance on centralized technical expertise and cultivation of knowledge and practice at the grassroots—were clearly at play.

The most important of the Mao-era agricultural models was Dazhai Brigade in Shaanxi Province. The national movement to "study Dazhai in agriculture" emerged in 1965, at the same time as the more general emphasis on models summarized in the slogan "Emulate, study, catch up to, assist, and surpass" (比学赶帮超). The idea was that local communities should look to models like Dazhai for inspiration and guidance, and then themselves become proficient, perhaps even rising to model status themselves. The Study Dazhai movement escalated in 1967 and soon rose to become the byword for advanced, revolutionary agriculture.[60] Throughout the Cultural Revolution, people all over China read about Dazhai or even traveled there to witness first-hand the transformation of agriculture in arid northern China. Just what lessons Dazhai was meant to teach changed over time. Sometimes it was a specific practice such as terracing, mechanization, the standardization of field shapes, or the use of work points to determine peasant income; sometimes it was a more abstract concept such as self-reliance or mass mobilization; and sometimes the phrase "study Dazhai" appeared to need no further elaboration, its significance lying in some more transcendent understanding of loyalty to socialism.[61] Tales of havoc wreaked by the inappropriate application of terracing and other practices promoted by Dazhai abound in postsocialist interview and text sources. However, as chapter 5 will show, Dazhai was above all celebrated for embodying the principle of self-reliance, so in a paradoxical way it also served as a model for localities seeking to resist the imposition of outside models.[62]

Though far and away the most famous, Dazhai was not the only model.

In 1969, Hunan's Huarong County created a "four-level agricultural scientific experiment network" (四级农业科学实验网). Huarong's system was
not fundamentally different from patterns that had emerged elsewhere, but
it was more elaborately codified and so served well as an example for others
to emulate. The top level of the network was the county agricultural science
institute. Within the county, every commune had an agricultural science station. Within the commune, every production brigade had an agricultural science brigade. And within the production brigade, every production team had
an agricultural science group, or scientific experiment group. By 1973, many
counties in Hunan and other provinces had established four-level networks
of their own. The goal was to train three to five technicians for every production team in the country, on the order of twenty million technicians; in 1974,
officials estimated that thirteen million people nationwide were already participating in four-level networks, which left them some way to go.[63]

In October that year, the Ministry of Agriculture and Forestry, together
with the Chinese Academy of Sciences, organized the National Conference
for the Exchange of Experiences in Four-Level Agricultural Science Experiment Networks in Huarong, which was attended by more than four hundred
people. Like almost everything that happened in 1974, the conference reflected
the politics of the Campaign to Criticize Lin Biao and Confucius, an effort by
Cultural Revolution radicals to beat back the moderates through ideological
mobilization.[64] The conference summary expounded: "Using the four-level
agricultural scientific experiment network to organize the masses to carry out
scientific experiment will not only raise the level of science and technology,
and promote the development of production, but will simultaneously play a
role in revolutionizing ideology."[65] But the strong political rhetoric deployed
by the revolutionary committees in charge of promoting the network should
not be taken to imply a lack of technical content: the four-level networks represented an attempt to conduct agricultural extension (typically a top-down,
technocratic program) within a radical paradigm of mass science.

The four-level networks were closely related to the point-to-plane system.
Both depended on, and at the same time reinforced, the nested structure that
linked locales with higher levels in the political-economic system. The networks not only facilitated the spread of new policies and technologies from
the top down but also provided a role for the grassroots in developing new
policies and technologies. So, for example, when hybrid rice arrived on the
scene in 1976, Cenxi County in Guangxi organized each commune to select a
few production teams with good conditions to serve as experimental points,
which cadres and members of the masses then observed to learn from their
experiences before extending the technology in other teams throughout the

county.[66] This was by all counts a highly effective system for spreading new agricultural technologies—but where local leaders had genuine commitments to local needs and local decision making, it could simultaneously serve as a check on the efforts of upper-level officials to push through big changes with minimal resistance. As will be more fully explored in chapter 5, the ideological emphasis placed on experience, practice, and local self-reliance—those earthy values associated with *tu* science—offered crucial support for local agents struggling to maximize the benefits and minimize the harms of state-imposed agricultural models.

The Supplanting of *Tu* Science

Chairman Mao died in September 1976. In October, four radical leaders (including Mao's wife, Jiang Qing) were arrested on the charge of attempting to usurp power from Mao's chosen successor, Hua Guofeng; henceforth they would be known as the "Gang of Four." Hua was in a tricky position: to justify his leadership, he needed to distinguish himself both from the reviled Gang of Four and from his chief rivals—Deng Xiaoping and the other "moderates," who were calling for increased emphasis on economic modernization and investments in science and technology. Hua attempted to straddle both positions by trumpeting his commitment to the Maoist political line (including mass science) even as he pushed for greater investments in elite education and professional science.[67] To bolster this position, he claimed credit for two scientific achievements in Hunan while he had been party secretary there—the four-level agricultural science network and the development of hybrid rice technology.[68]

In 1978, Deng Xiaoping wrested control away from Hua Guofeng and soon thereafter launched the process of decollectivization—that is, the dismantling of the communes and establishment of a family responsibility system for agricultural production. The 1981 trial of the Gang of Four and publication of the "Resolution on Chinese Communist Party History" officially discredited not only Jiang Qing and her colleagues but the radical politics of the Cultural Revolution itself. In the new political climate, *tu* science did not entirely disappear—as we will see, vestiges continue to make themselves felt today. However, its influence quickly paled in comparison with the extraordinary surge of investment in professional science when Deng Xiaoping picked up the "Four Modernizations" as the banner of his technocratic line. Across China, these changes spelled the end of the grassroots scientific experiment groups and the three-in-one epistemology they represented. Likewise, from this point forward scientific farming would take on an increasingly "green revolutionary" and decreasingly "red revolutionary" color.

The shift in state policy and rhetoric has had a profound effect on the way the Mao era is remembered. Today the history of agricultural science is expected to be a *yang* history evaluated in terms of the degree to which the state supported professional scientists and the ability of those scientists to accumulate research achievements so as to raise production levels, feeding hungry peasants and enriching the economy. The chapters that follow take a very different approach, in some ways more consistent with Mao-era perspectives on the relationship between science and politics, agricultural transformation and social revolution. By no means will I be repeating uncritically the claims of the Mao-era Chinese state. However, I will endeavor to put aside the assumptions of the current dominant paradigm in order to highlight questions once recognized as important, and so illuminate the political tensions and priorities that produced China's unique intersection of red and green revolutions.

Pu Zhelong: Making Socialist Science Work

One of the dominant ways the history of the Mao era has been publicly re-membered is through the accounts of the sufferings of intellectuals. Although to some extent protected by the state's need for their technical expertise, sci-entists were not immune from the violence and persecution experienced by their counterparts in the humanities. Narrowing the circle just to agricultural scientists, the examples are still easy to find: Plant geneticist Li Jingxiong had a PhD from Cornell and was known for his expertise in maize breeding, but was subjected to criticism during the Cultural Revolution when crops under his supervision, which had used Texas cytoplasm, succumbed to blight.[1] The agronomist Ye Duzhuang (whose daughter, incidentally, will appear in chap-ters 6 and 7 as an educated youth in the scientific experiment movement) was unable to turn his genuine commitments to China and to socialism into political safety, and endured two long stints in prison—first as a result of the Anti-rightist Movement and then again during the Cultural Revolution.[2] Then there was the entomologist Liu Chongle, who suffered "ruthless per-secution" before he died of illness in 1969.[3] And there are many, many more such examples.

But suffering is not the full story of science and scientists during the Mao era. Nor were Chinese scientists important only when they acted as political dissidents and bearers of the May Fourth standard of science and democ-racy.[4] Rather, some scientists not only had strong commitments to socialist ideals but were genuinely successful in tapping the possibilities offered by Maoist *tu* science—the populist, nativist counterpart to professional, trans-national *yang* science. It was a tip from an American scientist who visited China in 1973 with the radical US group Science for the People that first drew me to Pu Zhelong's story. He suggested I find out more about Pu Zhelong because Pu had struck him and the other delegates as the most obviously sincere in his belief not only in "serving the people" but also in the much more revolutionary concept of "learning from the peasants."[5] Thus, in choosing to focus this chapter on Pu Zhelong, I am by no means proposing to take him as representative of Chinese scientists more generally. Rather, I am suggesting that there were important exceptions to the familiar story of Chinese scientists in the Mao era, whose lives appear to follow a narrative arc beginning with hope followed by tragedy and then ending (for those who survived) with recovery or even triumph after the fall of the Cultural Revolution radicals. Pu Zhelong is one such exception: the beginning and end of his story fit the standard narrative, but the middle seems to have been taken from another tale altogether.

Though not generalizable, Pu Zhelong's story nonetheless offers much insight into the larger history of agricultural science, and especially the field of insect control, in socialist China. Seen through Pu's experience, science in socialist China was remarkably transnational and displayed strong continuities with both the prerevolutionary and postsocialist periods. At the same time, his work reflects the strong emphasis on nationalist self-reliance produced by economic constraints, China's odd position in Cold War geopolitics, and Maoist ideology.

Pu's life also provides a valuable window into the making of a socialist Chinese scientist and particularly the significance of the *tu/yang* binary in that process. Postsocialist biographies of Pu emphasize his professional achievements (*yang*) over his contributions to the self-reliance, mass mobilization, and nativism embodied in *tu* science. Recovering that *tu* side of Pu's life and work is of crucial importance in charting Pu's individual accomplishments and, more broadly, grasping what kind of science worked in socialist-era China. Pu's experience demonstrates that *tu* science could be a natural fit for agricultural scientists. At the same time, he skillfully navigated the tensions of the Cold War to serve as an effective agent of China's new form of transnationalism in the 1970s. Interestingly, China's renewal of relations with the United States did not result immediately in the rise of a more *yang* vision for science. Rather, Pu and others first mobilized their *yang* connections to promote the idea of a uniquely socialist Chinese *tu* science that China could offer to the world. Only after the end of the Mao era and the repudiation of radical politics did *yang* triumph over *tu*: the chapter concludes by charting this transformation in Pu's postsocialist biographies.

A Classic Tale of a Modern Chinese Scientist—with a Twist

The story of Pu's early introduction to science is in many ways a familiar trope of the young urban elite intellectual coming of age as the Chinese nation struggled to survive the trials of war and political disunity. Nor were his experiences abroad distinctive: many other Chinese students of his generation similarly journeyed to the United States, where they formed communities with other overseas Chinese students seeking knowledge of agricultural science and other subjects they hoped would help save China from war and poverty. Juxtaposing Pu's early experiences with those of his classmate at the University of Minnesota, Huai C. Chiang, will help illustrate these parallels and highlight where Pu's trajectory was more distinctive.

Pu and Chiang were both born into well-to-do families in the very early years of the new Chinese Republic—Pu in 1912 and Chiang in 1915. Both grew up in cosmopolitan cities—Pu in the southern city of Guangzhou and Chiang in northern Beijing. As a secondary school student, Pu had already had the experience of visiting rural areas around Guangzhou, where the contrast between the beautiful scenery and the poverty of the inhabitants is said to have instilled in him a desire to study nature and transform the backward Chinese countryside.[6] This is the classic tale of the young Chinese urbanite of the 1920s and will be very familiar to historians of the period.

Pu and Chiang both attended top-notch Chinese universities dominated by faculty who had received their doctorates in the United States.[7] Their mentors in China were trained in the United States and thoroughly a part of US-dominated transnational agricultural science. Chiang studied with Liu Chongle, who had a PhD from Cornell in 1926 and completed a dissertation on the "natural control of the eastern tent caterpillar." In 1935, Liu received funds to return to the United States to survey achievements in biological control (i.e., the use of natural enemies to control insect pests).[8] Pu's professor at Yanjing University, Chenfu F. Wu, received his PhD in entomology at Cornell in 1922, returned to China for a position at Yanjing University, and then later applied for a fellowship to study insect control in cotton, fruit trees, and cereal crops, focusing especially on "mechanical and cultural methods" in the United States ("cultural" here means "agricultural"—that is, it involves altering patterns of planting, plowing, harvesting, etc., to minimize pest damage).[9] Pu also studied with the up-and-coming geneticist Li Ruqi, who received his doctorate at Columbia University in 1926 under the guidance of the renowned (and in socialist countries, highly controversial) geneticist Thomas Morgan.

After the outbreak of full-scale war with Japan in 1937, Chiang and Pu joined the westward retreat—the largest wartime migration in history. Pu had by then finished his master's degree and was teaching at Sun Yat-sen University in Guangzhou; Chiang was still working on his BA at Qinghua University in Beijing. Both universities relocated to the southwestern province of Yunnan. Amid the harsh realities of war, Chinese scientists in every field sought practical applications for the knowledge they had cultivated in the academy: anything that protected China's supplies of food or raw materials was a noble national cause.[10] As entomologists, Chiang and Pu had little trouble finding useful projects: while in Yunnan, Chiang worked on the life cycles of ladybugs (an important biological control agent), on wax scale insects (used to produce candles and other items), and on insects damaging to pear crops; and Pu took up the study of forest insects, experimenting with the use of bacteria to control forest pests.[11]

The road from China to Minnesota had already been well prepared by the time Chiang and Pu began their doctoral studies in entomology. When Chiang enrolled in the graduate program in entomology at the University of Minnesota in 1945, he found a community of "old timer Chinese students" ready to help him buy warm clothing and otherwise acclimate to Minnesota.[12] Pu arrived in this community just a year later and found Minnesota an ideal place not only to pursue entomology but also to immerse himself in the study of foreign language (including French, Italian, and Japanese) and music.[13]

Chiang completed his PhD in 1948 on population dynamics of fruit flies. Pu finished the following year, with a dissertation on the taxonomy of Chinese moss beetles under the direction of Clarence Mickel.[14] Pu's wife, Li Cuiying, received a master's in entomology from Minnesota the same year. Figure 8 depicts the strong cohort of Chinese graduate students in the University of Minnesota entomology program in 1948—with Li and the other woman in stylish fur coats. Figure 9 captures Pu and Li at the time of their graduation, giving every appearance of being relaxed and in their element.

The communist victory in China in 1949 presented many Chinese people then living abroad with a tough choice. Chiang, like most of the other five thousand Chinese students and scientists in the United States in 1949, decided to stay in the United States.[15] But Pu and Li, like hundreds of others, opted to throw their lot in with the new socialist state; they arrived in the People's Republic less than a month after its founding. Thus Chiang and Pu represented the two different ways—remaining in the United States and returning to China—that Chinese-born scientists educated in the United States contributed to transnational science. As Wang Zuoyue has aptly expressed it, those who opted to remain in the United States after the communist revolution

FIGURE 8. Pu Zhelong (bottom row, far right), Li Cuiying (top row, second from right), and Huai C. Chiang (bottom row, middle) among their fellow Chinese students of entomology at the University of Minnesota posing for a picture in 1948. The third man from the left in the top row may be Alexander (Alec) Hodson, who advised a number of Chinese graduate students in entomology. Reproduced from Gu Dexiang, ed., *Pu Zhelong jinian yingji* (Pu Zhelong memorial album), 2002, 4.

FIGURE 9. Pu Zhelong and Li Cuiying after Pu's PhD graduation, 1949. Reproduced from Gu Dexiang, ed., *Pu Zhelong jinian yingji* (Pu Zhelong memorial album), 2002, 4.

of 1949 helped produce the "transnationalization of the American scientific community," while those who chose to return to China contributed to the "Americanization of international science."[16]

Entomologists suffered their share of difficulties and trauma during the Mao era. However, Pu weathered the political storms remarkably well. Pu's former student and colleague Gu Dexiang recalls that with the exception of a few "extreme actions" (过激行动, a standard phrase used today when describing Cultural Revolution violence), Pu's political life was without incident.[17] It would not be right to minimize the significance of these "few" events. Pu and Li lost a considerable number of valued possessions—more than the silver and copper coins or the collection of Qing dynasty stamps, the music albums, sheet music, and books must have been particularly painful losses for Pu as an intellectual and a lover of music.[18] Moreover, in 1967–1968, Pu oversaw a research project in Qianyang, Hunan, on silkworm breeding that was disrupted by the chaos of the early Cultural Revolution, such that much good work was wasted.[19] In 1969, when Pu and Li were sent down together to a May 7th Cadre School to participate in labor, Pu felt so discouraged that he reportedly even spoke of quitting his profession to teach English or suggested that it might be best to retire in a few years.[20] Yet, compared with the experiences of many other scientists, these political setbacks were mild, and Pu survived the upheavals notably unscathed.

To explain his good fortune, people who knew him point to his character: Pu's generous spirit and genuine desire to help others meant that few people bore him any grudges, and so he was unlikely to become a target during political campaigns.[21] But there was also another important reason for Pu's exceptional success through all the turmoil and despite the ever-shifting political winds. It was not just that Pu largely escaped trouble during the Cultural Revolution; he managed to do truly productive work that fulfilled both professional and political requirements, and that successfully bridged the *tu* and *yang* ideals of socialist Chinese science. Pu was without question a member of the intellectual elite, but he also had a deep respect for peasants and a willingness to embrace the Maoist priorities of eliminating the discrepancies between city and countryside and between mental and manual labor.

The post-Mao triumphs of professional science that receive so much attention in biographies of Chinese scientists thus represented, in Pu's case, not a rupture with the Cultural Revolution but rather the culmination of decades of continuously productive work. I will turn below to a deeper exploration of these themes, but will first step back briefly to explain the significance of Pu's research within the broader historical context of insect control science in socialist China.

The Transnational World of Socialist Chinese Insect Control

Prior to 1949, Chinese entomologists had been part of a larger world in which the United States played a disproportionate role; in that era biological control of insect pests was already an important research topic for both US and Chinese entomologists. In the 1950s, the United States officially became an enemy, and after 1960, China was on the outs with both Cold War superpowers. However, when we look beyond the relationships between nations and think about the actual people involved in scientific work, it quickly becomes obvious that Chinese science continued to be part of the same global network it enjoyed in the early twentieth century. And so it should not be surprising that though China and the United States experienced very different economic conditions that had strong shaping effects, their histories of insect control science proceeded along remarkably parallel trajectories.[22]

In later years, Pu's adoption of biological control as a research focus was celebrated by foreign and Chinese, scientific and popular accounts alike as a testament to his early awareness of the environmental and health consequences of toxic pesticides. But these were not the dominant concerns driving research on biological control in the Mao era. The struggle to attain self-sufficiency (again related to emphasizing *tu* over *yang*) was a consistent refrain in insect control as in every other economic sector in Mao-era China. According to *People's Daily*, Pu's own initial motivation for researching biological control arose from his anger over the refusal of "imperialists" to sell insecticides to China.[23] The Cold War geopolitics of agrochemicals was indeed a major structural force shaping agricultural science in China and other parts of the world: for example, in the early 1960s, requests from Cuba for chemical insecticides and fertilizer circulated through the top levels of the Chinese state where unfulfilled domestic needs competed with commitments to international socialist solidarity.[24] In short, whatever Pu's own concerns about the environment—and it does appear he was ahead of the curve on this front—the broader context for his focus on biological control at least in the 1950s was compensation for the unfortunate lack of chemical insecticides rather than environmentally conscious attempts to limit their use.

In fact, throughout this period China actively sought to increase its supply of insecticides. Nationwide, use of insecticides rose ten times from 1957 to 1965 and in 1974 reached twenty times the 1957 levels.[25] To meet this demand, Chinese industry steadily increased production levels of BHC (benzene hexachloride), DDT, and organophosphate insecticides. BHC was of particular interest: as a by-product of coke and soda manufacturing, it was relatively cheap to produce, and moreover it was highly effective. According to a 1954

People's Daily article, the main problem, it seemed, was actually convincing more people to use it.[26] This problem appears to have been quickly overcome: in 1955, a county agricultural department worker wrote to *People's Daily* to encourage education about the limits and dangers of BHC. People were treating it as a "cure-all" (万灵药), applying it directly on horses for skin ailments (which made the problem worse) and pouring it into rice paddies (which merely wasted it).[27] Despite increases in production, between 1950 and 1979 China imported more than nine hundred thousand tons of insecticides.[28] Even this was not enough: shortages of potent chemical insecticides encouraged local people to manufacture "indigenous insecticides" made from available plant material.[29]

The emergence of the concept of integrated control (which combines biological, chemical, and "cultural" methods) followed a strikingly similar pattern in China as in the United States. China's first published reference to "integrated control" came in a 1952 article in the key entomology journal *Kunchong xuebao* (Acta entomologica sinica). Later the same year, "integrated control" appeared in a US journal article that has since been recognized as a kind of locus classicus for the term.[30] Interestingly, the Chinese article's use of the term was much closer than the US article's to the meaning it soon came to carry in the United States as well as internationally.[31] However, rather than focusing on the integration of chemical and biological controls, as proponents of integrated control in the United States generally assumed, the Chinese article proposed that chemical insecticides should be used alongside cultural controls.[32] Moreover, according to an interview with the lead author in 1985, "The suggestion of integrated control was to emphasize the chemical control because pest control could not depend only on agricultural [i.e., cultural] control methods."[33]

Pu himself was by no means categorically opposed to the use of chemical insecticides: in 1961, he coauthored an article that expressed optimism about the use of BHC and DDT to control sugarcane weevils.[34] Still, in 1953, Pu began researching the cultivation of the parasitic wasp *Trichogramma*, a primary biological control agent internationally, to control sugarcane stem borer. He took this research into the field in 1956, and by 1958 he had reportedly achieved excellent results. Pu also pioneered the use of a less common type of ovoparasitic wasp, *Anastatus*, to control stinkbugs in lychee orchards.[35]

The 1950s in China was the period of "learning from the Soviet Union." In government ministries, research institutes, and universities around the country, about ten thousand Soviet advisers counseled, collaborated with, and often offended and exasperated their Chinese counterparts.[36] In some cases, however, the program resulted in genuine collaborations, with the visiting "experts" learning as much, and interested as much or more in pursuing the relationship, as their Chinese hosts.[37] Such was the case with Pu and the resident Soviet advisers in entomology at Sun Yat-sen University. The Soviets at that time had expertise in the use of *Trichogramma* wasps and other biological agents to control insect pests; Pu imported two types of ladybugs from the Soviet Union to control scale insect.[38] In its official report, a delegation of entomologists who visited China in 1975 speculated that the heavy use of *Trichogramma* in biological control regimens in China perhaps owed to Soviet influence.[39] In fact, *Trichogramma* studies in China began in the 1930s, and as with so many other areas in agricultural science, the dominant influence was clearly American.[40] Nevertheless, Soviet research on *Trichogramma* and other biological agents clearly also made an impact on China, especially after the revolution but also earlier on.[41] Articles in scientific journals and *People's Daily* allude to Soviet research, and Soviet books on insect control were available in Chinese translation.

More importantly, the Soviets were considered "advanced" in the area of insect ecology, and Pu and his colleagues benefited from their guidance in this area.[42] In 1957, Pu invited Professor Andreanova of Moscow University to spend a year teaching in the Biology Department; his senior seminars in insect ecology produced two cohorts of students trained in this important field. Upon Andreanova's departure, Pu invited Professor Grishin of Leningrad University, who oversaw the construction of China's first climate-controlled ecology laboratory.[43] Correspondence between Pu and Grishin testifies to Grishin's warm feelings for Pu, Li, and others in the department and his strong desire as late as 1961—after the "Sino-Soviet split" and the end of the resident adviser program—to continue the professional collaboration in insect ecology. Pu and his superiors were also interested in finding a way to facilitate the collaboration, but given the deteriorating relationship between the two nations, it is not surprising that their hopes did not materialize.[44]

Nineteen sixty-two was a big year for integrated control. In the United States, that was the year Rachel Carson's *Silent Spring* was published. In China, scientists at the Chinese Entomology Conference in January agreed that the heavy use of pesticides over more than a decade had caused resistance and thus larger insect pest populations in many countries. Some called for increased attention to biological control, while others argued strongly for greater use of chemical insecticides in the short term.[45] In April, two scientists, including one who had been a pioneer of biological control (and specifically use of *Trichogramma*) in Republican-era China, wrote an article for *Chinese Agricultural Science* promoting the "integration of chemical control and biological control." Consistent with the temporary waning of self-reliance rhetoric during Mao's post–Great Leap Forward retrenchment, the authors highlighted the international attention that had been brought to this subject, and they cited research from Canada, the United States, Germany, Australia, Egypt, the Soviet Union, and other countries.[46] Later that year, however, sixty-six scientists and technicians who attended a national conference on plant protection signed a report testifying to the sorry state of insect control and calling for increased attention to insecticide production and distribution—biological and cultural controls were not discussed, and the one mention of insect resistance was buried under an item related to the pricing of insecticides. The dominant perspective thus continued to be that the chief problem with insecticides was their short supply.[47]

In spring 1975, the Ministry of Agriculture and Forestry convened the National Plant Protection Conference where it was determined that "integrated control with prevention foremost" (预防为主，综合防治) would henceforth be national policy.[48] However, throughout most of China, labor-intensive cultural and manual methods continued to form the foundation for pest control, and "integrated control" typically meant maximizing the effectiveness of these methods in order to minimize applications of expensive chemical insecticides.[49] As in the United States, biological control in China was the exception rather than the rule. And though China did not have US-style chemical corporations weighting the scales in favor of increased dependence on chemical pesticides, local investments in pesticide manufacturing did sometimes create political pressure on scientists not to criticize pesticides too much.[50]

Pu Zhelong's most sustained and influential contribution to integrated control was his work in Big Sand, a commune about seventy kilometers from Sun Yat-sen University in Guangzhou. In the fall of 1972, Pu Zhelong accepted an invitation to visit Big Sand to give a lecture to peasants and cadres on insect control: he focused on biological control—specifically, the use of "bug-eat-bug" (以虫治虫) and microbial approaches.[51] With the enthusiastic support of a local cadre, Mai Baoxiang, Pu's team soon launched a pilot project.

In the midst of the Cultural Revolutionary Chinese countryside, Pu was still a *yang* scientist connected in deep intellectual ways to the international entomological community of which he had once been a part. In August 1973, Pu gave a lecture at Big Sand to more than seventy people; his arguments could easily have come from the lips of his counterparts in Western countries.[52] He emphasized that pesticides should not be considered silver bullets. When chemical pesticides were introduced in nineteenth-century Europe and the United States, some thought that insects would be eradicated. They even set about collecting specimens against that eventuality! But over the next century pesticides just became more and more poisonous: for example, one drop of some of the ones used at Big Sand could kill a person, while insects would not only survive but actually thrive. Second, it is not easy or necessary to eradicate insect pests. Instead, the commune should strive to control insect populations to prevent economic harm. Third, no single method of control is perfect. For this reason, the commune should follow an integrated control strategy that combined different methods.[53]

In late 1974, Pu and his younger colleague Gu Dexiang attended a national conference on integrated control of pests in rice paddies. They returned to Big Sand excited to tell their partners on the commune about the information they had shared on Big Sand and about what they had heard from others. This representation at the national level seems to have inspired commune members, since they resolved then and there to place the entire commune under integrated control beginning in 1975. Gu Dexiang remembers composing a verse to evoke the importance of the wider ecology for insect control: "By day swallows fly through the sky, in evening hear frogs and birds chirp, along the way spiders build their houses."[54] Commune members pledged to forbid the catching of frogs (beneficial in killing insects), to build a factory for microbial control and a station for rearing wasps, and to propagandize peasants and establish and train a technical team. And they determined to expand their program of raising ducks to control rice paddy pests.

This was by no means the first time ducks had been used for biological control in China. Peasants in 1930s Jiangsu used ducks in their battles with locusts, and many other examples must also exist.[55] Still, their massive employment in 1970s Big Sand—thirty-four thousand the first year and three hundred thousand the next—required careful planning and technical assistance. At first many became sick and died, prompting Pu to invite specialists from the agricultural institute to investigate and offer recommendations for changing feeding and management conditions. Their efforts appeared to pay off: one work unit reported that in 1975 nine thousand ducklings earned 18,000 yuan at the market, while expenditure on insecticide dropped from 2,560 yuan to just 32.80 yuan.[56] The ducks of Big Sand were not just good at catching bugs and raising money for the production teams. Much cuter and more delicious than wasps or bacteria, they captured the attention of visitors to the commune. With their help, Pu's university team and Big Sand Commune leaders put their pest control program on display to members of other Chinese communes, Communist Party officials, university leaders, newspaper reporters, and international delegations (figure 10). Because, of course,

FIGURE 10. Two stuffed ducks and a chart showing the effectiveness of ducks in reducing the need for chemical fertilizers at Big Sand Commune. From the Robert L. Metcalf Papers at the University of Illinois Archives.

the work at Big Sand by no means represented just self-reliance: rather, with the arrival of US entomologists, international exchange in insect control science had come full circle. The transnational story will resume below, after consideration of the *tu* side of the equation.

The Making of a *Tu* Scientist

When Pu and other young Chinese scientists returned from abroad, they helped establish a transnational foundation for the science of "new China." However, socialist Chinese science had another set of roots in the communist base area of Yan'an. There the experiences of war and revolution produced a lasting set of values, or to borrow Elizabeth Perry's phrase, a "revolutionary tradition" that could later be "mined" to bolster the *tu* side of the *tu/yang* binary in science.[57] Moving back to China required considerable reorientation. In the United States, no matter how committed agricultural scientists were to the farmers they served, they were inescapably part of a scientific culture and institutional economy that judged a scientist's worth according to his (or sometimes her) contributions to "basic" or "fundamental" science. Like many other entomologists, Pu had completed his dissertation research on a taxonomic subject, thus establishing his contribution to basic scientific research. In socialist China, scientists of all stripes would be expected to embrace the nativist, proletarian, application-oriented, earthy values of *tu* science. At the same time, scientists returning from the United States had an especially difficult road to navigate through the possibilities and perils associated with Cold War–era politics.

Given these challenges, Pu's political accomplishments were very impressive—and even more so considering that his official class background of "landlord" was very unfavorable and he was not a member of the Communist Party.[58] In 1956, he was awarded the national-level title of "advanced worker," and over the years he served in leadership roles in a number of scientific bodies and at Sun Yat-sen University (where he eventually served as vice president), as well as being a representative to the National People's Congress.[59] These honors were a mark not just of Pu's professional achievements but also of his ability and willingness to flow with the currents of science under Chinese socialism.

When Pu returned to China in 1949, he quickly shifted his research focus from taxonomy to pest control, a subject with far more direct social benefit and thus more in tune with science in a socialist system. To outside observers, such a switch may have suggested political pressure or even coercion. In a 1963 article in *Science*, a Chinese-American zoologist at Penn State, Tien-Hsi Cheng, referred to Pu as a "coleopterist" (i.e., beetle specialist) who "found it expedient to change from research in taxonomy to investigation of the biological control of sugar-cane borers."[60] But if Cheng worried about the political influence that drove this change, it is not clear that Pu shared his concern. After all, Pu's research prior to moving to Minnesota had been on insect control, and it is not surprising that a scientist who had opted to return to China immediately following the revolution should have been eager to tailor his research activities to the needs identified by the new state—needs that scientists themselves also often recognized. Pu's colleague Gu Dexiang notes that anyone whose research focused predominantly on theoretical issues would have faced criticism; however, he also emphasizes that Pu agreed with that position.[61]

Pu Zhelong's switch from basic to applied entomological research and his contributions to China's self-reliance in insect control might be explained by any number of factors; they do not necessarily indicate a deep commitment specifically to the Maoist vision of "mass science." More telling was Pu's ability to work with the tide of the Great Leap Forward's radical attempt to achieve full-fledged communism in just a few years by mobilizing the great revolutionary spirit of the masses. Pu emphasized mass participation in the testing of *Trichogramma* wasps, encouraging people on the ground to help modify the technology to fit local needs. For example, Pu reported that in Zhongshan County local people used bamboo stalks and other "materials at hand" (就地取材) to construct tubes for raising wasps, an improvement over the matchboxes that Pu had originally employed. He also credited peasants for determining that the tubes should face northeast to avoid wind and rain, and for employing thick paper to create a flexible seal on the tubes. In addition to encouraging innovation by "the masses," Pu also developed the means to release the wasps on the scale (areas over 1,000 *mu*) that Great Leap philosophy encouraged. By 1959, areas in Guangdong, Guangxi, and Fujian had reportedly adopted the methods he designed with good results.[62]

Pu was certainly not the only scientist to pepper his Great Leap–era scientific journal articles with references to mass participation and even mass innovation. But he was unusual in the lengths he went to synchronize with the radical politics of the day. Most strikingly, Pu took an active role in promoting a peasant, Li Shimei, who had developed an approach to controlling termites. Pu heard Li Shimei speak on termite control at the First Guangdong Provincial Conference on Scientific Work in April 1958, and subsequently invited him to Sun Yat-sen University so that he and Li Cuiying could help him complete a scientific paper on the subject. The Entomology Division of the Biology Department, led by Pu, then invited Li Shimei to accept the post of professor, a move covered on the front page of the 22 June issue of *People's Daily*.[63] In fact, during the Mao era Li Shimei appeared considerably more often in *People's Daily* articles than Pu himself: he perfectly embodied the radical emphasis on the mobilization of "native experts" in China's development. Without ever losing his own identity as a "foreign expert," sponsoring Li Shimei was one highly visible way that Pu could demonstrate his support for the *tu* agenda of mass science.[64]

Soon after Pu and Li were sent down to the May 7th Cadre School, they were recalled because their work on insect control was so closely tied to the needs of production. At the same time, the university was engaged in a "revolution in education" (教育革命). Pu was trusted in this area, and so he was sent with a group of students to Dongguan County to control insects and simultaneously carry out the revolution.[65] He was successful enough that in 1970 *People's Daily* published an article authored by the Sun Yat-sen University Biology Department's Revolution in Education Practice Team. The article noted that peasants in Dongguan County had reported problems of pest resistance to insecticides. Stinkbugs had become an increasing problem in lychee orchards; chemical insecticides had not been very effective, and worse yet had harmed pollinating bees. Teachers and students in the Biology Department determined that the ovoparasitic wasp *Anastatus* (平腹小蜂) would be an ideal means of combating the stinkbugs. Not only did they achieve a parasit-ism rate of 98 percent, but by embracing "crude and simple" (因陋就简) methods in the field, they reportedly broadened their own worldview beyond the confines of the university. On the basis of these experiences, the depart-ment established practical classes on the subject, and in the process was said to have created a "new entomology"—a victory for revolution in education.[66]

Because Pu's work resonated so effectively with the *tu* vision for science and other key aspects of Cultural Revolution politics, it won attention from the *People's Daily* in yet another feature article two years later. Here Pu him-self was the focus: the article celebrated this "old professor's new youth" and again highlighted the revolutionary aspects of his work. While Pu had reared *Anastatus* in the laboratory, that kind of experience was insufficient for large-scale operations in the field. So Pu took his research into the countryside and conducted experiments rearing *Anastatus* in a "crude shack" (简陋的茅棚) that he had been able to transform into a workable wasp station. He went on to train more than thirty technical workers to run the pest control opera-tion. The article further celebrated the half year Pu spent in the countryside sharing the peasants' life and labor, learning many things, and becoming "tempered" (锻炼, as steel is tempered in a furnace).[67] From this time on, Pu Zhelong "regularly entered the front lines of the production struggle and assimilated the rich experience of the masses in his research." For example, when he heard that peasants in Mei County had been getting good results with microbial agents, he traveled there with his students so that they could learn directly from the peasants. People often saw him in his lab or at home sitting with commune members and peasants from the countryside exchang-ing the results from their scientific experiments. Consequently, Pu's teaching and research reportedly underwent a profound shift to pay closer attention to the relationship between theory and practice. "This old professor's youth is reblooming," the article concluded, "as he pursues a new kind of scientific research into viral controls of insects and commits himself to making a new contribution to socialist construction in the fatherland."[68]

No doubt these official media depictions skew the picture in an effort to de-emphasize the *yang* and play up the *tu*. My main point in using these sources is not to capture true historical experience but rather to demonstrate the political significance of Pu's work within *tu* science. That said, interviews with people who worked with Pu during the Cultural Revolution confirm that Pu was exceptionally committed to working with peasants and willing to get his hands dirty in the countryside. So during the Mao era, while envi-ronmental and health concerns about the use of chemical insecticides grew only slowly, Pu's research on biological control held other, equally powerful political significance: it served to celebrate the "expert" bowing to the "red," the *yang* learning from the *tu.*

An official memo written to support Pu's application to leave the country on a professional delegation in 1972 captures in a nutshell Pu's political reliability in the eyes of the state. The memo praised Pu for taking direction from the party, loving the socialist fatherland, accepting new things, working hard, and being willing to serve socialist projects. It described his research projects, noting his work on insect taxonomy but especially highlighting insect control. In keeping with Cultural Revolution–era documents of this type, the memo highlighted Pu's prior susceptibility to "the individualist fame and profit mentality of the capitalist class" for which he had "received the masses' criticism and assistance," but applauded Pu's interest in the Cultural Revolution and his "attention to studying the thought of Mao and the party center." The memo writer had apparently heard that Pu was dissatisfied with his experience at the cadre school in 1969, for it reported on this, but emphasized that his participation in the revolution in education back on campus taught him that "he just needed to serve the people" and that "his own specialty and strengths were of use." Most importantly, the "poor and lower-middle peasants" of Dongguan had reportedly sung his praises, testifying, "It's great that Sun Yat-sen University has someone like Pu Zhelong. When we have problems, we can go to him at any time."[69]

As politically commendable as Pu's work was, in 1972 the authorities at his university decided he was still lacking in one key area: he had failed to "take grain as the key link"—a crucial priority in agriculture since the Great Leap Forward. It was not enough to control insects in sugarcane fields or lychee orchards; Pu needed to find a project that focused on rice.[70] The solution to his problem lay in Big Sand. There Pu succeeded in bringing the many priorities of Cultural Revolution science together with his own professional commitment to biological pest control.

In the early days, as Pu's younger colleague Gu Dexiang remembers, transportation was so limited that even scientists on official business had to purchase tickets three days in advance. The roads were dirt, and the bridge that now crosses the Beijiang River (between Sanshui and Big Sand) had not yet been built; parts of the journey had to be completed by ferry and bicycle, sometimes carrying boxes of parasitic wasps and other supplies. The road was so hard—six or seven hours, whereas now it takes just two—that the researchers often stayed for a month, and sometimes several months, at a time.[71]

Once there, Pu and his colleagues were immersed in commune life. Consistent with the *tu* vision of overcoming the barriers between mental and

manual labor, scientists engaged in physical work, attended meetings on production, and trained peasants to participate in research activities. The researchers had to build the facilities they required with the limited resources
available on the commune, and they slept with the workers in the factory they
built for producing microbial insect control agents. Ironically, they spent
their evenings in the hopeless pursuit of catching fleas, which nonetheless
bit them all night, leaving itchy welts. (China's current leader, Xi Jinping,
memorialized this common experience as the "flea test.")[72] They had rice to
eat, but not much to go with it. The difference between urban and rural conditions was so stark that they soon could not conjure up a memory of their
former lives.[73]

From all accounts, Big Sand peasants adored Pu and were proud to be
working with him. Local cadre Mai Baoxiang recorded in his journal that
when Pu first came to inspect the land designated for his experiments, he
was quickly surrounded by a number of young women workers excited to
tell him everything they had done to plant the field and prepare it for the
experiment.[74] One local participant especially recalls the effort Pu expended
organizing training classes and sending them books from the city: "He took
care of us agricultural technicians, really took care of us."[75] Here the personal and the political became tightly interwoven: Pu's commitment to the
peasants signified strongly in revolutionary terms, but it was simultaneously
experienced as human kindness expressed through personal relationships—
and this was not a contradiction.

In January 1974, Pu Zhelong and Gu Dexiang visited Big Sand to determine whether the commune leaders were interested in continuing the work.
Amid many courtesies and toasts—Pu ended with a toast to Mao—the leaders unsurprisingly elected to continue for a second year. In Mai's journal entries, Pu often seems to have had a politically appropriate comment on his
lips, and his commitment to the peasantry and the revolution appears strong.
However, he could also be lighthearted about the political campaign du jour.
Once he arrived in the midst of a meeting to criticize Confucianism, but instead of joining it, he requested that Mai accompany him on his inspection
of the fields, saying, "They're talking about struggles among human relationships; we'll go struggle with nature." Still, their tour provided an opportunity
to talk about some of the more enduring priorities of Chinese socialism, and
Mai recorded Pu's concerns about hardship in the countryside and his belief
that the government should do everything possible to help the peasants.[76]

Big Sand was an extraordinary example of the vision of science champi-
oned in Cultural Revolution China. Pu had successfully developed a project
that involved both urban and rural youth, integrated theory and practice,

transplanted laboratory research from the ivory tower to a "crude shack" in the countryside, and increased the self-sufficiency of Big Sand Commune by raising wasps, ducks, and other biological control agents on-site, all for the sake of serving the peasants. Moreover, reports on the research were often coauthored by the commune's revolutionary committee, the county science and technology department, and Sun Yat-sen University's Biology Department, providing further institutional validation for the slogan "Bringing together *tu* and *yang*."[77] Big Sand had crystallized Pu's reputation as a Cultural Revolutionary scientist of the first water.

The Meaning of *Tu* in a Transnational World

If success in Mao-era science had been just about cultivating a *tu* orientation, we would look for exemplars among "native experts" like Li Shimei rather than among professionals like Pu Zhelong. But *tu* science was never meant to stay in the Chinese villages. Even when China was isolated from the two Cold War superpowers, it was striving to become the leader of the Third World. And when China began the process of renewing relations with the United States through the early 1970s scientific and cultural exchanges, *tu* science became an important contribution China could claim to make in the international scientific arena. Native experts served important purposes, but at the end of the day China could not do without people like Pu Zhelong: scientists with *tu* credibility, *yang* connections, and the ability to navigate the very turbulent political waters of Cold War politics.

In February 1951, Pu Zhelong and Li Cuiying wrote to Li's adviser, Minnesota professor A. Glenn Richards, and his wife:

> We heard that the price of food and commodity in the States is going up but we don't think that yours [*sic*] living would be menaced by such happening. Our living is still hard, but the salary has been raised every four or five months. . . . Probably, you have been hearing lots of news of China from newspaper or radio. We must [tell] you honestly that China is in progress. There is not any force that suppresses the people and the farm tenants are really liberated from the oppression of landlords who are real enemies of modernization of China.

They also asked after Li's master's thesis, which had been published as an article in *Biological Bulletin*, and which involved a histological study of the insemination reaction in house flies.[78] Perhaps this was the real impetus for their letter, and the political statements were meant to please the censors. However, given the other evidence as to Pu and Li's commitments, I interpret these as genuine statements reflecting the couple's beliefs at the time. Like

many other intellectuals of this era, Pu and Li were excited about the prospects of building the new China and optimistic about the leadership of the
Chinese Communist Party in this endeavor.

It is unclear whether Pu and Li continued to have correspondence with
their American teachers and colleagues after this point. Certainly, it would
have been risky to make their connections with the United States conspicu-
ous while the two countries were at war. In 1952, their fellow student of entomology at the University of Minnesota, Ma Shijun, joined other scientists
in publicly testifying to what he had heard in US scientific circles that allegedly lent weight to accusations the United States had employed germ warfare
against China during the Korean War, going so far as to name specific individuals.[79] Perhaps Ma and the others truly believed the charges were justified;
perhaps they simply sought to decrease the vulnerability associated with their
American PhDs.[80] Pu, however, appears to have had little trouble fitting into
the new political order, and I have not found evidence that he resorted to at-
tacking his former mentors and colleagues in the United States to curry favor
with political leaders.[81] As we have seen, Pu was also remarkably successful
in navigating China's shifting relationship with the other side of the Cold
War axis.

In 1970, when state officials sought politically trustworthy intellectuals to
carry out the "revolution in education," they chose Pu. Following ping-pong
diplomacy in 1971, Pu was an even more obvious choice for state officials
seeking competent and reliable hosts for international scientific delegations.
Pu's international connections, his outgoing personality, and his sincere commitment to many of the most fundamental tenets of socialist Chinese science
made him an ideal representative. Because he communicated so effectively
in English, Pu was a more convincing interview subject than many who had
to work through a translator. Also, he was comfortable around Westerners
from the years he had spent in Minnesota. And he came with a partner, Li
Cuiying, who shared these qualities and added a domestic touch appreciated
by delegates who visited their home. Of course, Pu acted under close supervision. When a former student of Pu's, Lü Mingheng, returned to China in
1973 and sought to visit Pu, approval from the party work team at Sun Yat-
sen University was required.[82] Similar approval was necessary even for Pu to
send a book on insect control to the Iraqi ambassador stationed in Beijing.[83]
Still, to the extent that anyone—especially any intellectual—was trusted, Pu
certainly was, and it is no surprise that so many foreign visitors stopped to see
Pu Zhelong on their carefully orchestrated trips through China.[84]

Not all the Chinese hosts had such skill in translating the politics of the
Cultural Revolution for their foreign guests as Pu Zhelong. When "father" of

the green revolution Norman Borlaug visited Sun Yat-sen University in 1974, the obligatory "short introduction" delivered by the chairman of the revolutionary committee contained what for Borlaug was an unpalatable amount of enthusiasm for the early years of the Cultural Revolution, when "workers marched into this school" to "rub out" Lin Biao and the "cultural bougesues [*sic*, probably bourgeois] philosophy," sending professors out to work in fields and factories, and bringing peasants into the university to teach. After a few pages of such notes, Borlaug apparently took some pleasure in writing, "Chairman of Rev Comm here a nut!!" and "Head of Rev Comm here a 'hot dog' showboat."[85]

Meeting Pu Zhelong at his university laboratories, as well as visiting his research site at Big Sand, even partaking of roast duck in the commune cafeteria, was another experience entirely. Political tensions were not completely absent even from these encounters. When Huai C. Chiang arrived with the 1975 US insect control delegation, Pu refrained from recognizing their "special relationship" as classmates back at the University of Minnesota, and Chiang tactfully respected Pu's distance.[86] But such dissonant notes were a muted counterpoint to the dominant and much brighter theme of foreign admiration for Pu's accomplishments. Westerners were especially excited to be able to use Big Sand and other Chinese models of integrated control as a stick with which to prod the scientific and political leaders in their own countries, where the environmental consequences of chemical insecticides were of growing concern.[87] One British delegate reportedly told his hosts at Big Sand, "In Western countries people talk a lot about integrated control but do very little of it. You do so much work; you are our model."[88] The official report of the Swedish delegation similarly posited the relative backwardness of biological control in Sweden and suggested that knowledge should be sought in China, where biological methods and integrated pest control were more developed.[89] Indeed, so effective was the demonstration at Big Sand that the UN Food and Agriculture Organization, drawing on the US delegation's enthusiastic report, presented Big Sand as *the* example of advanced integrated control in its 1979 *Guidelines for Integrated Control of Rice Insect Pests*.

We might expect the dramatic change in China's global position beginning with ping-pong diplomacy to have resulted in an equally dramatic fall in the emphasis on self-reliance in socialist Chinese science and a corresponding shift from *tu* to *yang*. Significantly, this was far from the case. Even as the Chinese state hosted foreign delegations of scientists and sent its own scientists abroad, it maintained a strong commitment to anti-imperialist self-reliance that showed up in its continued efforts to lead the Third World in socialist revolution and in its narrative of China's own history. And in fact, China's emergence into a more prominent global position in some ways increased the desirability of emphasizing the *tu* aspects of its science. Chinese research institutions were unmistakably underequipped, and Chinese scientists had far fewer accomplishments to trumpet than their Western counterparts. China could not compete in *yang* science, but *tu* could be promoted as the basis for a uniquely socialist Chinese style of science from which other countries could learn. This was the extension into the détente era of China's desire to present a "third way" to the world, an alternative—not only for the Third World, but also for potential European allies like Sweden—to the options offered by the two superpowers.[90] And so for example, when the International Rice Research Institute delegation visited Big Sand in 1976, its members heard from Pu Zhelong not only about the integration of theory and practice, and the way students and teachers worked with peasants in the communes, but also that "all teaching materials are indigenous" and "foreign teaching material is used only if alternative material is not available within the country."[91]

In addition to hosting foreign delegations, in 1975 Pu traveled to Sweden and Canada, where he strongly articulated the vision of Cultural Revolution science and the historical narrative then dominant. In a publication from the conference he attended in Sweden, his essay emphasized that Confucianism and feudalism had retarded scientific progress in China, but the "Great Proletarian Culture [*sic*] Revolution" brought a "step forward," and now scientific personnel cooperated with peasants to conduct research based on practical problems and devise solutions directly in the people's communes.[92]

That same year, China's chief zoology journal published an article attributed to Pu that spoke just a bit about Pu's own research projects (specifically at Dongguan and Big Sand) and was packed instead with perspectives on the politics of science as they stood in 1975. The history of insect control science apparently illustrated perfectly the struggle to resist the cultural domination of Western practices and create a uniquely socialist Chinese style of science:

> We researchers in the natural sciences, having been influenced in the past by a mentality of slavishness to the West (洋奴哲学), frequently could not distinguish between good and bad experience, to the extent that some people thought that all foreign experience was good. For example, in the past, some of our comrades were totally enthralled by foreign agrochemical products. Now the once-fashionable DDT has been discovered to pose a definite danger to humans, while insects have to different degrees developed resistance against it. Over the past few years, we have engaged in biological control, taking China's concrete reality as our starting point, and now we have quite a bit of our own experience. Thus, we must follow the great leader Chairman Mao's teachings, "Overturn the Western slave mentality and bury dogmatism" and

"Through self-reliance and hard work, eliminate superstition and liberate thought," in order to integrate [science] with China's industrial and agricultural production, diligently synthesize experience, and carve out our own road for scientific development."[93]

Of course, Pu was himself intimately aware that "foreign experience" in insect control included not just agrochemicals but also biological control. Pu's Chinese professors of entomology had studied biological control in the United States, and Pu himself had been instrumental in bringing Soviet work on insect ecology and beneficial parasites to China. It was ironic that just as Chinese scientists were linking back up to the transnational networks in which they had begun their careers, the *yang* side of socialist Chinese science was being de-emphasized, and those connections elided. But if it was ironic, it was also fitting: China's new, more prominent role on the global stage sharpened the need for a unique contribution that socialist China could make to science. And Pu Zhelong, with his *tu* credibility and his *yang* connections, was an ideal spokesman for this project.

The Post-Mao Fate of *Yang* and *Tu*

Mao-era narratives—including those articulated by Pu himself—highlighted the contributions of peasants and workers, the need for science to suit local conditions and serve production, and the primary goal of self-reliance. They spoke of scientists and university students embedding themselves in the countryside, where they conducted research alongside peasant technicians, learned from the rich experience of the masses, made do with limited resources, and participated fully in production and political life. They emphasized revolutionary mass science over professional science, *tu* over *yang*. This narrative was to change dramatically in the post-Mao era.

Beginning in the 1980s, biographical articles and books on intellectuals, including scientists, proliferated in China. In some ways, the accounts of Pu's life are consistent with what can be fairly characterized as a genre—as authors and the reading public became deeply familiar with the specific form such stories were expected to take.[94] Thus, postsocialist biographical accounts typically emphasize the mounting accomplishments of scientific research produced by a top-notch university scientist through laboratory and field experiments, and validated through publication in scientific journals and engagement in the international scientific arena. This emphasis reflects the postsocialist valuing of professional science (*yang*) over revolutionary

FIGURE 11. Pu Zhelong playing the violin. Reproduced from Gu Dexiang, ed., *Pu Zhelong jinian yingji* (Pu Zhelong memorial album), 2002, 65.

mass science (*tu*). Through photographs and descriptions of his musical accomplishments on the violin and piano (figure 11), such accounts further highlight Pu's intellectualism. In postsocialist writings we also see newer, environmentalist values coming to the fore: they often celebrate Pu as a rare far-sighted intellectual who called attention to the environmental consequences of increased use of chemical insecticides, and offer the promise of a more ecologically sound future based on his research on biological control of insect pests. They paint Pu as a "pioneering environmentalist" whose personal character was like a "lofty mountain" and whose "noble deeds will live forever."[95]

The narratives typically begin with Pu's early experiences witnessing rural poverty and becoming inspired to help Chinese peasants, a story told of many Chinese intellectuals of Pu's generation. The accounts further highlight Pu's return to China at the birth of the People's Republic as evidence of his patriotism: "In October 1949, as the five-starred red flag was flying over Tiananmen Square and new China had just been born, this pair of kindred spirits [Pu and his wife, Li Cuiying] resolutely determined to leave the United States and return to the bosom of their ancestral country."[96] Despite such similarities, in other ways Pu Zhelong is remembered today in ways strikingly different from the biographies of most Chinese scientists who lived and worked during the Mao era.

Unlike the "scar stories" that quickly became the template for the narratives of intellectuals' experiences during the Cultural Revolution, Pu's biographies rarely discuss political persecution or even obstacles to his research. The longest such work, a memorial volume published in 2012 to mark the one-hundredth anniversary of Pu's birth, does detail the demoralizing destruction of his silkworm research in Qianyang, but otherwise such accounts are free of traumas and grievances. Indeed, one biographical article strikes precisely the opposite note. Speaking of his work up until 1972, the author writes, "As Pu Zhelong's scientific research won victory after victory, he received honor after honor, but he never let that stop him from moving forward. Instead, he saw each achievement as a new starting place, and forever scaled new heights."[97] And just one of the biographies considered here references the historical signposts of the fall of the "Gang of Four" and rise of Deng Xiaoping at the Eleventh Plenum, which are nearly ubiquitous in other scientists' biographies. (In an interesting echo of the 1972 *People's Daily* article, which credited the Cultural Revolution for inspiring a "new youth" in Pu, the author claimed that after the Eleventh Plenum, Pu "shone with a new youthfulness.") Thus, while the biographies of most scientists emphasize historical rupture during the Cultural Revolution and renewal with the coming of Deng Xiaoping's reforms, in Pu Zhelong's story we see continuity—a sense that what was built in the early years of the People's Republic flourished in the Cultural Revolution and continued uninterrupted into the post-Mao era.

Biographies of Pu further highlight his extraordinary leadership abilities, a characteristic that those who knew him frequently emphasize in casual conversation. One employee I chatted with at Sun Yat-sen University told me that Pu was such a wonderful administrator because he was always moving out of the way to encourage younger colleagues to succeed. His biographers stress his "skill in bringing people together" and his commitment to cultivating the next generation of scientific talent, such that his "peaches and plums" (that is, his students—the fruits of his intellectual life) "filled the world."[98] They also highlight his work organizing technicians and peasants in Dongguan.

It is thus striking, if not exactly surprising, that Pu's long list of recognized accomplishments does not include the one that received the most media attention and continues to be the best known today outside entomology circles: the recruitment and cultivation of "*tu* expert" Li Shimei. Recognizing this young laborer's potential and bringing him to the university was certainly an extraordinary example of Pu's commitment to cultivating talent, and it especially helps us understand why Pu was so successful in making the most of the *tu* style of science promoted during the Mao era. But it is an example just too close to the most radical, and thus most discredited, aspects of Maoist science. In postsocialist biographies of Pu, we do not find celebrations of "native experts" or of peasant ingenuity: where peasants appear, they are for the most part merely "welcoming" Pu's efforts to introduce new pest control technologies rather than teaching Pu the value of reliance on the masses or playing an important role of their own.

One aspect of Maoism that does appear in two of these stories is the integration of theory and practice.[99] This is consistent with the post-Mao refashioning of "socialism with Chinese characteristics," which preserves specific, relatively safe elements of Mao's philosophy—with the two classic essays "On Practice" and "On Contradiction" in prominent positions—and sidelines much of the rest of what had been canonized as "Mao Zedong Thought." We can also detect a certain flavor of *tu* science in postsocialist biographies of Pu; however, the emphasis is on Pu's own character rather than the peasants he mobilized or the larger political philosophy of socialism. Perhaps the most "*tu*" thing about Pu that comes across in these stories is his willingness to engage in menial work. As one article explains, "He had none of the arrogance typical of professors, but rather did much of the scut work, which deeply moved local peasants and his fellow teachers and students."[100] This same quality is captured visually in figure 12, which appears in two memorial volumes documenting Pu's life and work. The photograph depicts Pu bracing

FIGURE 12. Pu Zhelong's wristwatch clearly marked him as a man of position, but he was not above engaging in manual labor in this photograph of construction at Big Sand's new bacterial fertilizer factory (c. 1975). Reproduced from Gu Dexiang, ed., *Pu Zhelong jinian yingji* (Pu Zhelong memorial album), 2002, 25.

a bare foot against a board as he participates in construction at the factory that produced bacteria for biological control at Big Sand.

But these echoes of the Mao era aside, the dominant narrative found in these biographies is clearly one of *yang* professional science. More prominent than the photograph of Pu with his shoe removed are the pictures of Pu with his violin or at the piano. Both instruments are a standard trope of intellectual life. (Here China scholars may think of Richard Kraus's book *Pianos and Politics.* Others may be familiar with Dai Sijie's novel and film *Balzac and the Little Chinese Seamstress,* which feature an urban, educated youth who brought his violin with him when he was sent down to the countryside during the Cultural Revolution; the violin symbolized his intellectual status.) And of course, every biography of Pu highlights his impressive list of professional achievements and the recognition his research won from foreign observers. As in the 1970s, the effort in these post-Mao biographies is to establish Chinese scientists and Chinese science as worthy on the world stage, but now instead of offering a challenge to the Western standard of *yang* science, the goal is to embrace and excel in it. The bottom line in remembering Pu today is that he "passed on to our country's scientific establishment a precious wealth."[101]

The next chapter turns to another agricultural scientist who straddled *yang* and *tu* identities. Like the accounts of Pu Zhelong's work, accounts of Yuan Longping's work tended in the Mao era to highlight *tu* over *yang*, and in the postsocialist era to highlight *yang* over *tu*. In other ways, however, their lives and the stories told about them have been remarkably different.

Yuan Longping: "Intellectual Peasant"

The life of the renowned rice breeder Yuan Longping (袁隆平, 1930–) was in many ways very different from that of Pu Zhelong. To begin with, Yuan was born almost twenty years after Pu and belonged to the generation of Pu's younger colleague Gu Dexiang. By the time Yuan and Gu were of age to begin college, the communists were just months away from victory in the civil war: there would be no opportunities for them to travel to the United States for graduate school. When Yuan graduated from the newly established Southwestern Agricultural University in 1953, he was assigned to teach at Anjiang Agricultural School (also known as Qianyang Agricultural School) in the remote hills of western Hunan Province, where many of his students were young peasants bound to return after graduation to their hometowns.[1] Yuan rose from this humble starting point to become the most famous agricultural scientist in China (figure 13).

In the postsocialist era, his name is a household word; he is far more famous in China than the "father" of the green revolution, Norman Borlaug, is in the United States. Yet even though his claim to fame—hybrid rice—was a product of the Mao era, he is almost invisible in Mao-era sources. His celebrity status is a product of post-Mao publicity, and it was by no means a forgone conclusion. The public knows nothing about who "invented" hybrid sorghum, though it predated hybrid rice and involved the same technology.[2] Nor have the many other people involved in the invention of hybrid rice received acclaim of the kind Yuan has been afforded.

Thus, despite the important differences between Pu and Yuan, the starting point must be the same: an emphasis on the need to think carefully about the different types of narratives that emerge from different historical periods. This chapter tracks the changing ways that the story of hybrid rice has been

FIGURE 13. A commonly reproduced photograph of Yuan Longping (squatting). I collected this example at Panjiayuan antiques market in Beijing when I asked people for "materials related to agriculture." As further testament to Yuan's fame, a Chinese graduate student in the History Department at UMass who visited my office immediately recognized Yuan from the photograph and expressed surprise when I was impressed by his knowledge: "of course" he would recognize Yuan Longping!

told, paying special attention to the role of the *tu/yang* binary in the narratives. Although Yuan's homegrown education and assignment to a rural agricultural school fit the values of earthy, nativist *tu* science far better than Pu Zhelong's ivory-tower pedigree, Yuan was ironically not famous until the post-Mao era, when professional, transnational *yang* values rose to the fore. As Pu had in the Mao era, Yuan came to embody both *tu* and *yang*, but if Pu was a *yang* scientist cultivating a *tu* persona, Yuan and his biographers have had to work hard on both sides of the binary—to lay claim on the one hand to the *yang* credentials of a professional scientist and on the other hand to the, now nostalgic, *tu* values that Yuan's humble background inevitably invokes.

Yuan Longping through Mao-Era Sources

Yuan authored only one article during the Mao era. In the April 1966 issue of *Chinese Science Bulletin*, Yuan reported on his discovery of mutant male-sterile rice plants, the first critical step on the road to tapping the benefits of hybrid vigor in rice.[3] Hybrid corn had long since been put into production around the world, but the challenges with self-pollinating plants such as rice and sorghum were much greater. Rice in particular has a very low rate of cross-pollination—more than 95 percent of the time rice seeds are produced by the male and female parts of a single rice plant. Thus, to hybridize two rice plants, one of the plants must first have the male parts removed so that it will not fertilize itself. It can then be exposed to the pollen from another plant so that its seeds will contain the genetic material from both plants. This was already a standard practice in China, as in many other places, in the production of new varieties of rice. However, the phenomenon of "hybrid vigor" (or heterosis) applies only to the first generation (called F_1) of plants produced through this crossing. So the new variety may have many useful qualities and be worth stabilizing and putting into production as an "improved variety," but those first-generation hybrid plants will typically perform significantly better than the later generations. The trick with "hybrid rice," as with "hybrid corn," was to find a convenient way to repeat the hybridization process on a large scale every year, so that farmers could be provided with large quantities of seed that would grow into first-generation hybrid plants. For rice, this meant finding plants that were already male-sterile so that each individual plant would not have to be sterilized by hand. Yuan's discovery of a male-sterile plant was thus worthy of attention from China's most important science journal.[4]

After this one article, published on the eve of the Cultural Revolution, Yuan's name disappeared from print until several months after Mao's death and the fall of the Cultural Revolution radicals. However, an article published in 1972 in *Agricultural Science and Technology News* under the name of the Qianyang Agricultural School Scientific Research Group was undoubtedly his creation. By this time, Yuan and his colleagues had given up on the male-sterile line they had identified in 1964 among the cultivars in their fields, but after a painstaking search, they had found on the island of Hainan a male-sterile plant of a wild rice variety, which has ever since been known as "wild abortive" (WA). In the 1972 article, Yuan described the next challenge: identifying the second genetic strain needed for the three-line method of producing hybrid seed. This was the so-called maintainer line—the strain that when crossed with the male-sterile line would preserve the male-sterile trait and so

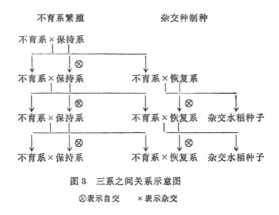

图 8　三系之间关系示意图

⊗表示自交　　×表示杂交

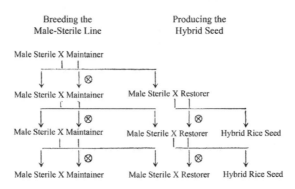

Relations among the Three Lines

⊗ Self-fertilization　　X Hybridization

FIGURE 14. Diagram showing the three-line method of producing hybrid rice. The left column shows the process of maintaining a line of male-sterile rice. When crossed with the male-sterile line, the maintainer line preserves the gene for male sterility, thus providing a continuing supply of the male-sterile plants for further breeding work. To produce the hybrid seed, the male-sterile line must be crossed with another line to restore fertility (as shown in the right column). It is this cross that produces the F_1 seeds possessing hybrid vigor (heterosis): the F_1 seeds grow into plants that produce more of the grain (i.e., the seeds) that are eaten as rice. In the meantime, the maintainer and restorer lines are perpetuated through ordinary self-fertilization. The original Chinese chart is reproduced from Hunan sheng nongye ju and Hunan sheng nongye kexueyuan, eds., *Zajiao shuidao shengchan jishu wenda* (Questions and answers on technology in hybrid rice production) (Beijing: Nongye chubanshe, 1977), 9.

produce more male-sterile plants that could then be crossed with the restorer line to generate the desired F_1 hybrid seeds (figure 14). Drawing on Mao's contribution to dialectical materialism, Yuan reasoned that the emergence of the male-sterile line was "the result of the movement of contradiction," and that the "maintainer line and sterile line were two sides of the same contradiction."[5] In framing his research questions this way, Yuan was following common practice not only in scientific research in general but also specifically in hybridization work. Indeed, an article in 1971 applied the theory of contradiction to the three-line method of producing F_1 hybrid sorghum, and just a month before the publication of Yuan's own article, a group in Anhui published an article that applied the same theory to the exploration of a two-line method in hybrid rice.[6]

One other article associated with Yuan dates from the Mao era, but as with many other aspects of Yuan's Mao-era work, it involves a few levels of mystery. The article, ostensibly dating from 1974, appears in a recently published volume of Yuan's collected works and is said to have originated in *Hunan Agricultural Sciences* (Hunan nongye kexue). However, the article cannot be found there, or in any available database, or anywhere on the web other than in the collected volume itself. But whether or not the article in fact was published in a scientific journal, the political phrases used in the article's introduction do strongly suggest that the text was written in or around 1974.

The second and more important oddity involves the article's claim that research into hybrid rice had been stymied by the "traditional" theory that self-pollinating plants such as rice do not display hybrid vigor. This appears to be the first account of the alleged conflict, which has been widely repeated in biographies about Yuan in the postsocialist era. Unlike many later references, in this article Yuan provides a source for this "traditional" theory: a "foreign" book titled *Principles of Genetics*. However, I have not found any articles in Chinese science journals that suggest that this theory was commonly referenced in, much less had a paralyzing effect on, Chinese breeding theory or practice. Moreover, by the mid-1960s, hybrid sorghum—another self-pollinating plant—was already being widely popularized. I will explore this mystery further below; for now, what is worth noting is that in 1974 political culture, it was useful to be able to identify a Western antagonist against which to set Yuan's research achievements.

Until late 1976, the research on hybrid rice received scant attention from popular news sources such as *People's Daily* or *Scientific Experiment*, and it appeared surprisingly seldom even in the national scientific journals. Moreover, although by 1974 the production of first-generation hybrid rice seed was already at the evaluation stage, no mention was made to the US Plant Studies Delegation that visited that year. However, provincial-level journals and archival materials make clear that hybrid rice research was proceeding rapidly and involved research institutes and extension systems around the country. Articles published in 1972–1973 discussed progress made in Hunan, Beijing, Sichuan, Guangdong, Heilongjiang, Anhui, Liaoning, Fujian, and Guizhou.[7] A document in the Guangdong provincial archives from 1973 emphasized the priority placed on two new important agricultural technologies: biological control of insect pests and hybrid rice.[8] Even in the far northwestern province of Xinjiang, seeds from male-sterile rice plants had been imported from Hunan in 1971, and by 1974 researchers in Xinjiang had succeeded in producing F_1 hybrid seeds through the three-line method.[9] And research activity was not limited to provincial-level institutes: for example, in 1975 Zhao'an County in Fujian reported the development of a new variety of early-ripening hybrid rice produced through the three-line method.[10] The involvement of research institutes representing not only numerous provinces but also county-level organizations was highly characteristic of the specific style of scientific research developed during the Mao era, which involved the massive mobilization of human resources.[11]

Where was Yuan in all of this? Although he was not mentioned by name, his school was routinely recognized for its importance in pioneering the research on hybrid rice, and there is no doubt that Hunan's special place in the record owes first and foremost to his efforts.[12] Still, the story that emerges from the Mao-era documents when considered on their own is not that of the lone heroic scientist that we will encounter when we turn to Yuan's reform-era biographies. True to the ideals of Maoist *tu* science, the Mao-era story avoids celebrating individual scientists—though occasionally Yuan's peasant-student assistant Li Bihu might be mentioned by name—and emphasizes instead the collaborative nature of the work and the large number of people participating.[13] Even in 2012, when I interviewed peasants about the history of hybrid rice, Yuan Longping's name did not emerge until I brought it up. They recognized the name because of his later fame, but it was clear that he was not an important part of the story of hybrid rice when they were participating in it.[14]

As a 1972 article in the *Bulletin of Agricultural Science and Technology* emphasized, the development of effective hybrid rice technology was conceived as part of a larger "mass hybrid breeding scientific experiment movement" that had already made great strides in corn and sorghum. Noting the many provinces already engaged in the work, the article paid special attention to Hunan, which by then had more than one hundred research sites including at the brigade level with more than eight hundred people participating in fourteen thousand experiments on four hundred varieties of rice.[15] In 1976, a book produced by a government office in Zhejiang titled *Actively Test and Extend Hybrid Rice* made a special effort to spread the credit across three provinces in its summary of the research achievements to date: "It was Qianyang Agricultural School that raised the research topic, Pingxiang Agricultural Science Institute [in Jiangxi Province] that created the infertile line, and

Guangxi Academy of Agricultural Sciences that found a more ideal restorer line, thus realizing the three-line complete set."[16] Further characteristic of Mao-era agricultural science was the integration of research and extension. The articles published in this early period covered a range of issues, from identifying and crossing lines in order to breed suitable varieties, to producing seed, testing the varieties in different places, and developing effective cultivation practices.

And of course, if the story were told from the perspective of Mao-era sources, the production of hybrid rice would be as much a political struggle as a technical achievement. Some materials placed hybrid rice research and extension thoroughly in the context of ongoing political campaigns. Like Pu Zhelong's 1975 polemic (see chapter 2), an article from Guangxi in 1975 used hybrid rice to forward the Campaign to Criticize Lin Biao and Confucius, accusing Confucians of having inhibited science, and celebrating the support for science shown by their rivals the Legalists. It went on to criticize scientists who went through the "cold door" (i.e., the isolated, elitist path) of science in picking esoteric research subjects like frog vocalizations or eight-legged pigs. In the past, the authors said, science courses followed the foreign model in selecting lilies as the premier example for teaching plant reproduction. Now some teachers were choosing instead an example relevant to production, the three-line method of hybrid rice.[17] A 1975 publication, *How to Plant Hybrid Rice*, packed full of slogans from every conceivable political campaign, ended in no uncertain terms with the warning that class enemies would certainly attempt to disrupt hybrid rice research and production.[18] A special issue of *Guangxi Agricultural Sciences* devoted to a conference on hybrid rice similarly called attention to the "two-line struggle," alleging that people who complained about the extra trouble involved in hybrid rice production were following the capitalist road. The solution was to intensify political education to show the relationship between hybrid rice on one hand and, on the other, the proletarian mass line, the criticism of Lin Biao and Confucius, the fight against the theory of innate genius and the Western slave mentality, the need to squash superstition and liberate thought, and the struggle against the cowardly and lazy worldview of the diehard conservatives.[19] When that same year Deng Xiaoping fell from power for the second time, he became another target in political struggles across the country, including the struggle for hybrid rice.[20]

As told from Mao-era sources, the story of hybrid rice was a perfect example of green revolution cultivated in red revolutionary soil. *Tu* science overshadowed *yang*: the heroic achievements of individuals, especially intellectual individuals, had no place there. Rather, science was collaborative, involving peasants as well as scientists and integrating theory and practice, research and extension, breeding and cultivation. Moreover, contrary to the technocratic vision of the US green revolution, hybrid rice was in these Mao-era sources far more than just a technical achievement. Instead, it was part and parcel of every other aspect of the ongoing social and political revolutions that defined Chinese socialism under Mao.

Chairman Hua Hitches His Star to Yuan Longping's

After Mao died in September 1976, politics in China underwent a series of profound shifts that transformed the story that could be told about the invention of hybrid rice. The first change was the emerging emphasis on Yuan Longping himself. Yuan's rise to stardom was extraordinarily sudden and rapid, and responsibility for this lies with Hua Guofeng. When Mao died and left Hua Guofeng in command, Hua faced enormous challenges: not only did he have very large revolutionary shoes to fill, but he had competitors, led by Deng Xiaoping, waving the attractive flag of science and technology. Hua is typically remembered as the chief of the "whateverist" faction—clutching at power for two tenuous years by promising to uphold "whatever" Mao said and did at a time when people all over China clamored for change. Yet some scholars have pointed out that the policy differences between Hua Guofeng and Deng Xiaoping were minimal; Deng triumphed over Hua not because Hua lacked vision, but because as a relative latecomer to the party, he lacked Deng's political connections.[21] There is no doubt that in education, culture, and economic policy Hua embraced the same commitments to modernizing reform for which Deng and other moderates had been advocating since the early 1960s. But agriculture, where Hua's experience was particularly strong, offered the potential for an interesting distinction: Hua was in a perfect position to grasp what was most modernizing about the revolution and what was most revolutionary about modernization.

Hua thus scrambled to present himself as a champion of both Mao's revolutionary line and scientific modernization, upholding mass science alongside the Four Modernizations for which Deng and others had been arguing since the early 1960s. Hua had been party secretary of Hunan from 1970 to 1976, and he had been given responsibility nationwide for agricultural development. The story of hybrid rice was one of a handful of exciting Hunanese examples of agricultural mass science that Hua could use to bolster his image. In this new story, Yuan Longping became an ideal hero, with the "Gang of Four" as villains and Hua Guofeng as a kind of party godfather. Yet even with the introduction of the heroic scientist, Hua-era publicity on hybrid rice painted the research carefully within the lines of Maoist *tu* science, upholding the principles of mass mobilization and collectivism.

Yuan's introduction to the public came in a December 1976 article in *People's Daily* titled "How Hybrid Rice Was Successfully Cultivated"—the same moment that Hua himself became associated with the story of hybrid rice.[22] Typical of articles during this period, it opened with a denunciation of the Gang of Four and a warm welcome to Chairman Hua. The story of hybrid rice began in 1964, when Yuan Longping, his student Li Bihu, and others, responding to the pressing needs of the revolution and building on the success of hybrid corn and hybrid sorghum, determined to research hybrid rice. (The article continued to refer to "Yuan and the others" throughout—Yuan had not yet become an isolated genius.) Their decision "immediately gave rise to a debate in the world of agricultural science," as some people "brought out the foreigners' genetics, quoting chapter and verse," and said things like "Rice is a self-pollinating crop, so hybridizing it won't produce vigor. Proposing this topic reflects ignorance about genetics." But "Yuan and the others" were undaunted by this appeal to the "authorities," drawing inspiration instead from Mao's essays "On Contradiction" and "On Practice." Nor did they lose heart when faced with the ravaging effects of typhoons, disparaging comments by conservative-minded people, or sabotage of their experiment fields by class enemies. Instead, they continued their grueling search across several provinces for a male-sterile plant. "With the impetus of the Great Proletarian Cultural Revolution, uniting with a large number of breeding workers and the broad masses of poor and lower-middle peasants from many provinces and regions, working morning and night, relying on their own strength, using our country's plentiful rice resources, they collected thousands of rice varieties and materials, combined them in millions of ways, and conducted innumerable hybridization experiments." In 1970, Li Bihu discovered a wild male-sterile plant, which provided the basis for many more experiments in provinces all over China, and in 1972–1973 research teams in Jiangxi, Hunan, and Guangxi successively bred male-sterile lines and paired them with maintainer and restorer lines to create a successful three-line technology. The rest of the article turned away from "Yuan and others," emphasizing instead how quickly the research progressed because of the "guidance of the party" and "socialist collectivism," as manifested in the establishment of "three-in-one" research groups involving cadres, masses, and technicians. The key was to move the research from a small project involving just a few people to a massive coordinated effort, where "one place's research results and breeding material very quickly became everybody's shared resource." Hybrid rice research became a "mass scientific experiment movement." And of course the authors highlighted the support hybrid rice had received in the early 1970s from "the party center, state council, and Comrade Hua Guofeng."[23]

By the time *People's Daily* next published on the subject the following April, Hua's role in the story had expanded and the political drama had heightened. The article began with a flourish: "Chairman Mao's line steered the course; Chairman Hua will not lead the ship astray. Satellites ascend to the heavens with red flags waving; scientific experiment begins a new symphony." The masses of Hunan were reportedly recalling Hua's long support of science and technology and celebrating his rise to national leadership. Hua was credited for supporting the "four-level agricultural scientific experiment network" perfected in Huarong County and the successful production of bacterial fertilizer using "local methods" in the city of Changde. And he became a main character in the triumph of hybrid rice: At a conference in Hunan in 1971 (*sic*, 1970), Hua reportedly saw an exhibit about the results of Yuan's research and personally awarded a prize to Yuan's research team, after which he directed that the research should proceed according to the spirit of Chairman Mao's teaching that "in all work, we must organize mass movements." Thus did the entire province mobilize breeding technicians together with cadres and peasants in a mass movement to tackle the remaining problems. They reportedly made great progress before Lin Biao and the Gang of Four began "blowing a cold wind," saying that scientific research was the "white-expert road." But Hua was said to have stood up for Mao's revolutionary line, and the National Conference on Agricultural and Forestry Science and Technology in May 1972 decided that hybrid rice research would be a national research focus.[24] An even more elaborate account appeared the following year. In that story, the "black storm" unleashed by Lin Biao and the Gang of Four descended on Yuan's school, and he became the target of political struggle as a representative of the "white-expert road" and as a "capitalist technology swindler." Some people even tore up the plants he had painstakingly cultivated over the years. But in 1970 (here the date has been corrected), Hua saw the exhibit on Yuan's research at the conference. When Yuan mounted the stage and received the award from Hua, the audience erupted in "thunderous applause."[25]

With each source drawing from its predecessors, the story became more highly elaborated and certain key elements cemented in place. The 1977 celebratory book *Chairman Hua in Hunan* devoted an entire chapter to hybrid rice authored by the Hunan Academy of Agricultural Sciences. Consistent with Cultural Revolution politics, it explained that while Yuan's research began in 1964, at that time the influence of Liu Shaoqi's "revisionist line" was still strong. This meant that only a few people were pursuing the research in a "cold and sterile" way, and progress was slow. The Cultural Revolution swept away all that "poison" and allowed science to "bloom" again. When Hua got behind the research in 1970, the province was able to mobilize all available resources, including the four-level agricultural scientific experiment network, the agricultural science academy, the agricultural schools, and most importantly the masses of poor and lower-middle peasants. Class enemies "quoted chapter and verse" from foreign genetics authorities to discredit the work, but all over the country cadres, masses, and technicians worked together to move the research forward. The success of hybrid rice research won the attention not only of Hua but also of Chen Yonggui, the champion of the supreme agricultural model, Dazhai Brigade. Hua stood up again to the Gang of Four by holding a big national conference in Guangzhou on hybrid rice. With "capitalist scholars" still debating the merits of hybrid rice, China put it into production, leaving the foreigners in the dust.[26]

In the reporting on a 1977 national conference on hybrid rice, we see clearly the distinctively Hua-era rhetoric of upholding the mass line while scapegoating the Gang of Four and waving the Four Modernizations flag. Beginning with his efforts to support Yuan Longping in 1970, Hua is said to have "encouraged science and technology personnel to give three-line rice research over to the masses to do, to develop this research from a small number of specialists' experiments to a new phase of a thousand armies and ten thousand horses." The article claimed:

> Science and technology personnel and the broad masses have cultivated a lofty ambition to speedily achieve the Four Modernizations. They are committed to "self reliance and energetic efforts to enrich the country" and to "squashing superstition and liberating thought." And they have criticized the Western-slavish and conservative mentality that assumed "if foreign countries could work on hybrid rice for years without success, we will not be able to succeed either." Using Chairman Mao's philosophy as their guide, they are committed to practice over theory, arduously scaling scientific peaks, and using the clear facts of hybrid vigor in rice to vigorously refute the outdated genetics theory that "rice is a self-pollinating crop, and so hybridizing it will not produce vigor." . . . Hybrid rice research and extension is a powerful criticism of the "Gang of Four's" counterrevolutionary crimes of meddling with the science and technology world; wrecking the revolution, scientific research, and production; and bringing chaos to the party ranks and to ideology.[27]

Yuan himself authored a few articles during the Hua era, and these too credited Mao's line for the success of the research. In an article he coauthored with his two peasant-student assistants, Li Bihu and Yin Huaqi, Yuan claimed to have been inspired by Mao's famous essay "On Contradiction" to understand the "internal cause" of male sterility in rice; this helped him overcome any doubts raised by foreign geneticists, whose assumptions that rice would not display hybrid vigor resulted in their failure to inquire into the internal nature, rather than merely the outward appearance, of the organism. Yuan credited the research with reflecting just about every Maoist value in the book: mass mobilization, self-reliance, socialist cooperation, and courage in going their own road (an old slogan associated with the split from Soviet revisionism). And the new rice had reportedly won the support of that crucial segment of the population, the poor and lower-middle peasants, who dubbed it "revolution rice."[28] How much this rhetoric reflects Yuan's own perspective at the time is impossible to know for certain, but a few clues suggest that his interests rested far more in the technical details of the research and less in its political significance. The article he published in 1966 was devoid of political rhetoric, and in another 1977 article he authored, we find Maoist language about mass mobilization attached somewhat clumsily to the beginning and end, while the meat of the article presents no evidence that the research involved the "masses" at all.[29] However, as we will soon see, Yuan has much more recently testified to the inspiration he found in Mao's "On Contradiction"; this ideal of the Mao era has survived the political transformations that have changed so many other elements of the story of hybrid rice.

Rewriting the History of Hybrid Rice in Postsocialist China

The materials on hybrid rice published since the rise of Deng Xiaoping have a complicated dance to perform. When Hua's brief administration ended, Deng Xiaoping did not modestly claim to be keeping China's "ship" from going astray; instead, he steered it in a dramatically new direction, to a place where the values of *tu* science would be far more muted. The story of hybrid rice had to become one of triumph that brings glory to the Chinese nation and the Chinese Communist Party, without seeming to give credit to the utterly discredited Cultural Revolution.

The vessel for this new story is Yuan Longping himself, as presented in a veritable industry of Yuan Longping biographies that borrow liberally from one another and so create a strong dominant narrative about the invention of hybrid rice. The biographies have retained some elements of the Hua-era story, jettisoned others, and added a few new ones to tell a tale in tune with the new political and scientific values of postsocialist China. Yuan is the key plot element, while the network of three-in-one organizations are of little significance, and the "masses" figure mostly as impoverished peasants whose hunger inspires Yuan to help China produce more rice for its people.

Marxism and some of the politically safer strands of Mao's philosophy are claimed as key influences on Yuan's Mao-era research, and credit is given to the party as a whole for its invaluable support. At the same time, the story of hybrid rice follows the template of Cultural Revolution sagas—with heroes and villains in all the expected places, and strikingly little mention of the vast agricultural science network that supported hybrid rice—such that success appears to come in spite of, rather than because of, Cultural Revolution–era approaches to the organization of agricultural science.

The inspirational role played by Marxist philosophy, especially Mao Zedong Thought, represents a key continuity between the Hua-era and post-Hua stories of hybrid rice: these are the crucial narrative elements that ensure criticisms of the Cultural Revolution will not undermine the legitimacy of the socialist state itself. One biographer who has written three books on Yuan has in each case included an entire chapter on "the nurturing influence of party thought" or on Mao Zedong Thought more specifically. He characterized Yuan as "not just a famous breeding expert, but also a natural philosopher who brings Marxism-Leninism-Mao Zedong Thought closely together with hybrid rice research."[30] Another biographer highlighted the influence of "On Practice" and "On Contradiction," along with Engels's *Dialectics of Nature.*[31] And yet another book includes an appendix with an essay by the Hunan Hybrid Rice Research Center party committee secretary on the use of contradiction theory in Yuan's research.[32] We have seen already that in his Mao-era and Hua-era writings, Yuan framed the search for the maintainer line in terms of contradiction. Whatever else of Yuan's political rhetoric was necessitated by the demands of political propagandists, in this case—if we are to believe his oral history—Yuan appears to have been sincere. He sees himself as "less politically sharp than other people" but still greatly influenced by Mao's "On Contradiction" and "On Practice." He explained, "With respect to 'On Contradiction,' internal contradictions actuate the motive force in the development of all things. Hybrid vigor is simply the hybridization of two varieties that are genetically different. Only when there is contradiction is there vigor. . . . In addition, with respect to whether rice has hybrid vigor, it was demonstrated through practice that it has vigor, and then we increased [our understanding of this] on the theoretical level, and then used that theory again to guide practice. This is the thought method of 'On Practice.' "[33] His testimony on the influence of "On Practice" is somewhat less convincing than for "On Contradiction"—in part because there are no corresponding Mao- or Hua-era materials on "On Practice," and in part because the reasoning Yuan provides seems vague and far-fetched. But the continuity of emphasis on "On Contradiction" is striking: it remains a part of the story because it probably was truthfully an early inspiration for Yuan, and also because in postsocialist China it represents Maoist philosophy in a way that shores up the party's legacy without threatening the new political-economic order.

Despite the emphasis on Maoist philosophy in these materials, the story of hybrid rice reinforces the dominant narrative on the failures of the Cultural Revolution. In postsocialist accounts, what saved Yuan and his hybrid rice from Cultural Revolution radicalism was the leadership of good party officials. In a 2001 *Guangming Daily* article cited in one popular biography, Yuan put it this way: "Some people cannot understand how China could make a breakthrough in hybrid rice in such a time as the 'Cultural Revolution.' There are numerous reasons, I think. For example, sincerity and solidarity between colleagues, close collaboration between people in various areas. However, what I want to emphasize is how the party nurtured this wonderful flower— hybrid rice."[34] These are the bright threads that run through each of the accounts. Yuan became the target of criticism in the early days of the Cultural Revolution because he had published in a professional journal and because he had once dared to suggest that Mao's famous "Eight-Character Charter" for agriculture needed one more character: timing. But he was saved from the "cowsheds" that served as prisons for so many other intellectuals by the intervention of an official in Beijing who had read his article and recognized the value of his research, ordering not only that Yuan should be left alone but even that he should have some funding.[35] A few years later, Yuan was sent to work in a coal mine to "temper" himself and reform his thought. He was saved again when Li Bihu and Yin Huaqi—who came from poor peasant backgrounds and thus were politically secure—sent a telegram directly to the State Science and Technology Commission, which sent a representative to western Hunan to investigate. The students reportedly served him dinner and tearfully told of Yuan's great virtues and sufferings. The result was that Yuan was quickly able to return to his research. Soon after, the provincial authorities began taking more notice and moved Yuan and his assistants up to the more central and well-supplied Hunan Academy of Agricultural Sciences.[36]

The focus on party officials as the stewards of the hybrid rice project represents a continuity with the Hua-era narrative, but with a key difference. One of the most politically transparent changes to the story is the waning role of Hua Guofeng and the celebration instead of Deng Xiaoping. It seems that the chief person who continues to recognize Hua rather than Deng is Yuan himself. Hua was called to write the preface to Yuan's oral history in 2008, just a few months before he died. Among Yuan's numerous appreciative comments about Hua, including a touching account of their reunion in 2006, Yuan states in no uncertain terms: "If it weren't for Hua Guofeng's support, the extension of hybrid rice would not have been so fast. Over all these years, I have held deep feelings of gratitude for Hua Lao [a term of address reflecting affection and respect]."[37] In contrast, Yuan's biographers tend to refer to Hua without fanfare or even to neglect him entirely. For example, a 1990 biography told the story of Hua's encouragement of Yuan in 1970 without crediting Hua at all, instead just referring to "a provincial leader." But the biographer favored Deng and his technocratic platform with reference to a little ditty attributed to "a peasant in Hunan": "In resolving our food problem, we have relied on 'two Pings.' We rely on Deng Xiaoping (for the responsibility system) and we rely on Yuan Longping (for hybrid rice)." The parentheticals were added by the biographer, who further explained, "This means that first they rely on policy and second on science."[38] The quintessentially technocratic "two Pings" formulation appears frequently in Yuan biographies. Biographers also tell of a peasant in Hunan who worshiped Yuan, built a statue to honor him, and hung a couplet at his house that read, "To get rich, rely on Deng Xiaoping / For a bountiful harvest, rely on Yuan Longping." The former revolutionary secretary of the Hunan Hybrid Rice Research Center has further extended the more common rhyme "In revolutionary transformation don't forget Chairman Mao, in getting rich don't forget Deng Xiaoping" to read, "In getting rich don't forget Deng Xiaoping, in eating rice remember Yuan Longping."[39]

Along with the shift to emphasizing Deng has come a shift to downplaying or even criticizing collectivism. While the Yuan biographies do periodically recognize the contributions to hybrid rice research by other people and in other provinces, they often take pains to emphasize that the work was done under Yuan's direction, or that the others knew nothing about hybrid rice until Yuan taught them.[40] Some of this may be attributed to the stylistic expectations of the biography genre, which inevitably emphasizes the achievements of the individual over group efforts or structural forces. However, in the specific context of postsocialist China, the story of hybrid rice has a much higher political charge: to uphold the rightness of Deng Xiaoping's new course for the Chinese political economy. Thus, it is not surprising to find negative comments about collectivism sprinkled through these sources and very few moments of recognition for the scientific experiment movement and the agricultural extension system, which were structurally rooted in the collectivist system.[41] For example, the term "collective" (集体) appears only twice in the 2002 book *Yuan Longping*: in one instance, Yuan was compelled to eat in the peasants' "collective dining hall," where the food was "fit only for pigs"; and in the other, he was sent down to live with peasants while mining coal and leading "a militarized collective life."[42] While collectivism is blamed for causing a lack of motivation among the peasants, the reforms of the late 1970s are said to have "unleashed peasants' positive action and liberated production power."[43] (Here again Yuan's own account is somewhat more faithful to the Hua-era reporting: in his oral history, he acknowledges the research advances made by other provinces and even by other countries, and he emphasizes that China's success in hybrid rice relied on "collective strength" and was established on a foundation of "socialist cooperation.")[44]

In postsocialist materials, the mass scientific organizations that appeared so frequently in Mao-era sources are now sidelined or downright disparaged. In 1967, Yuan is said to have urged his young assistants not to join any of the Red Guard groups then forming; he pointed out that they already belonged to a "mass organization," Anjiang's Male-Sterile Rice Research Group, which by this time had provincial-level sponsorship.[45] In another biography, Yuan is even said to have told Li Bihu that since theirs was a government organization with science funding, what need would there be for "some kind of mass organization"?[46] In the Mao era, such a comment would have been traitorous; it is striking to see how much the political values have changed. And in fact, the biographies could easily have characterized Yuan's group in Cultural Revolution terms: the involvement of youth in experiment groups was standard practice during those years. In the postsocialist biographies, however, there is no sense that Yuan's experiment group was part of a larger "revolutionary movement" then sweeping the country.

Finally, while the Mao- and Hua-era sources emphasize the thorough integration of research and extension, the postsocialist biographies rarely mention extension. There is one moment in Yuan's oral history where he expresses enthusiasm for the revolutionary energy involved in the extension work for hybrid rice. He says, "It was really a mass movement. The cadres had their demonstration fields, soldiers had 'battle readiness' fields, women had March 8th fields, youth had team fields [跟班田], old peasants had 'hand down the classics' fields [传经田], students had 'study agriculture' fields, agricultural science organizations had model fields."[47] However, such discussions of extension are few and far between, and like the biographies, Yuan's oral history devotes far more space to criticizing Cultural Revolutionary approaches to organizing science, especially those closely tied to extension work. For example, he complains about the system of "point-squatting" (蹲点, see chapter 5), which required scientists to spend long hours into the night talking politics with the local production team, and the pressure he received from local cadres to research sweet potatoes instead of rice.[48] Through a combination of omission and criticism, the key structures and principles of Cultural Revolution–era agricultural science have lost their role in the history of hybrid rice.

Between Geneticists and Lysenkoists

Historians of science will be especially interested in the way Yuan's story intersects with the infamous conflict between geneticists and Lysenkoists in China. Here again we see the impact of shifting political contexts on the narrative of scientific discovery and academic debate: the enduring quality of some of the struggles in the Mao era, along with how their stakes have changed in recent decades. Most Yuan biographies include some version of the episode in which Yuan was chastised for being "ignorant about genetics" based on the verdict of foreign authorities that self-pollinating crops such as rice do not display heterosis when hybridized. As one biographer wrote, "Yuan threw down a gauntlet in front of famous international authorities and their conclusions. The famous American geneticists Sinnott and Dunn's 1950s–1960s American college textbook *Principles of Genetics* asserted that rice is a self-pollinating crop and so has no heterosis when hybridized. . . . This was forbidden territory, a kind of 'Thunder Lake' [雷池, a famous lake that people historically did not dare to pass]. But Yuan Longping believed that his own theory was proven scientifically on the foundation of scientific knowledge and possessed strict internal logic. He would break through that forbidden territory, pass that Thunder Lake."[49] According to another biographer, Yuan's courage in this area represented something of the Maoist revolutionary spirit against the ivory tower: "It is just as Chairman Mao said, youth are the least conservative in their thinking. . . . Yuan Longping resolved to break through the taboo erected by the academists [学院派]."[50]

And it is not just the biographers who continue to foreground the significance of this classic *tu/yang* conflict. In Yuan's oral history, he again pointed specifically to the "famous American geneticists Sinnott, Dunn, and Dobzhansky's *Principles of Genetics*," and the book even included a photograph of Yuan's copy of the book in Chinese translation with the key passage underlined.[51] From this one passage, Yuan concluded, "the academic world of crop genetics generally held an attitude of denial about the phenomenon of hybrid vigor with respect to the rigidly self-pollinating crop that is rice." He went on, as he had previously, to refer more broadly to the allegedly stifling "theoretical perspective of 'no heterosis.'"[52]

It is a bit suspect that what is repeatedly painted as a widely accepted assumption among geneticists is represented by just one example. While influential, the Sinnott, Dunn, and Dobzhansky textbook did not reflect a consensus on heterosis in self-pollinating plants among Western geneticists.[53] To what extent this notion was so powerful in China that it impeded research on hybrid rice is hard to say. The book was certainly a key influence on at least some very important Chinese geneticists who received their educations in the 1930s, 1940s, and 1950s.[54] However, aside from Yuan's own writings, the journal literature does not turn up concern over limited heterosis in self-pollinating plants; even in 1962, at one of the high points for Western genetics in China, Chinese scientists published on the successful hybridization of japonica and indica strains of rice with resulting heterosis.[55] Moreover, sorghum is a self-pollinating plant, and by 1962 Chinese researchers were reporting on their development, building on research done internationally, of a three-line method of hybridizing sorghum; not long thereafter, the state began aggressively extending the technology throughout sorghum-producing provinces.[56] No hagiographies have emerged to celebrate the "inventor" of hybrid sorghum; nor do we hear stories about people opposing such research. A 1971 article on the "mass movement of hybrid breeding" explained that for a long time it had been difficult to get hybrid vigor out of self-pollinating crops, but recently some people had found male-sterile sorghum plants: the process of hybridization resulted in a 30 to 50 percent increase in production. Moreover, Xin County in Shanxi had in the past year mobilized fifty thousand poor and lower-middle peasants and had found male-sterile stalks of wheat, millet, and other plants. The article then mentioned the Qianyang School in Hunan along with a few other places that had made contributions to achieving hybrid vigor in rice.[57] Nothing in the article suggested that the prejudices of Western-trained geneticists had stifled this research.

Clearly, the national climate was not hostile to researchers seeking methods of producing hybrid seed from self-pollinating plants, and hybrid rice research was at the time perceived as one part of a larger research effort in this area. Thus the idea that hybrid rice research was "forbidden territory" is undoubtedly exaggerated. Still, it seems likely that some geneticists who used the Sinnott, Dunn, and Dobzhansky textbook in school could have taken it as a kind of bible. If in fact a few such people "quoted chapter and verse" in discouraging Yuan's research, this would lend some support to the criticism—very much emphasized during the Mao era, but generally pooh-poohed today—that Chinese scientists were apt to idolize Western genetics texts in a way that inhibited research progress.

If all we knew of Yuan was that he had overcome his colleagues' slavish mentality with respect to Western genetics, we might associate him with the Lysenkoist school. Lysenkoism is without question the most reviled of all efforts to transform science in a communist country—so much so that historians of science have had to struggle to wrest attention away from the negative example of Lysenkoism to arrive at a fuller and fairer understanding of science in the Soviet Union.[58] Lysenko and the man who inspired his research, Michurin, maintained the neo-Lamarckian position that individual organisms adapt to their environments and then pass on their new traits to their offspring. In contrast, scientists in the West had by the 1920s revived Gregor Mendel's research into heredity and were increasingly confident that an individual organism could not pass on acquired traits: species underwent change through selection (natural, sexual, or artificial) of individuals with the most advantageous or most preferred traits. The Lysenkoists had the upper hand in socialist China until 1956 and continued throughout the Mao era to pose a serious political challenge to the Mendelian geneticists.[59] All subjects related to breeding presented opportunities for debate between Lysenkoists and Mendelians, and research on hybrid vigor in crops was no different, as evidenced in many journal articles from the period.

But since the end of the Cultural Revolution, Lysenkoism has been roundly criticized in China as elsewhere; it now symbolizes the bad, old scientific politics of the Mao era. And so a new narrative is woven into the story of hybrid rice, in which Yuan originally encouraged his students to pursue Lysenkoist experiments but came to doubt the validity of Lysenkoism when the new varieties failed to live up to expectations. In 1962, Yuan reportedly read about Watson and Crick's work on DNA in the party publication *Reference Information* (参考消息). Demonstrating his intellectual courage and independent spirit, Yuan is said to have traveled all the way to Beijing to consult the geneticist Bao Wenkui, who explained the problems with Lysenkoism to him and encouraged him in his pursuit of hybrid rice.[60] Later, Yuan had the opportunity to pass this knowledge on to a local party official back in Hunan; Yuan reportedly defended Mendel and Morgan, arguing that it was unfair to slap "capitalist hats" on them.[61]

The triumphal disproving of an assumption about hybridization associated with Western geneticists—from Harvard, no less—served Yuan well in the *tu* political context of Mao-era China. In the postsocialist era, antielitism has largely disappeared from discourse on science, but nationalist pride continues to make Yuan's victory over Sinnott and company worth retelling over and over. However, to underscore Yuan's opposition to discredited aspects of Mao-era science, biographical materials highlight his encounters with the failures of Lysenkoism and his personal efforts on the side of the geneticists. In the process, as we saw with Pu Zhelong's biographies, *yang* has once again trumped *tu*.

The "Intellectual Peasant"

The overall picture painted in biographies of Yuan is clearly one of *yang* science. We learn that Yuan was the child of intellectuals, and that even in the re-

FIGURE 15. Yuan Longping posing with his violin in a rice field. Xu Jingsong, "Yuan Longping ling yige wenya de aihao: La xiaotiqin" (Another of Yuan Longping's cultured pastimes: Playing violin), *Yangshi guoji*, 22 May 2007.

mote hills of western Hunan, he retained a sense of himself as an intellectual. Like Pu Zhelong's memorials, every book-length biography cited here describes Yuan as an avid violin player; he brought his violin to western Hunan with him and is said to have especially loved to play romantic melodies such as Schumann's *Fantasia* (figure 15).[62] In keeping with the heroic individualist narratives familiar in "bourgeois-scientific" societies around the world, biographies consistently characterize Yuan as a lone scholar in the wilderness, boldly going into uncharted waters. In the face of much doubt and hostility, he makes his discoveries by himself or in the company of his few supporters.

As we saw in Pu Zhelong's biographies, with the "masses" and their "mass scientific movement" no longer holding water, where earthy nativist *tu* values appear, they are embodied in Yuan himself. So, for example, Yuan is said to have emphasized experiment and study in the fields rather than classroom learning. He made do with meager equipment and even committed to "learning from practice and learning from peasants."[63] One biography reports that in his early years of research at Anjiang, he used discarded earthenware pots from a kiln factory to grow his seedlings—a specific example of "self-reliance" and "frugality" commonly seen in propaganda materials from the time. However, the biography does not mention these or any other Mao-era slogans; rather, the episode is reframed to fit better with postsocialist values. Resonating with Deng-era injunctions to "lay down a family fortune," Yuan's use of the pots is depicted as helping his family by avoiding dipping too far into his savings to support his research. In another episode, Yuan's colleagues encourage him to spend more time on his family's private vegetable garden and less on his breeding work. Consistent with the post-Mao narrative on the stifling of science during the Cultural Revolution, their cautionary advice is attributed to fear of the political consequences of conspicuously pursuing scientific research. Yet the comments sound strikingly like those often encountered in Cultural Revolution–era writings, where they would have been said to represent the efforts of "class enemies" to undermine mass science and further the "capitalist road."[64]

Another example of *tu* science that makes its way into the biographies is Yuan's employment of many simple "native methods" (*tu banfa*, 土办法) in hybrid seed production.[65] Echoing decades of mass-science rhetoric, Yuan is said to have maintained, "We technicians should learn from farmers, for they have rich farming experience." One biography suggests that Yuan greatly admired the peasant technique of placing a stick in the ground to use as a sundial while working in the fields. More importantly, Yuan adopted a fertilization method from local peasants: two people holding a rope between them would walk across a rice paddy, allowing the rope to graze the ears of the rice to release the pollen.[66]

Consistent with a strong Mao-era expectation of scientists, biographies frequently praise Yuan for his willingness to get down and dirty in the field.[67] One biography applauded his willingness to walk barefoot through manure-laden fields. And in a touching moment, it reconstructed a conversation between Yuan and a peasant. The peasant asks him why "cultured" people like Yuan would want to live alongside "us muddy-legs" (泥腿子). ("Muddy-legs" was a common Mao-era term for peasants, a way of taking an elitist epithet and reclaiming it as a badge of honor.) Yuan replied that if it weren't for "us muddy-legs," the world would surely starve.[68]

However, Yuan is by no means painted as just a peasant. His earthy *tu* side is always balanced by examples that highlight his status as an intellectual. One biography characterized him as an "intellectual who grew up in a big city but embraced a simple life deep in the desolate mountains."[69] Another called him "an intellectual who still retains a peasant's manner," explaining, "He did not put on airs or seek special privileges."[70] Yuan affirms he is an "intellectual peasant" (有知识的农民).[71] A 2002 biography captures this image well in a scene said to take place in 1973, when Yuan was just beginning to see the results of hybrid vigor in his test plots. One night he pulled out his violin, brought it out to the fields, and gave a performance, not "under the bright lights of the stage," but "in front of the boundless golden grain," and in front of the "muddy-legs." He is said to have felt that his music "belonged to the 'muddy-legs,' belonged to the living grain he loved so well."[72] The violin marks him unambiguously as an intellectual, but the contrast between the "big city" and the "desolate mountains," the "bright lights of the stage" and the "muddy-legs," recalls the classic Mao-era demand for scientists to descend from the ivory towers and get dirty in the fields.

Despite Yuan's own cultivation of a "peasant" persona, some present-day Chinese Maoists have objected to the glorification of Yuan, and they have sought to highlight instead the contributions of "pure peasants" (纯正的农民). For example, on a Maoist website, Li Zhensheng is said to have had no education, and so could not write scientific articles, but nonetheless succeeded in hybridizing rice in 1967, eight years earlier than Yuan Longping. And, as a reader of the website commented, there was also the "native expert" of hybrid rice, Lin Ruoshan.[73] Subsequent chapters will discuss such stories more fully. Here suffice it to say that their research, worthy or not, did not solve the problem of mass production of first-generation hybrid rice seed. Others had used hybridization to create new varieties of rice in the past, but this did not allow for the wide-scale tapping of hybrid vigor.

Maoist critics of Yuan's celebrity are on firmer ground when they highlight the very real mass-science base of Yuan's own work. Yuan's two long-time assistants, Li Bihu and Yin Huaqi, did not just materialize from thin air to play their supporting roles. Like many of the other students at Anjiang, they came through a nationwide program for rural youth intended to raise the educational level of rural communities. Despite its many destructive aspects, the Cultural Revolution presented rural youth with greater educational opportunities than they had ever seen.[74] Another of Yuan's most important collaborators, Luo Xiaohe, was also a peasant. According to one biography of Yuan Longping, Luo had always loved farming and never thought it a hardship. He spoke to Yuan of his childhood reveries: "I thought of the land as a big bed. When I was tired from working, I could just lie down on this bed and take a big nap, surrounded by crickets playing their music for me. So beautiful!"[75] After completing secondary school, Luo took the exam for the provincial agricultural academy, whence he graduated in 1962. He went on to win many awards in the postsocialist era for his work with Yuan. Interestingly, Luo Xiaohe also refers to himself as an "intellectual peasant"—the movement in his case is the reverse, since he came from a peasant family into the ranks of science.[76]

Questions about what kind of scientist Yuan is have emerged also in the controversy over whether Yuan deserves election to the Chinese Academy of Sciences. Yuan was nominated three times, but each time he was turned down. Instead, he was elected in 1995 to the Chinese Academy of Engineering. This has angered some of Yuan's supporters, who feel that his accomplishments have not been properly appreciated. However, the larger question here is what constitutes "science." In Mao-era China, there would have been little question that developing a technology for hybridizing rice counted as science. But with the embrace of more international standards, the criteria have turned from practice to theory, and from *tu* to *yang*. Hence the continued emphasis on Yuan's triumph over Sinnott: it suggests Yuan has made a contribution to basic genetics research and not just a technical innovation. This is the kind of claim scientists need to be able to make if they are to hold their own in the realm of *yang* science.

Tu Science Meets Pedal Steel

Unlike Pu Zhelong, Yuan Longping did not participate, as either host or guest, in the Mao-era international scientific exchanges. The 1974 Plant Studies Delegation from the United States came and went without ever hearing of hybrid rice research, although by that time Yuan and others had already achieved significant results with the breeding process he had developed. Norman Borlaug was later under the impression—and I know of nothing to the contrary—that the first time foreigners learned of the scientific breakthrough was when he returned in 1977, this time with the CIMMYT wheat studies delegation. Although the journals of the delegates do contain notes from a presentation they heard on the subject at the beginning of their trip, it appeared not to have made a particularly sensational impression on them at the time. And in any case, Yuan himself was apparently not highlighted, though he did travel to the Philippines that year to discuss hybrid rice at the International Rice Research Institute.[77]

When Yuan entered the world of international scientific exchange, the arena was undergoing a dramatic transformation. In May 1979, US industry giant Armand Hammer took fifteen executives from his corporation Occidental Petroleum to China. They discovered through "casual conversation" that China had developed a method for producing hybrid rice. This led to an exchange agreement between Occidental's agricultural subsidiary Ring Around Products and the newly formed China National Seed Corporation: China would get hybrid cotton, and the United States would get hybrid rice.

The episode represents a quirky intersection of communist and capitalist forces characteristic of Cold War history. As China was transitioning to "market socialism" and beginning to open the doors to US capitalist inter-ests, who better to make this early overture than Armand Hammer—son of a prominent American communist, named for the emblem ("arm and ham-mer") of the Socialist Labor Party, and one-time resident American capital-ist in the Soviet Union under Lenin's New Economic Policy.[78] Hammer met Deng Xiaoping at a rodeo in Houston during Deng's famous introduction to the United States in 1978, and within months Hammer had booked his China trip. China's embarkment on the Four Modernizations seemed to Hammer to herald opportunities for American business like those of the Soviet Union under Lenin in the 1920s.[79]

The exchange was so exciting to Ring Around Products that they made a film about it, with footage of Yuan Longping at home and work in Hunan. As with so many other Western visitors to China then and throughout the cen-turies, the filmmakers saw themselves as voyaging into a strange and exotic land. Titled *From the Garden of the Middle Kingdom*, the film begins by plac-ing rice research in a quintessentially traditional Chinese context—panning across old Chinese paintings with zither music in the background as the narrator quotes classical Chinese poetry and speaks of the role of "religious practice and mysticism" in rice farming. However, its treatment of Yuan's ex-periences in the Cultural Revolution reflects the politics of 1982, which were starkly different from those in play in the 1970s when Pu Zhelong received delegations of foreign scientists to witness China's revolutionary mass-based approach to science: the narrative is entirely textbook, with clips of intellec-tuals in dunce caps resolved by the trial of Jiang Qing. The film then explains Ring Around Products' discovery that China had mastered the production of hybrid rice and presents footage of Armand Hammer and his colleagues at a Chinese banquet, where their hosts offer toasts to "friendship" and "world peace." The next segment depicts two Chinese rice breeders who came to the United States to help launch Ring Around Products' hybrid rice proj-ect. Footage of a major US metropolis accompanies a lively pedal steel music track; then we see two men in classic blue suits of Mao-era China riding bi-cycles through a quintessentially US landscape. Their existence in the United States is meant to appear hilariously incongruous. The Chinese in their blue suits are presented as rubes—naive babes just entering the modern, capital-ist world (figure 16). And in some ways this was accurate: new to intellectual property law, China did not get the best of the deal with Ring Around Prod-ucts, which gained the right not only to grow hybrid rice in the United States but also to market it internationally.[80]

David Livingstone has argued that the "regional geographies of scientific endeavors" are typically erased in the process of claiming the universal valid-

FIGURE 16. Three stills from the film *From the Garden of the Middle Kingdom*, dir. Kenneth Locker (Armand Hammer Productions, 1982).

ity of the knowledge they produce.[81] But this film suggests something quite different. We see socialist Chinese science both marked as "local" through the emphasis on quaint national-cultural characteristics and also portrayed as traveling very well—its inescapable, endearing Chineseness apparently does not prevent the knowledge from being brought across the ocean intact to serve "world peace" (and capitalism). Chinese scientists wear Mao jackets and ride bicycles even in the land of steel guitars and pickup trucks—but their science is sound. As a Berkeley scientist interviewed in the film recognized, they "know everything about rice that there is to know." However, according to this 1982 portrayal, their knowledge comes not from the sociopolitical context of Mao-era China—which the film portrays as solely destructive of science—but from a kind of "ancient Chinese wisdom" that transcends history. We see no mass mobilization, no recruitment of peasant technicians, no revolutionary self-reliance, no revolutionary meaning of science whatsoever. The significance of the sociopolitical context for the production of knowledge in socialist-era China has been erased.

This is *tu* science through a thoroughly colonialist gaze, appropriated by the corporate United States, missing the revolutionary politics that made *tu* a challenge to capitalist science. How quickly, tragically *tu* became disempowered in the hands of global capitalism — disempowered both at the symbolic level (the "cuteness" no longer has any power to provoke or inspire) and at the practical level (China got the short end of the stick in the exchange with Ring Around Products). Perhaps this fate was not the fault of *tu* science itself, but merely a small part of the larger victory of liberal capitalist technocracy. But I suspect that in its replication of the binaries of colonialism and modernity, the *tu*/*yang* configuration of science was doomed to be co-opted by wielders of global economic power, whose position the twentieth-century revolutions in the end had utterly failed to vanquish.

Conclusion

Like Pu Zhelong, Yuan Longping embodied the tensions of postcolonial subjectivity, manifested particularly in the struggle to be simultaneously *tu* and *yang*. Pu Zhelong's connections to US science were profoundly significant for both his entomological practice and his sense of personal identity; yet as a specifically agricultural scientist committed to socialist values of practical application and mass mobilization, Pu cut a convincing figure when he took off his shoes and adopted a more *tu* persona. Yuan had no opportunity to study abroad and so missed the chance to form the kinds of transnational networks that Pu enjoyed. But despite his very different experiences, his status as a professional scientist meant that he too was *yang* enough in the Mao era to incur the kinds of attacks so often directed against intellectuals. When the tides turned in the post-Mao era, it was no longer clear whether Yuan was *yang* enough — the focus in biographical materials on Yuan's violin playing (which we see also in post-Mao depictions of Pu) and on his supposed contributions to genetics theory suggest an effort to make up for gaps in Yuan's intellectual profile. At the same time, his biographies frequently evoke the nostalgia many feel for certain aspects of the Mao era — including the humbleness, sincerity, humanity, and Chineseness of *tu* science. The positive epithet "intellectual peasant" is a clear echo of the Mao-era call to "raise *tu* and *yang* together."

As it has been constructed in the post-Mao era, the history of Yuan Longping and hybrid rice is in some ways a testament to the lasting resonance of the *tu*/*yang* concept. But there is a difference between celebrating *tu*/*yang* as symbolized by the "intellectual peasant" Yuan Longping and acknowledging the importance of *tu* science more broadly—the integration of research and extension, the three-in-one model of grassroots research, the emphasis on mass mobilization—in the development of hybrid rice technology. The latter is far harder to find in postsocialist accounts. The next two chapters will pursue a critical reading of socialist-era documents and publications to explore scientific farming as it was experienced in rural communities.

Chinese Peasants: "Experience" and "Backwardness"

No social group was as important to Mao and the Chinese Communist Party as the peasantry, which defied the predictions of more orthodox Marxists to propel the Chinese communist revolution to victory in 1949. The Chinese word for peasant (农民, literally "rural person") denotes not just an occupation, but rather a social class and identity.[1] Official reports and propaganda accounts from the 1960s and 1970s typically emphasized this by referring not simply to "peasants" but more specifically to "poor and lower-middle-class peasants": these were the people with the political credentials on which the state sought to capitalize. The term also has something of a deprecatory feel to it in Chinese that the English word "peasant" captures well. The sense of pejorative was reinforced by the occasional substitution of other names, such as "muddy-legs" and "hick" (泥腿子 and 大老粗). Propagandists thus tried less to erase the stigma of being a peasant and more to claim that stigma as power.[2]

Scientific and political elites themselves also often benefited from claiming peasant status. For example, during the Cultural Revolution, foreign visitors learned of the "peasants" Hong Qunying and Hong Chunli who in 1956 developed China's first variety of dwarf rice. In fact, these "peasants" were the party secretary of the production brigade and an agricultural technician.[3] And as the previous two chapters showed, scientists like Pu Zhelong and Yuan Longping had to work hard to prove that their feet were as dirty—and thus their claims to authority as unimpeachable—as those of their peasant colleagues. Such claims worked in two directions: every time someone identified as a "peasant" (even if he was in fact a cadre or scientist) achieved something important, he helped buttress state propaganda about the "genius" of the peasantry; at the same time, claiming peasant status helped protect cadres and intellectuals against accusations of being "white experts."

In addition to its class connotations, the term "peasant" was associated with old age. Like everyone else, peasants come in all ages, but young peasants who had attended secondary school lost something of their peasant identity. The contrast between "educated youth" and "old peasants" (who need not be so very aged in actual years) helped reinforce the idea that the peasantry was in essence "old." The term "old peasants" (老农, *laonong*) was used already in the late imperial era to designate farmers with valuable experience, and literati bent on agricultural improvement knew enough to consult such people.[4] Figure 17 represents a Cultural Revolution–era version of this theme.

On its surface, the intent in including peasants in scientific experiment groups was simple enough: the state sought to emphasize that peasants could

大学办到咱山村

FIGURE 17. In this propaganda poster, a university professor and students consult with an old peasant, identifiable as such by his head covering and waist sash. Note the clearly dominant position of the old peasant, who stands squarely facing forward, while the professor and students are clustered around him. The sign stuck amid the rice plants says "Experiment Field." Hong Tao, "Daxue ban dao zan shancun" (The university has moved to our mountain village) (Beijing: Renmin chubanshe, November 1976). Stefan R. Landsberger Collection, International Institute of Social History, Netherlands, http://chineseposters.net.

and should contribute to agricultural science. But there were a number of distinct ways that this could be configured. To some extent the idea was that old peasants, by virtue of their age and class position, already had expertise that could be tapped to promote scientific farming. Sometimes this even extended to the notion that old peasants possessed a body of knowledge that had been handed down over the generations, but more often old peasants were said to have personal experience that could be brought to bear. Even this was often insufficient if the goal was to bring new technologies rapidly into production. In such cases, peasants were more likely to be applauded for quickly mastering new technologies developed in research centers, and sometimes they were celebrated as innovators of such new technologies.

These were all very different ways of recognizing peasant participation in agricultural science, and they had different implications for the politics of knowledge. At stake here were some of the most potent and politically risky questions the new state faced in mobilizing peasants; they were questions deeply embedded in the decolonial and class politics of Maoist *tu* science. How should "traditional" knowledge be viewed? What kind of *wenhua* (文化, literally "culture," but more broadly "education" and "knowledge") did peasants possess? And how should peasant society be transformed? These dilemmas and their consequences for peasants and rural Chinese society are the subjects of this chapter. Exploring these topics will further shed light on a central scholarly concern about modernization and knowledge: Does the introduction of new agricultural technologies result in "skilling" or "deskilling" of farmers?

The Construction of "Old Peasants" and Their Knowledge

Beginning in the early 1950s, articles relating the experiences of "old peasants" appeared in Chinese scientific journals; often the experiences were synthesized at conferences where old peasants came together to discuss specific subjects, from expanding sunflower oil production, to managing late-ripening wheat, to preventing frost damage in rapeseed plant.[5] In 1960, *People's Daily* reported on a county in Jilin Province that had developed an "old peasant adviser department" to help cadres benefit from peasant advice whenever the need arose. Throughout the county, 7,250 poor and lower-middle peasants— most of them old, but some young—served as advisers to help cadres determine the best way to implement new agricultural directives from above and overcome any technical trouble encountered. Individual "old peasants" could be consulted on the spur of the moment, or a number of them could be called together for a conference on a more difficult problem.[6]

In 1965, in a speech at the Jiangsu Provincial Agricultural Science and Technology Work Conference, Jiangsu party secretary Xu Jiatun declared that peasants had "lots of valuable experience coming from many years of struggling with nature," such that they possessed a "fierce fighting character." He went on, "Their generalist experience is great in combination with specialist scientific research. Their experience suits today's level of production; it is good for getting the most out of small investments." Moreover, he considered peasant knowledge to be "suited to local conditions." Xu recognized a "difference" between peasant experience and science but insisted there was no "Great Wall" that divided them and indeed declared their experience necessary to scientific advance.[7]

Sometimes, though perhaps not as often as we might expect, old peasants were consulted not just for their personal experience but as conduits for "traditional" knowledge. During the mid-1950s, the state called upon researchers to study China's "agricultural heritage" (农业遗产). Over the next ten years, they republished imperial-era agricultural treatises and gathered traditional "agricultural maxims" (农谚) for interpretation and application to current farming conditions.[8] The content of these texts and oral transmissions presented some sticky problems for the researchers, and they were careful to place them in an appropriate political context, which included articulating what it meant to consider them for their scientific value. For example, a 1957 volume titled *Annotated Agricultural Maxims* proclaimed the need to "verify and apply agricultural maxims that accord with scientific rationality, criticize and refute those that do not accord with scientific rationality, and study the form of agricultural maxims so as to express new agricultural scientific knowledge in that form."[9] A 1963 collection of maxims from Guangzhou celebrated agricultural maxims as "a kind of living agricultural science textbook with practical significance and reference value for the broad peasantry (especially young peasants and all the rural cadres who directly lead agricultural production) in critically carrying on traditional agricultural experience"; moreover, it maintained that agricultural maxims constituted a "folk oral literature" whose study could provide the intellectual and aesthetic education necessary for creating new maxims.[10]

The 1957 volume further emphasized that agricultural maxims are bounded by their class character, geographical place, and historical time. It highlighted, for example, the maxim "Raising pigs does not turn a profit, but wait a bit and look to the fields" (养猪不赚钱, 回头望望田), which suggests (in keeping with widespread practice) that the chief importance of pigs is the benefit their manure brings to the fields. The authors offered the following analysis: "At first glance, this agricultural maxim is full of rationality, but it nonetheless represents oppressive experience spit from the mouths of the landlord class, because landlords don't themselves participate in labor. In order to raise pigs, they must hire a laborer, and adding up the wages and feed, raising pigs makes no money. Only if you add the fertilizer collected and calculate it all, does it become profitable. So how do peasants talk about it? Peasants say: 'Poor people are never far from pigs, rich people are never far from books' [贫不离猪，富不离书] and 'Raising pigs is neither costly nor difficult, with spare change you can piece together whole money' [养猪不费难，零钱凑整钱]. . . . To raise pigs you need to buy a bit of feed with your spare change; when the pig grows big you sell it, and then you can get a chunk of money, which in practical terms means putting money into savings [储蓄]."[11]

Place dependence amounted to little more than differences in timing based on climate (e.g., wheat ripened thirty-three days after the frogs started calling in Shandong, but it was forty-five days in Anhui). However, changes over time also carried political meaning—in particular as they related to the collectivization of the economy and the advance of agricultural science and technology. So, for example, one maxim advised, "If you select early seeds, the next year you'll feed the birds" (选种选得早，来年饲雀鸟). The authors explained that in the old days under the private economy, if someone selected the seeds for the next year's planting early in the season, their crop would ripen earlier than their neighbors' and so would become the target for birds. However, in the era of collectivization and the policy of eliminating the four pests, this was no longer a problem.[12] The adage "Plant wheat close together, but for cotton leave room for an ox to lie down" (麦子宜稠，棉花地里卧下牛) was similarly outdated: close planting of cotton required more labor than was available in the private economy, but this was no longer an issue under collectivization. And now that there were effective means of killing insect pests, peasants no longer had to follow the maxim "If last year locusts swarmed, this year plant black beans and cotton" (上年蝗虫闹成灾，今年多把黑豆棉花栽).[13]

A 1965 book published by the China Youth Press, *Scientific Rationality in Agricultural Maxims*, similarly warned that, while some agricultural maxims were completely in accord with science, others were no longer applicable because of changes in the social system and the development of science and technology. It used the maxims as a means of introducing youth to concepts in agricultural science: each maxim served as a springboard for a more general discussion of a topic. For example, the method of using natural enemies to control insect pests was conveniently introduced with the adage "One animal conquers another; praying mantis conquers poisonous animals [i.e., pests]" (一物降一物，螳螂降毒物). Similarly, the maxim "When corn loses its head, it becomes strong as an ox" (玉米去了头，力气大如牛) was a perfect introduction to the use of male sterility in the production of hybrid vigor. And the book offered a fuller discussion of the maxim "If you select early seeds, the next year you'll feed the birds": it explained that early ripening should generally be adopted as a worthy goal when selecting seeds, though weather, soil, and other factors would also need to be considered according to the principle of "suiting local conditions" (因地制宜).[14]

When the Cultural Revolution shook up professional scientific activities, agricultural maxims continued to be of interest, though sometimes in different ways. In 1974, Cultural Revolution radicals mobilized the entire country in a campaign against Confucius. In contrast to the Confucians, the rival Legalist school was praised for having emphasized the collection of agricultural maxims and thereby promoted the development of agricultural science and production.[15] During the 1970s and even as late as 1981 popular media published articles on youth who collected agricultural maxims from old peasants—a demonstration of the value of combining the experience of old peasants and the scientific advancements of educated youth.[16] One such case appeared in a 1974 article in the popular magazine *Scientific Experiment.* When a rural youth graduated from middle school and returned to the countryside, he was assigned to a weather station where poor and lower-middle-class peasants observed leech behavior to forecast the weather. At first he had a negative attitude and did not realize how much he could learn from the peasants. Then came a day when the youth carelessly lost his leeches. He found a new leech, but his next forecast failed. A peasant explained that there are three kinds of leeches, and he had collected the wrong kind. The youth then realized that old peasants had a wealth of experience watching weather patterns: he visited more than eighty old peasants to collect their knowledge of observing animals to predict weather.[17] The enthusiasm for youth learning agricultural lore from peasants was so strong, the author of a *People's Daily* article once rashly assumed that the publication of excerpts from a six-hundred-year-old agricultural text, titled *The Farmer's Five Phases* (田家五行), arose instead from the work of an educated youth who had learned from old peasants, collected agricultural maxims, and then made his own observations in the fields.[18]

A few specific technologies rooted in long-standing peasant practice proved especially interesting to Mao-era agricultural researchers. I will save the most important, fertilization, for in-depth discussion below, but several others deserve mention here. In 1960 and 1961, the state made a big push to promote intercropping—the practice of planting multiple crops in a single plot. Newspaper articles emphasized intercropping as a system long used by

Chinese peasants to promote soil fertility, withstand natural disasters, and achieve high yields. Scientists sought to "synthesize" peasant experience in this area, both "traditional" and deriving from more recent experiences; they further consulted imperial-era Chinese agricultural texts for insight into past precedent.[19] The 1965 volume *Scientific Rationality in Agricultural Maxims* introduced intercropping with the adage "If you want to be rich, make your crops into a variety store" (若要富，庄稼开个杂货铺): it dutifully noted the offensive landlord-class character of the saying, but upheld the scientific validity of mixing different crops together.[20] The interest in intercropping lasted through the Cultural Revolution and was a key element of socialist Chinese agriculture displayed to foreign visitors in the 1970s (during a 1977 trip to China, "father" of the green revolution Norman Borlaug lay awake one night at 2:30 a.m. worrying about how intercropping could work in a modernized agricultural system).[21]

Mao-era agricultural extension agents I interviewed in Qinzhou and western Guangxi recall inviting old peasants from the Chaoshan region of Guangdong every year to impart their expertise in "scientific farming."[22] Chaoshan was—and remains—famous for its style of "intensive cultivation" (精耕细作), which Mao had promoted in his 1957 essay "Be Activists in Promoting the Revolution."[23] In 1965, *People's Daily* reported that twelve thousand "old peasants from Chaoshan" had been invited to communes all over Guangdong to set up "Chaoshan-style demonstration fields," in which the intensive cultivation methods were modified to suit local conditions.[24] Another interview subject encountered the old peasants of Chaoshan when she was a sent-down youth during the Cultural Revolution. They were promoting a weeding technology that involved both hands and feet: by going on all fours, they would first loosen the clay with their toes then use their hands to pluck out the weeds. It was arduous, but the promise was increased yields to the tune of 100 to 200 extra *jin* per *mu* (about 600 to 1,200 pounds per acre).[25] Chaoshan peasants also inspired a critical innovation in rice breeding technology when a scientist, Huang Yaoxiang, witnessed "old peasants" from Chaoshan adopting unusual cultivation and fertilization practices to stunt the plants' growth and so prevent "lodging" (i.e., falling over). This inspired Huang to pursue the breeding of a semi-dwarf variety of rice.[26] Today, celebrations of Chaoshan's "sustainable" approach to farming focus on the rich cultural traditions and clan networks that undergird "intensive cultivation" practices.[27] In 1965, however, *People's Daily* was careful to refer not to traditional cultural forms but rather to an "accumulated set of relatively complete experiences" that integrated sowing, transplanting, fertilization, and irrigation practices.[28]

Discomfort with the idea that peasants might contribute "traditional" forms of knowledge to scientific farming was not limited to the framers of state ideology. In his visit to China with the US Plant Studies Delegation in 1974, China historian Philip Kuhn expressed relief that peasants were not being tapped out of some faith in "the formalized and sacrosanct body of rural lore."[29] For agricultural technicians sent to the countryside, concern with the "backwardness" of traditional knowledge was of far more immediate and personal concern. Indeed, interviews conducted today make clear that the celebration of "old peasants" as experienced and knowledgeable operated in very narrow constraints and was outweighed by far more negative perceptions on the part of state agents and even peasants themselves. When I interviewed Mao-era agricultural technicians and cadres, I always asked about the participation of old peasants. Typically, they would dismiss the possibility that old peasants were involved in scientific farming and emphasize that peasants did not have the *wenhua* (that is, education) to be useful in this way; but when pressed, some would identify one or two specific areas in which old peasants contributed to scientific farming.

For example, one former production team leader declared, "most of us didn't have much faith in" local, folk (*tu*) veterinarians. But then he acceded that because horses had been used for transportation in that area for "thousands of years," on this "one subject" alone local veterinarians could be trusted. Otherwise, "only Western veterinarians would do, local veterinarians would not do." At another point in the interview, when pressed about the supposed involvement of "old peasants" in three-in-one teams, he said, "Most were young peasants. Old peasants didn't usually participate. . . . Every year when we made the production plan we needed a lot of old peasants to come and do it with us. But as far as spreading pesticides and chemical fertilizer, old peasants didn't participate in these advanced technologies. . . . In planning old peasants played a key role because they knew how to plant specific plots, which plots had lots of water, which were dry, which varieties were better in dry land, etc. At those times the role of old peasants was very big. But they didn't participate in the agricultural science group."[30]

At another site in northwestern Guangxi, I conducted a group interview with seven Mao-era agricultural extension specialists. When asked whether local peasants had any traditional knowledge that they contributed to scientific farming or whether there were any *tu* experts (that is, peasants recognized for their expertise) in the area, they said, "Not around here." They agreed that in more "developed" places—they specifically cited the Chaoshan region of Guangdong—old peasants might have valuable knowledge, but not in "backward" areas, especially places like northwestern Guangxi with large minority nationality populations. I heard similar ideas expressed many times over the course of my travels in Guangxi. Taken together, such comments help explain why it is so difficult for interview subjects to identify peasant contributions to the scientific farming movement. Knowledge and technology were (and remain) understood to be by definition the antithesis of the traditional and the peasant. Peasants, especially those of minority nationalities, were (and remain) seen as "backward." The ubiquity and shrillness of Mao-era propaganda insisting that "the lowliest are the smartest" should be taken as evidence of just how pervasive the opposite perspective in fact was.

On the other hand, when my collaborator Cao Xingsui and I were eating dinner at a rural restaurant, Cao pointed to some clusters of tangled thorny branches hung from the roof of one of the kitchen buildings. He asked the proprietors about them and learned that they served to discourage rats. Cao asked his assistant to take a picture of it, suggesting it should be documented for the National Agricultural Museum in Beijing. He observed that it was a good method because using poison was "inhumane" (不人道). This display of appreciation for local knowledge emerged when I did not ask directly about it: direct questions tend to trigger automatic responses about backwardness, a phenomenon common when interviewing on topics where dominant perspectives are strongly cemented.

But what of the traditional agricultural maxims that the state so encouraged youth to collect from old peasants? Cao Xingsui remembers the great emphasis on the maxims and even participated in the project himself. However, he emphasizes that it was not the agricultural science group but rather the culture group (文化组) that organized the collection of agricultural maxims.[31] Members of the agricultural science group got information on the weather, when to plant, and other important questions from the technicians at the commune agricultural station, not from the sayings of old peasants. Members of the cultural group put on performances, taught peasants to read, and collected the sayings to document the historical accumulation of peasant culture. Today, an eight-year-old girl I met in eastern Guangxi tells me she has learned about agricultural maxims about weather prediction in her course in humanities (or "culture," 文科).

If the inclusion of agricultural maxims during the Cultural Revolution in *wenhua* groups and more recently in humanities courses represents state efforts to recognize peasant "culture," this priority has been overwhelmed by the more powerful conviction that peasants are backward and peasant knowledge either deficient or downright suspect.[32] It is hard to know whether peasants during the Mao era perceived themselves as "experienced" or "backward," but today it is striking to hear how widely it is accepted among peasants that they lack the *wenhua* so essential to science and modernity. One former production team leader told me about the time that the government installed a water pump for irrigation. At first, he and the other peasants had little interest in it. They did not know how to use it and figured the government "might as well take it and give it to someone else." Only after an agricultural technician came and showed them how to use it did they realize it was much more efficient than the water wheel they had been using before.[33] His wife also spoke of how much assistance peasants needed from agricultural experts. In the past, she said, they operated just by slowly figuring things out on their own. Before Liberation only rich and middle peasants could go to school, so poor peasants like them had no education. Agricultural experts brought much-needed knowledge. This is a common kind of exchange in rural China today; it is evidence of how accepted, and how powerful, the narrative of backwardness and development is. Movements now spreading globally to recognize farmers' "indigenous knowledge" were not tangible in the exchanges I had in Guangxi in 2012, though as the epilogue will explain, they are deeply influential in other Guangxi circles.

"Peasants Grasping Agricultural Science"

While it requires some effort to dig up examples of state interest in the existing forms of agricultural knowledge possessed by peasants, no such effort is needed to find examples in propaganda materials of peasants, young and old, learning new technologies and thereby contributing to agricultural science. In 1975, party leaders in Huarong summarized the overarching attitude well. They warned about the dangers of assuming that "peasants know how to farm, so all leaders really need to do is take a little interest." This represented a relaxing of party leadership in agriculture and opposition to the extension of advanced technologies. As an example of the need for intervention, the "four-level network" in Huarong worked to "overturn conventional practices" in rice seedling cultivation, introducing in their place new technologies that worked much more effectively to prevent the shoots from rotting.[34]

The political mandate to celebrate peasant contributions to science thus did not mean embracing traditional methods and adopting a laissez-faire attitude toward farming. Rather, it meant actively cultivating peasants to transform them into "peasant technicians," "peasant breeders," and other kinds of "new peasants" with the necessary scientific knowledge to build the "new socialist countryside." This was accomplished through education, extension, and participation in scientific experiment groups—and in rarer cases also through interactions with scientists. Through such activities, peasants were

FIGURE 18. Wang Tianjie, "Kexue zhongtian" (Scientific farming), ink on paper, 1961. The colophon indicates that it was painted at Beijing Art Institute (北京艺院). Here an old peasant (identifiable from his beard and head scarf) looks through a microscope while a young woman takes notes and a young man stands by.

expected to learn not only specific new technologies but also an attitude toward agricultural modernization and science more generally (figure 18).[35] At the same time, some older forms of knowledge were undoubtedly lost to the new generations or demeaned to the point of abandonment.

Even during the populist high tide of the Great Leap Forward, peasant contributions to agricultural science were framed far more in terms of peasants grasping new scientific knowledge and producing new scientific innovations, and much less in terms of what peasants had learned from their forebears. In 1958, a collection titled *Short Biographies of Worker and Peasant*

Innovators: "Native Experts" Rival "Foreign Experts" included several articles on Pu Zhelong's protégé, the termite expert Li Shimei, along with articles on workers and peasants developing new tools, a factory worker cum "rat-catching expert" (we will meet another such expert, dubbed the "Rat King," in chapter 5), and the "new style of a new peasant" who made many advances using innovative fertilizers and testing new varieties.[36]

In 1966, a *People's Daily* article on mass agricultural science activities in a commune in Fujian Province announced, "The era of peasants consciously grasping agricultural science has begun." It recounted that in spring 1962, a member of the commune's "old peasant advisory committee," Zhang Xiang-zao, attended a county-wide agricultural conference and obtained from a peasant breeder seeds from a new variety of dwarf rice, which performed very well compared with the rice the commune usually planted. The commune began involving old peasants as "backbones" in scientific experiment activities, and Zhang Xiangzao went from "old peasant adviser" to "peasant breeder." As old peasants became more involved in scientific experiment activities, their attitudes also reportedly underwent transformation. In the past, the article claimed, old peasants viewed science as mysterious. They looked at technological cadres and thought, "You are *yang* [possessing foreign, modern scientific knowledge], and I am *tu* [possessing only native, traditional Chinese forms of knowledge]." But by 1965, old peasants and technological cadres alike had joined the new agricultural studies group. "In the past old peasants discussed only old experiences, but now . . . they can also talk about new things. This gradually clarified our understanding of the question of whether peasants can practice science."[37] A report from a 1965 conference in Beijing similarly claimed, "Many old peasants have broadened the scope of their knowledge by participating in scientific experiment activities." Peasants who had accumulated enough scientific knowledge to become experts in breeding, pest control, or other areas reportedly earned the laudatory title "farmer-scientists" (亦农亦科) and "technicians able in both knowledge and action" (能文能武的技术员).[38]

Documents on the scientific experiment movement are full of references to formal education programs established to provide peasants with enough agricultural science knowledge to engage meaningfully in grassroots scientific extension and experiment work. During the Cultural Revolution, efforts to provide technical training for peasants only increased, through programs for young peasants at regional agricultural schools (such as the one where Yuan Longping taught), local night classes and short training courses scheduled during slack periods in the agricultural year, and technology tutoring networks to provide more ad hoc educational opportunities.[39] One agricultural expert in Qinzhou who rose through the ranks to become agricultural bureau chief testified that during the Cultural Revolution peasants were brought to agricultural stations and provided free housing and food while they studied the new technologies. Some peasants, he said, were so successful that they were promoted multiple times and ended up becoming officials.[40] Another agricultural expert in Qinzhou reinforced this, saying, "We selected local peasant technicians and paid them well to guide [the other peasants]. This was a mass movement. They had more practical experience than we did, and we had more theory."[41]

Documents produced during the Cultural Revolution display a more radical class politics than that found in the reminiscences of interview subjects today. For example, a 1975 account traced the history of agricultural scientific experiment activities beginning in 1964. In this story, the party secretary of the production brigade was said to have had the "spirit of Dazhai" and so in spring of 1964 created an agricultural secondary school.[42] However, because of the influence of Liu Shaoqi and Lin Biao, the school was allegedly controlled not by the "poor and lower-middle peasants," but by "capitalist intellectuals." The school "wore new shoes but walked the old road and practiced closed-door education, planting cotton on blackboards and running machinery on stage, so that when the students went to the experiment fields they couldn't distinguish between fruiting branches and spindling branches, between seven-spotted ladybugs [a beneficial 'natural enemy' of insect pests] and 'fart bugs' [a colloquial term for stinkbugs, an insect pest]." So in 1966 more radical-minded locals transformed the school, offering courses in politics, culture, technology, military, and labor in three types of classrooms (society, field, and indoors) and with three methods (theory linking with practice, less is more, and democratic teaching). The school reportedly graduated 1,700 students who became the backbone for agricultural scientific experiment activities for the whole county.[43] Such accounts testify to the Cultural Revolution's greater emphasis on politics relative to technical knowledge, and practice relative to theory.

Sometimes peasants were further encouraged to use practice as a means of understanding more general patterns in science. Following intensive state efforts to promote cotton cultivation and especially to encourage women's participation in it, a group of girls in Shaanxi formed an experiment group focusing on cotton production.[44] A 1966 report on their achievements explained, "In the beginning we didn't believe that we could become experts in cotton technology, or even more that we could investigate and grasp cotton growth patterns. We thought this was for science and technology personnel to do and that we would be fine doing just planting, managing, and harvesting. The possibility that we could produce innovations was even further from our minds. But when we had studied the Chairman's writings, our thinking was liberated and we realized that all we needed was to study through practice. By whole-heartedly studying from the masses, we would not only become proficient at technology but would also be able to grasp the life patterns of cotton."[45]

Some peasants greatly benefited from the opportunity to interact closely with scientists, and in some cases, these interactions resulted in peasants rising to participate directly in professional scientific activities or even become scientists themselves. Chapter 2 presented the case of Li Shimei, who came from a peasant family and taught himself to become a termite expert; during the populist tide of the Great Leap Forward he gained the patronage of Pu Zhelong and joined Sun Yat-sen University as a professor. Li Shimei was but one of many such famous "native experts" who helped demonstrate the capacity of peasants to make meaningful contributions to science. From composting to breeding to rat catching, no element of the Eight-Character Charter was imagined to lie beyond the ken of model peasants.

One of the earliest peasants celebrated for his contributions to agricultural science was Chen Yongkang. Born in 1907, Chen was middle-aged when the revolution came. By 1951, he had already earned national recognition for his bumper harvests of rice, which media attributed to his careful management of plant density and irrigation, his use of manure and nitrogen-fixing milk vetch for fertilizer, and his efforts year after year to select the best progeny from a plant he originally found in a neighbor's field.[46] The 1955 film *Seedling Cultivation* introduced Chen's "advanced experience" to rural audiences eager for such entertaining and useful media experiences.[47] In 1958, he brought a new method, called "three black and three yellow" (or, variably, "three yellow and three black"), to a national conference on increasing rice yields. This was not just a simple technique but rather the result of close observation of the life cycle of the rice plant to provide clues to proper management. Chen found that the rice plants went through clear stages in their development, and that by charting these shifts the farmer could provide just the right fertilizer and water conditions at the right times to promote maximum yield. After this, he earned the status of "special researcher" in the Jiangsu branch of the Chinese Academy of Sciences. He authored several books, and his research reached an international audience in 1964 when representatives from countries in Asia, Africa, Latin America, and Oceania attended a scientific conference held in Beijing, and again when the International Rice Research Institute visited in 1976.[48]

After the landmark National Conference on Agricultural Science and Technology Work in 1963, *People's Daily* published accounts of scientists who had been inspired by meeting Chen Yongkang. For example, the soil scientist Chen Yuping, already known among Sichuan peasants as a "dirt doctor" (泥巴医生), went down to the countryside to help a peasant-born soil specialist, Li Sifu, write up his ideas in the form of scientific articles. At the subsequent annual meeting of the Sichuan Soil Society, Chen Yuping further encouraged more peasants to become experts in this field.[49]

A later example of a "native expert" is Li Zhensheng, a peasant of Korean ethnicity from Jilin Province celebrated for proving that "hicks" (大老粗) could master science. He came to prominence in 1975 when a series of articles in newspapers and journals reported on his successful hybridization of corn and rice. Li was very much a star of his time: in 1975, the Cultural Revolution radicals had gone into high gear in their attacks on the more technocratic party leaders like Zhou Enlai and Deng Xiaoping. Li's story was thus one of many illustrating the need for vigilance against "rightists" and "capitalists" bent on stealing science back from the masses. As a peasant with only three years of schooling, Li purportedly demonstrated that book learning was less important to agricultural science than good old-fashioned farming experience.[50] The results of his research were published in an article he authored for the Chinese *Journal of Genetics*, titled "Mao Zedong Philosophical Thought Was My Golden Key in Breeding Corn-Rice."[51]

Such examples are, of course, very rare. And in the case of Li Zhensheng, there are serious doubts about the legitimacy of the "invention." Of greater importance are the effects of the scientific experiment movement on the much larger numbers of rural people who participated in scientific experiment groups and other organizations that taught peasants new technologies or even basic scientific methodology—for example, the Ningwu County Improved Variety Breeding Farm in Shanxi Province, where twenty-three "old hicks" bred sweet potatoes for local suitability and disease resistance.[52] As discussed in chapter 3, the early success of the hybrid rice program required very rapid training in seed production for vast numbers of peasants. Peasants today remember the production of hybrid rice seed as a complicated technology that was difficult to learn; often one young member of the scientific experiment group was selected to go to Hainan for intensive training and then returned to the village to lead the work. The greatest challenge was caring for the male and female plants separately and timing their development so they would flower simultaneously, thus making hybridization possible.

Even the number of peasants participating in scientific experiment groups was no more than a few percent of China's vast peasant population. And so the need for experiment groups to educate and involve other commune members became a focus in some propaganda materials. As reported from Yangchun County, Henan, in 1966, "Under the guidance of the scientific experiment group, all commune members obtain a handle on scientific knowledge so that they are not just performing agricultural tasks but also observing, and then reporting any growth abnormalities or pests to the scientific experiment group."[53] Thus, the scientific experiment movement was ideally meant to involve every peasant, transforming an entire population into scientifically competent members of the new socialist society.

The Ambiguity of Technological Transformation: Old Skills or New?

The relationship between "traditional" and "advanced" forms of knowledge was far more ambiguous than the political rhetoric implied. Chen Yongkang's "three black and three yellow" theory was in fact common knowledge in many places, though with understandable local variation and called by different names.[54] The choice to identify a single peasant to celebrate as its innovator, to codify the knowledge, and to count it as science reflected a very specific attitude toward knowledge. Under other political circumstances, "three black and three yellow" could have been identified as what Kuhn called a "formalized and sacrosanct body of rural lore," but this would have been giving too much credit to "tradition"—to the cultural fabric of the bad, old "feudal" days. Focusing instead on what Kuhn called "the innate inventiveness of the ordinary peasant" allowed for a class-based standpoint epistemology without the danger of appearing to valorize "tradition."[55]

Fertilization was perhaps the most obvious area where the wisdom of the ages could come to the fore, yet here more than anywhere else the distinction between old and new was very blurry. On one hand, China was rapidly building chemical fertilizer plants in an effort to make this technology more widely available. Historical documents and interviews with former cadres, agricultural extension agents, and sent-down youth all point to the tremendous effort expended in teaching peasants to use chemical fertilizers. Everyone seems to have a story—sometimes funny, sometimes downright scary—of peasants misunderstanding how to apply the chemicals. For example, one agricultural technician told me, "Peasants didn't understand chemical fertilizers. They fed the ammonia to the cows, because they thought the cows would grow quickly [as the fertilized plants did], but the cows died. Chinese peasants are like this. They saw where the ammonia was stored and since no one was watching it, they stole it to boil vegetables because they thought it was salt."[56] Agricultural experts saw spreading chemical fertilizer to be an "advanced technology" out of the reach of most old peasants. As Cao told me, "It was the agricultural science group's responsibility to take the tiny amount of chemical fertilizer supplied by the government and use it well. Because if used badly, it has side effects. If used at the wrong time, it actually decreases production. This is an extremely complicated scientific question."[57]

On the other hand, right alongside the efforts to expand the use of chemical fertilizers was a very explicit state commitment to organic fertilizers. During the Great Leap Forward, Mao had offered his strong support for organic fertilizers, especially pig manure, for several reasons: the destructive effects of chemical fertilizers; the other economic and nutritional benefits that come with extending pig husbandry; and his fundamental commitment to mechanization over chemicalization, since the former was far more closely tied to the move to collectivizing agriculture on a large scale.[58] This led to a national policy of "relying mainly on farmers' fertilizers [i.e., organic fertilizer], and secondarily on chemical fertilizers" (以农家肥为主，以化肥为辅), which lasted through the Cultural Revolution and remains a touchstone for sensible agricultural practice today.[59] For example, a 1974 collection of materials on the scientific experiment movement used the experience of Dazhai as an example to "criticize the erroneous thinking of only relying on chemical fertilizer." Among the benefits of "farmers' fertilizer" were its ability "to improve the soil structure, turn dead soil into living soil, and preserve moisture"; the way it helped crops absorb nutrients and promote plant growth; and the low cost and high benefits.[60]

The use of the term "farmers' fertilizers" suggests a recognition that composting is a body of knowledge possessed by farmers. What is harder to find is an explicit articulation of that idea. Scattered references exist, especially during the early 1960s. For example, in 1960 Fujian, peasants were celebrated for having over a long period of time developed a system of fertilizing that was both economical and scientific. The system, particularly suited to places with much land and little labor power, was based on an old maxim, "Fertilizer packed tightly brings perfection, fertilizer scattered easily runs off."[61] In 1961, a scientist from the Soil Institute authored a scientific article on the need for more attention to fertilizer in the drive to increase agricultural production: "In this respect, Chinese peasants have over a very long period of production practice accumulated extraordinarily rich experience in economical use of fertilizer." He reported on experiments demonstrating the effectiveness of farmers' fertilizer for adding phosphate to the soil.[62] A 1961 book on agricultural chemistry stated, "For a long time rural China has had the custom of collecting and using farmers' fertilizers, fully using many kinds of organic materials to participate in the biological cycle and thereby increase agricultural production. This is the most important characteristic of China's agricultural production." It continued, "Farmers' fertilizers have several thousand years of history in China, and Chinese peasants have deep knowledge and a wealth of experience with them."[63] In 1964, *Zhejiang Agricultural Science* published an article supporting the "traditional excellent fertilizer" *Azolla* (绿萍, water fern).[64] A 1965 *People's Daily* article praised an agricultural school in Harbin for combining study of "advanced scientific knowledge" along with "local traditional experience," highlighting as a specific example the discussion of chemical fertilizer alongside "local farmers' fertilizers."[65] That same year, another article appeared praising the Song Dynasty work *On Rejuvenating Fertility*.[66] During most of the Cultural Revolution, people were careful not to allude to "tradition" in their discussion of fertilization practices, but in 1975 an article in *People's Daily* on Dazhai's fertilizing experience spoke of "Chinese laboring people's excellent tradition of 'enriching the fields with lots of manure' and 'rejuvenating fertility'"—here they borrowed the title of the classic Song Dynasty work—which made possible the simultaneous use and enrichment of the soil.[67]

Although the "traditional" (or at least "long-time") roots of organic fertilizer technologies were sometimes acknowledged, they were more often treated within the context of "scientific farming" as technologies in need of deliberate extension by the state. In some cases, the technologies were specific to locales and so peasants in other communities required convincing and education before the technology could be extended. A clear case in point was the promotion of various types of "green manure"—plants grown specifically to be turned into the soil to raise nitrogen levels. Green manure technologies appeared in Chinese texts as early as the sixth century AD (and in fact still earlier in Greece and Rome).[68] The practice of cultivating milk vetch to enrich the soil gained the notice, and applause, of US agricultural scientist F. H. King when he toured China in 1907 and documented his observations in the highly influential celebration of Asian agriculture *Farmers of Forty Centuries*.[69] However, the technology was by no means universally adopted, and for many peasants in the Mao era, it required just as much "extension" as chemical fertilizer did. For example, in 1965 Jiang Qizhang—a member of a scientific experiment group in Guangdong—attended a conference at the commune's agricultural science station. Jiang heard about the use of the nitrogen-fixing cover crop milk vetch for fertilizer and listened to an "old peasant's" experience with the technology. Jiang encouraged the experiment group to buy milk vetch seeds to start a local trial, but met opposition from some commune members who doubted the usefulness of milk vetch and advocated using the money to buy chemical fertilizer instead. The experiment group reportedly worked to educate the commune members and in the end convinced them.[70]

Historical documents and interviews record other types of "traditional" technologies extended as "scientific farming" during the Mao era. For example, in a group interview in Qinzhou, Guangxi, interview subjects raised the practice of "sun drying the fields" (晒田) as such a technology. (Discussed as early as the seventeenth century, the practice of draining off the water in the rice paddies and allowing the sun to dry and warm the soil is currently credited with increasing oxygen levels and strengthening the rice shoots so that they resist lodging.)[71] A peasant from a poorer area of Guangxi similarly referred to sun drying the fields as a "scientific" practice, and he blamed low yields from experiments with close planting in 1965 on the failure to understand sun drying the fields.[72] Agricultural technicians in Qinzhou also highlighted the technology of mixing manure with dredged pond water, then baking it in the sun and smashing it into pieces to use as fertilizer.[73] This too may have been long-standing practice in some places but brand new in others. And in 1976, the Jilin-based journal *Rural Scientific Experiment* published a number of articles on high-temperature methods of composting. They did not celebrate composting as a "traditional" form of knowledge but rather emphasized the need for better methods of producing more effective fertilizers.

The ambiguous character of technologies promoted through agricultural extension appeared again in an interview I conducted with a former production team leader and a younger man in his team. I asked when they began to use "chemical fertilizer," and they answered that at first it was *tu* (or "local") chemical fertilizer (土化肥). Because purchasing chemical fertilizers was cost-prohibitive, they were encouraged to mine the nearby limestone caves. The "local chemical fertilizer" technology was extended in the mid-1960s. "At that time," he said, "we didn't know anything, so it was the technicians from the extension station and the Department of Agriculture who came and guided us. It was very slow work, digging it all out." I asked if people thought it was too arduous and not worth mining. "It was arduous," he replied, "and we also had to cut a lot of grass and wood in order to fire the lime—as much as building a house. At that time, things were really hard. . . . At that time, it was also possible to buy commercial fertilizer, but it was too expensive for us. . . . Production was very backward, and we peasants didn't even have enough rice to eat."[74] In this case, just because a technology was *tu* did not mean it was rooted in any local practices; rather it was as much of an outside imposition as the introduction of any other new technology, and it was more unwelcome than most.[75]

Some technologies, however, could hardly have been seen as "new" no matter where they were promoted. For example, the use of manure collected under the pig sty would not have been new, but when the March 8 Agricultural Science Group (discussed in the introduction) used this type of fertilizer it was celebrated as forwarding "scientific farming." At a breakfast meeting with agricultural specialists in Qinzhou, I asked them why using manure or compost counted as scientific farming during the Mao era. They answered that although these methods were "traditional," people were reluctant to use them because they required much effort, and so the government was compelled to emphasize them a lot in extension efforts.[76] Critics of the Mao-era political economy might attribute this reluctance to an incentive problem produced by collectivism—that is, people were unwilling to do the dirty work when the benefits would not come directly to their own families. However, peasants today continue to report both the conviction that organic fertilizers are superior and the observation that most peasants avoid the work involved in using organic fertilizers if possible.[77] This suggests another possible incentive issue: people sometimes value rest more than they value the extra yield or higher quality that comes with greater investment of labor. And that was precisely the kind of cultural orientation that China's modernizers—both radical and technocratic—sought to combat.

Transforming Rural Communities

We have seen that part of what made the green revolution so different in China compared with other places was the insistence that science and technology not be divorced from social and cultural revolution. Scientific experiment groups were supposed to be actively transforming rural society and culture at the same time that they were transforming agriculture. This did happen: scientific experiment groups and other institutions provided girls and women new opportunities to gain scientific knowledge and become recognized authorities in agriculture, while also helping push out certain stubborn elements of the agricultural market economy. But rural society and culture also exerted a force, and sometimes the state found it more expedient to work with rather than against this grain.

Other than class struggle, transforming gender relationships is the most significant aspect of social revolution that emerges from the documents on the rural scientific experiment movement.[78] Moreover, discourse on gender was more likely than discourse on class struggle to reflect actual social and cultural patterns, since by the 1960s class labels no longer directly reflected current power relationships. Despite frequent claims about class antagonism over scientific farming, it is hard to believe that by the 1960s former "landlords" and "rich peasants" really had a class-based interest in blocking the

introduction of new technologies. However, it is very believable that girls
and women who joined scientific experiment groups or otherwise began participating in new agricultural practices faced hostility from fellow commune
members attached to existing gendered divisions of labor. Of course, accounts of women practicing scientific farming served state propaganda goals
very well. Not only did they provide opportunities to trumpet social revolution, but they also helped promote women's greater participation in farming,
which was essential to the collectivist economy.[79] Given their propaganda
value, we must read such accounts extra critically. (The repetition of certain
patterns in the accounts, especially the sexist remarks attributed to "conservative" elements, raises flags about the heavy hands of the propaganda spinners.) However, the amount of effort expended on this in propaganda suggests that sexist ideas about women's participation in certain types of farming
were very well entrenched, and anecdotal evidence further suggests that the
rural scientific experiment movement really did open avenues for girls and
women. This would not be surprising, since we know from very different
kinds of scenarios that the emergence of new types of organizations—for example, in peasant rebellions—have often provided opportunities for women
and others marginalized within existing power structures.[80]

No matter how fervently state officials sought to target sexism in rural
culture, to a large extent they had to work with the terms familiar to rural
people. We see this very literally in the term "old peasants." Although technically the term is gender-neutral, in practice it was gendered male; women
would more likely be called "old ladies" (老太太). On one hand, the distinction between the terms suggests that members of the older generation with
the social capital to participate in scientific experiment groups as "old peasants" were usually men. Another way of looking at this is that old men were
in a far better position to resist new technologies than old women, so it was
old men that the state most sought to bring on board via participation in the
experiment groups. On the other hand, the intentional use of the term "old
ladies" reflected the extra political significance old women carried: when they
participated in scientific farming, they helped celebrate state-sponsored efforts to overturn not only classist but also sexist assumptions about science
and expertise.

Girls and women often participated in groups composed specifically
of girls and women. A document from the 1965 Conference of Activists in
Beijing Municipal Rural Scientific Experiment Groups offers an example of
this—and is, moreover, a rich text for interpreting state ideological investments in women's participation in scientific experiment. It tells the story of

an "old lady" with bound feet named Dan Liangyu who was selected to be a people's representative to her county in 1960. The next year at the county-level People's Congress, she reportedly heard that the state was encouraging people to plant experiment fields. When she returned she found two other women and together they planted a field of sorghum. For two years, they had poor results, and people reportedly began saying the kinds of sexist remarks so often found in such accounts: "Women planting experiment fields, that's like a toad wanting to eat swan meat!" and "With women wanting to plant on good land, then what will we need all the men for?" The other two women encouraged her to give up, and one even quit, using her small child as an excuse. Dan encouraged the other woman to persevere, saying, "Let's show them what women are worth!" Reflecting on their experiences to date and heeding the advice of "old peasants" (presumably men), the two women experimented with different approaches, including the interplanting of yams, corn, and soybeans. Some in the community doubted these methods, but the "facts schooled those cold-wind-blowing people." Their success reportedly encouraged many other women, and in 1965 with the encouragement of the local party organization, Dan formed a science and technology group with seven women. In this way, Dan earned the name "iron-foot" because even though her feet were bound, she could work like a man.[81]

Other times, women participated in scientific groups alongside men. In 1973, a women's association meeting at a production brigade in Guangdong addressed the question of how to encourage women to participate in agricultural scientific experiment. The brigade had fourteen scientific experiment groups, all of which included women and men. Women made up 53 percent of the total members, with three women serving as group leaders and six women as vice-leaders. To support these women, the meeting emphasized the need to resolve problems preventing women from participating fully. For example, old people could help watch children to free women from this domestic responsibility. Consistent with pervasive ideas about women and labor, it also highlighted the need to accommodate the "four special times" for women: menstruation, pregnancy, birth, and lactation. When menstruating, women should do "dry work" (following an idea that it was unhealthy for women to stand in water during menstruation); during pregnancy, they should be assigned light and indoor work; and during lactation, they should work close to home so they can nurse their children conveniently. Although the document noted that the women had rich experience in production, they lacked scientific knowledge. Thus, the women's association asked technical staff and native experts to teach women about various agricultural techniques—for example, identification and control of insects. The meeting also noted that when women married (which entailed leaving their villages), they would need help "carrying on the revolution" in their new communities.[82]

Whether forming their own scientific groups or participating alongside men, women practicing "scientific farming" challenged long-standing gender norms. However, the above examples also hint at ways in which cultural expectations about women's bodies and social roles continued to influence the implementation of state policy on the ground. We gain still clearer insight into the complex dance between state and rural cultures in a celebrated example of girls' participation in science from Shaanxi. In the early 1960s, a group of nine girls, ages twelve to fifteen, began working (with considerable guidance from the local party secretary) on increasing cotton yields. (These efforts were part of a long-running program, from 1956 through the early 1980s, of "Silver Flower Contests" that promoted women's participation in cotton production and led to a relatively small number of women earning recognition or even fame, while exhausting a great many more through what amounted to "badly remunerated production drives.")[83] After much reported struggle against class enemies and nature itself, they succeeded. The songs they wrote to summarize their experiences were published in a volume titled *Songs of the Nine Girls Planting Cotton* (九女植棉歌).[84]

Celebrating the achievements of these girls served a number of revolutionary agendas, both red and green.[85] Still more interesting, however, were the ways their story was channeled to satisfy cultural expectations about girls and marriage. In 1964, *People's Daily* highlighted this aspect of their work in an article titled "Marrying Out the 'Silver Flowers,'" and it was further emphasized in a 1966 collection of articles on the agricultural scientific experiment movement as a whole. According to the stories, after establishing themselves as cotton experts, the original participants went on to cultivate fifty-two more "silver-flower girls" possessed of red thought, technological capability, and the ability to labor. As the girls married into different production brigades, they each became the leader of a cotton production group. Before a girl left, the whole group would evaluate her and point out strong and weak points. They also prepared six kinds of dowry presents: a hoe, a packet of cotton seeds, a volume of selected readings from Chairman Mao, a copy of the summary of the plan, and a glass plaque with an inscription about listening to Mao's words and preserving the glorious name of the Nine Girl Group. Sometimes at key moments in the cotton production cycle, the group would call the "married silver flowers" back to their "natal families." The women reportedly always first returned to the group, then to their families; they first came to see the cotton, then to see their mothers. In articulating their work as

cultivating girls for marriage, propaganda about the Nine Girl Group was ful-
filling time-honored cultural expectations about gender roles. The attention
to "dowries" was especially telling, since the party had long campaigned vigorously against dowries as part of the larger effort to overturn patriarchal and
otherwise "feudal" aspects of the traditional marriage system. And yet with
the dowries, as with the Eight-Character Charter, this apparent inconsistency
in fact represented a familiar pattern in which the party sought to co-opt
and simultaneously transform, rather than outright battle, popular customs.
Dowries had persisted despite state opposition;[86] the Nine Girl Group's form
of dowry was at least revolutionary and further reinforced state priorities in
both the cultural sphere (attempting to replace family loyalties with loyalties
to revolutionary groups) and the realm of economics (promoting collective
cultivation of cotton necessary for the textile industry).

Further evidence on the relationship between the scientific experiment
movement and rural gender relations comes from an interview I conducted
with a woman agricultural expert who got her start as a sent-down youth in
Guangxi. The production team leader in her village assigned relatively light
tasks to the girls, and heavier tasks such as dredging the river for ammonia-
rich sediment went to the boys. When I mentioned the Mao-era sources I had
seen that described such attitudes as sexist or even "feudal," she strongly dis-
agreed. The leader, she said, was like other Chinese men who sought to "protect their women comrades." She noted that women were also not assigned to
spray pesticides because they might be lactating. But a few minutes later she
acknowledged that the leader also might not have wanted to assign women
such tasks out of fear they would leave to get married or have children—"Sex
discrimination," she said with a chuckle. The greatest impact of the scientific
farming movement on gender relations in her village came about in a much
more structural way. Before dwarf varieties were introduced, it was a poor,
rice-importing village (i.e., they did not have enough rice to feed the population throughout the year). Women would not marry into such a village, and
so there were always old men without wives. Once the new varieties raised
production, women flocked to get married in the village. "Old people said
that was the biggest change."[87]

Gender norms were not the only elements of rural society in tension with
the scientific experiment movement. Cao Xingsui (who also began his work
in agriculture as a sent-down youth) told me that in his area of northwestern
Guangxi, peasants were unwilling to eat the dwarf varieties, preferring traditional varieties—including glutinous rice, black rice, and keng rice (粳米)
that local peasants had been eating for generations. Local peasants insisted
that at least one-third of their grain ration consist of these varieties so that

there would be something good to cook for New Year celebrations and when entertaining guests. At these times, "if the family cooked the government rice [i.e., the kind the government promoted], it was so coarse, people would think the family lacked hospitality or was very poor." Another time the "government rice" absolutely would not do was when a woman had a new baby. Local people thought the government rice was not nutritious enough; only if she had keng rice would the mother and child both be healthy, and would the mother make milk to feed the baby. If she ate the government rice, her child would not thrive.[88] This kind of tension between local needs and government mandates required careful management by skilled cadres.

The historical documents also occasionally betray evidence that state efforts to transform agricultural practice rubbed up against local interests. It is easy to miss these moments. Internal reports and propaganda stories alike are so full of ritualistic language about "class enemies" that we may overlook the very real conflicts happening beneath the rhetorical level. For example, a 1965 account claimed that at one point the Nine Girl Group wanted to merge the land they had been given for growing cotton with the commune members' personal plots (自留地), but some "middle and rich peasants" allegedly objected, saying it was their "lifeline" (命根子) and that they were not willing to use it for selling cotton to the nation.[89] Whether the objections actually came only from people in these class categories is questionable. However, the account does strongly suggest that the state actively sought to promote cotton production for the national economy, and that this put pressure on peasants to cede the land that had been returned to them during the experiments with family-based farming after the Great Leap Forward.

A further example comes in an account of yet another group of women working to introduce scientific farming. In 1961, a community in Jiangxi Province experienced a major outbreak of swine disease. There were very few veterinarians available, and so some folk vets stepped forward, but they charged very high prices and the pigs still died. A local "housewife," Wu Lanxian, went to the party secretary to discuss learning to provide veterinary services herself. The party secretary encouraged her along with three other housewives to study at the county veterinary station for several months. When they came back, they opened a clinic in an unused building and hung a red paper outside with their name on it: Four Sisters Vet Station. Even before they started work, people allegedly began spreading slander and making sexist comments. Wu's mother-in-law even forbade her to come back in the house. Especially problematic was their decision to keep a boar that they could take to sows around the community for breeding. Wu's father said that it disgraced the family for three generations to have women doing that kind of work. But the real resistance came from the locals who had been performing this service in the past. The local boar keepers charged three yuan for their services, while the Four Sisters Vet Station charged only one yuan. And so the Four Sisters came into competition not only with folk vets but also with the local boar keepers, who began threatening sow owners, noting that the Four Sisters kept only one boar, and that if that boar died, sow owners who had switched allegiances could not expect to resume business relations with the locals again. The appearance of folk vets and local boar keepers in this document offers a fleeting glimpse of rural society outside of state-promoted programs. Even as some people, including women, found opportunities to learn new skills and acquire scientific knowledge, others found their livelihoods threatened. These were people living in tension with state socialism, people who continued to pursue small for-profit business by selling their services to their neighbors.[90] The interests of such groups were often at odds with those of the modernizing, revolutionary state. Such conflicting interests—along with the slippage between discourses of peasant backwardness and peasant experience—provided the foundation for the kinds of local resistance, and state responses to resistance, that will emerge in more detail in chapter 5.

Conclusion

The evidence discussed in this chapter speaks to the growing literature on the role of introduced technologies in the "skilling" or "deskilling" of rural people. The debate originates in an argument advanced by Harry Braverman that the introduction of new industrial technologies under capitalism results in the loss of skills previously possessed by laborers.[91] Historians of agriculture have made similar claims about the consequences of modern agricultural technologies. For example, Deborah Fitzgerald has argued that the introduction of hybrid corn supplanted farmers' existing knowledge of seed selection for the improvement and maintenance of preferred varieties.[92] Turning to insect control, Ann Vandeman has argued that the adoption of chemical pesticides has similarly led to the loss of more intricate ecological knowledge that supports more sustainable forms of agriculture.[93] In response, other scholars have emphasized instead the acquisition of new skills—what has elsewhere been called "reskilling."[94]

More recently, Glenn Davis Stone has applied the literature on deskilling to consider the introduction of genetically modified cotton in India. Although he finds considerable evidence of deskilling, he notes that "there is nothing intrinsically deskilling" about Bt cotton; rather, the "deskilling effects depend on local conditions" and "where conditions are different, there

are intriguing hints that genetically modified seeds may mitigate deskilling."[95]
This is a useful intervention, especially when we turn to socialist-era China,
where political, economic, and social conditions were certainly very different from those at play in the contexts where arguments about agricultural
deskilling have been developed. Indeed, arguments about deskilling have
been especially important to Marxists critiquing capitalism: What does this
suggest about the possibilities of deskilling in socialist contexts?

Jacob Eyferth has convincingly argued that the introduction of modern
papermaking technologies by the socialist Chinese state resulted in dramatic
deskilling in communities with long histories of the craft.[96] The hostility of
the socialist state to anything classed as "tradition" and its eagerness to modernize in industry and agriculture made attacks on long-standing knowledge
forms almost inevitable, and this was as true in agriculture as in papermaking. However, the politics of mass science simultaneously encouraged the
development of new skills among a large number of "peasant technicians,"
encouraged a smaller number of peasant innovators (the *tu zhuanjia*), and
even encouraged some existing knowledge forms associated with the "experience" of old peasants.

The question of whether scientific farming resulted in the "skilling" or
"deskilling" of Chinese peasants is far from simple. Chemical insecticides introduced as part of the technological transformation of agriculture have been
blamed for causing the disappearance of the historic Chinese practice of using ants to control pests in citrus orchards. On the other hand, Pu Zhelong
and other advocates of biological control practices taught a great many peasants how to breed wasps to manage insect pests. Which of these is ultimately
more important in cultural or ecological terms is difficult to say, and impossible to generalize about.

Looking only at the vision projected in state propaganda, we see a great
deal of skilling. Peasants received specialized training in a variety of new technologies, some of them quite complex—as when they learned to breed biological control agents or traveled to Hainan Island to learn the complicated
processes involved in producing three-line hybrid rice. The 1965 Beijing conference proceedings highlighted the role played by the "three fields" system
in allowing scientific experiment groups to build their skills. For example,
one group reportedly used a demonstration field to master technologies for
growing wheat from seed to harvest; moreover, they learned to distinguish
more than ten kinds of wheat and understand their characteristics and needs.
Crucially, this knowledge was said to have given them the ability to manage
the growing of improved varieties independently from technical advisers at
the extension stations.[97] A smaller number of peasants were lauded for their

own innovations and were brought to other communities to impart their knowledge.

In some cases, the disseminated technologies were rooted in long-standing practices that in other political-historical climates would be celebrated as "traditional" and/or "sustainable." However, the vitriol with which Mao-era radicals attacked anything tinged with what they called "feudalism" made highlighting the traditional roots of such practices challenging to say the least. This marks a crucial difference between the Maoist concept of mass, nativist *tu* science and the decolonialist concept of indigenous knowledge.

Skilling is what state propaganda celebrated, and it did actually exist. However, some of the luster of that vision fades when we poke below the surface and discover how many people in how many communities did not experience any of these things, how much this vision competed with a developmentalist narrative often profoundly disrespectful of existing knowledge forms, and how quickly more skill-rich techniques were replaced by the use of seeds and chemicals imported from outside the community once economics made such substitutions possible.

Whether we bemoan the loss of skills acquired over centuries within rural society or celebrate the gain of new skills through state-organized education may ultimately depend on what kind of knowledge we consider more valuable. And because, as Jacob Eyferth has shown in Chinese papermaking communities, the acquisition of skill is embedded in social organization and cultural practice,[98] our preference may also relate to our political orientation. Do we favor "traditional" societies with long histories, and so mourn the loss of skills acquired through relationships steeped in long-standing cultural practices? Or do we pin our hopes on knowledge produced through modernizing and/or revolutionary transformations? Do we see in traditional communities the vestiges of old power structures of patriarchy and class oppression, or resources for combating new power structures of global capitalism and state authoritarianism?

Radicals in Mao-era China and elsewhere have shared a hope not only that an old society can be ripped down and a new one based on more egalitarian values erected in its place, but also that the knowledge produced in that new society will be better as a result.[99] In what must be one of her most often-quoted statements, Donna Haraway says, "It is a matter for struggle. I do not know what life science would be like if the historical structure of our lives minimized domination. I do know that the history of biology convinces me that basic knowledge would reflect and reproduce the new world, just as it has participated in maintaining the old."[100] Many leftist observers saw in 1970s China the potential for a reinvention of science, but to suggest that Maoist science "minimized domination" (of nature or of fellow humans) would have been a tremendous stretch for even the most sympathetic eyes.[101] It was a "matter for struggle," certainly, but not a nonviolent one. Nor were Chinese politics stable enough, or the economy strong enough, to produce a coherent "successor science" (to borrow Sandra Harding's term) based on radical ideals. Perhaps no society complex enough to be real could. Socialist Chinese agriculture's patchwork of old and new, ecological and chemical, arduous and labor-saving technologies "reflected and reproduced" the complicated political and economic realities of the society from which it emerged.

Seeing Like a State Agent

In a 1953 *Foreign Affairs* article, international relations scholar John Kerry King wrote in relation to Chinese agriculture: "Totalitarian methods sometimes bring impressive short-term results. A Communist does not concern himself with how to induce a rice farmer to adopt new varieties of seed, or to use different methods of production, milling, storage and transport. Under a Communist regime individual preferences make little or no difference. The individual does what he is ordered to do."[1] Today such faith (as it were) in totalitarianism seems almost laughably naive, for of course it was simply not true. The Chinese state could not merely order something and expect people to obey. Between the policies of the upper leadership and the "masses" of peasants whose cooperation the state desperately needed lay the local cadres and agricultural technicians: the success or failure of newly introduced technologies depended first and foremost on their efforts to convince skeptical (or at times even resistant) peasants to get on board.

This chapter draws from documents and interviews to explore more fully the roles of local cadres and technicians in China's green revolution. Although the two groups had different responsibilities (and each of the groups was itself diverse in composition), both represented low-level state agents charged with carrying out the policies of agricultural modernization within the larger revolutionary political movements that continually swept through the country. The chapter's title plays on the title of James Scott's influential book *Seeing Like a State: How Certain Schemes to Improve the Human Condition Have Failed.* Scott focuses on exposing the weaknesses of the "high-modernist" vision that states around the world have imposed on communities in their efforts to bring complex societies and ecologies into line with scientific planning—a critique highly relevant to the Chinese state's insistence that local communities emulate national models even when local conditions made those models inappropriate. In shifting the inquiry to "state agents," I am moving away from the abstract concept of "the state" per se, and highlighting instead the somewhat different and more conflicted gaze of the people on the ground who were accountable both to their local communities and to the state apparatus above them.[2]

Local state agents sat at the hinge of the contradiction between green and red revolutions—between technocracy and class struggle, between top-down and bottom-up approaches to change. At times their position placed them in a type of "patron-client" relationship with peasants, but they also faced checks on their power from below.[3] And they encountered practical conundrums on a regular basis. Not infrequently, the tasks the state assigned simply could not work: they did not suit the local environmental conditions, or they clashed with deeply held local customs and values. To succeed, state agents had to pay close attention to cultivating personal relationships and building good will among diverse players. And they often had to be skillful in negotiating between the needs of state and society, sometimes to the extent of actively but covertly resisting state mandates.[4] Their efforts in managing poor policies from above and resistance from below help illuminate just what the "scientific experiment movement" was actually testing. The question of whether new technologies would work in a given place involved not just biology, chemistry, and physics, but also what might be considered social sciences. And of course, given that science and technology are inextricably embedded in society, it should come as no surprise that "scientific experiment" should extend to include social, political, cultural, and economic relationships. A return to Big Sand Commune will provide a capsule example to launch this exploration.

Big Sand Commune: View from the Grassroots

As we saw in chapter 2, Big Sand Commune's achievements in integrated pest control owed much to the vision of Pu Zhelong. Yet just as the story of hybrid rice involves far more than the lone scientist Yuan Longping, insect control at Big Sand involved many diverse members of an intricate agricultural science network. Chief among these was local cadre Mai Baoxiang.

Mai Baoxiang had not leapt at the chance to oversee agriculture at Big Sand when the post was first offered to him in 1969.[5] He had been "struggled against" in the early years of the Cultural Revolution and had been demoted from his position as deputy director of nearby Qingkuang Commune to become merely a worker. He was still bitter, and when the call came he stayed home for three months, refusing even to consider the transfer. But the higher-ups continued to see him as the right person for the job: he had considerable experience on the ground in agricultural extension and from 1958 to 1963 had served as the director of the Sihui County Agricultural Science Research Institute. Finally, a county-level agricultural leader came to his house by bicycle and insisted that Mai at least visit Big Sand and meet with the people there. And so Mai ended up taking on the task.

This was a job few people would relish. Big Sand had big problems. It had too much low-lying terrain amid four rivers, and the soil was too sandy to absorb it all; the high water levels—reaching more than ten feet in the winter—tended to swamp average varieties of rice. Its crops also suffered unusually from pests of all kinds—weeds, rats, and insects. (The dismal situation was well captured in a Cantonese adage "In eliminating bugs there's no good solution, so just get an early start" [除虫有乜巧，总要早动手].)[6] To make matters worse, agricultural extension was terribly underdeveloped in Big Sand, though in surrounding areas it worked quite well. Big Sand peasants simply refused to participate in the activities promoted by extension officials. And for good reason. At the time, the extension stations were following Mao's directive in advocating "intensive cultivation." But Mao had been talking broadly about strategies for feeding a large population with relatively little arable land. Big Sand's situation was different: relatively few people worked a relatively large area. Any intensification programs would have increased labor inputs intolerably.

Even today, decades removed from the political priorities of the Mao era, Mai speaks of his work at Big Sand in Maoist terms. As he puts it, he "went down to synthesize the production experience of the peasants there" and came up with "one sentence: breed a new variety based on the characteristics of the water." That is, Big Sand needed a new variety of rice that could handle the local water conditions.

He also discovered that peasants were willing to accept new agricultural technologies tailored to local conditions. They had not been willing to exert effort reducing weeds when this meant manually pulling them as part of "intensive cultivation." There was too much land, too spread out, for this kind of work. However, when Mai and his colleagues introduced the use of a fermented substance to kill weeds quickly over large areas, peasants were happy to participate. Big Sand became a leader in this practice, holding conferences that brought representatives from other areas. So Mai did not succeed in Big Sand by following the specific Mao quotation being promoted by high-level officials. Nonetheless, in seeing his role as "synthesizing peasant experience" to derive solutions to local problems, Mai was acting in concert with core principles of the Maoist vision for agricultural science.

True to the perspective on knowledge promoted by Maoist *tu* science, Mai often sought out those then known as "native experts." Wherever Mai went, he would collect "agricultural maxims" from such experts and "synthesize" their knowledge. He recalls many kinds of native experts emerging during this period, but the most vivid in his memory was one nicknamed "Rat King" because of his remarkable abilities to track and catch rats. The Rat King could tell from their footprints how big the rats were and where they had gone; wherever they went, he would follow until he had caught them.

In the summer of 1972, all conversation at Big Sand seemed to revolve around the problem of eliminating insect pests, but the more people fought the insects, the more people seemed to be spinning their wheels, and they still endured heavy losses.[7] It was then that Mai learned from the Guangdong official who had recruited him of Pu Zhelong's ability to control insects. Pu Zhelong was a *yang* (professional) science version of the Rat King—an individual possessed of a special skill, in this case "catching insects" rather than "catching rats." And so on 24 July, Mai and a local technician named Qin Yunfeng traveled all the way to Guangzhou, to Pu's home on the Sun Yat-sen University campus.

Mai and Qin related their troubles to the sympathetic ears of Pu Zhelong and his wife, Li Cuiying. The class struggle of the Cultural Revolution had not erased Mai's profound sense of the social gap between himself (as a "low-level cadre, and a peasant to boot") and these two esteemed scientists. Today he marvels at his own daring in appealing directly to Pu; but as he says, the magnitude of his problems left him little choice. He was deeply moved that Li insisted on having their housekeeper prepare a dinner for the travelers, and that at the end of the visit Pu and Li both accompanied them downstairs and out of the gate to see them safely on their way.

Adding Mai Baoxiang to the story improves our understanding of the rich social texture of agricultural science in the Mao era. However, Mai was still just one of the many people involved. In 1970, Big Sand had only sixteen science technicians, one for each brigade. This severely limited the ability to conduct experiments and manage extension work in weed control, one of Mai's first targets. By 1972, this figure had dramatically increased to 1,270 people involved in a "scientific experiment network." Each brigade had a science and technology central group, and each production team had a science and technology small group. Strikingly similar to what was soon to become the famous "four-level network" of Huarong County, Hunan, Big Sand had created a system where "in every layer there was organization and at every level someone was working on it [抓]."[8] This was the system in place when Mai and Pu forged their alliance.

At the planning level, Mai and Pu participated in a leaders group made up of agricultural officials and party officials from the county and commune levels. The party committee made integrated control part of its official daily business.[9] Policies were devised to encourage brigade-level plant protection workers (植保员), letting them "work for a long time in peace" by having them arrange remuneration with brigade comrades, and giving them power over use of insecticides—no insecticides could be purchased without their order.[10] With the brigade as the chief accounting unit, land was taxed to support the purchase of insecticides, the completion of surveys, and other insect control requirements. And they implemented a plan of "using *tu* and *yang* together" (土洋并举, see chapter 1) to produce borer-killing bacteria: they imported bacillus powder for the starter and then grew the bacteria in a medium made from locally available materials—husks, sand, mud, ash, and so on.[11]

It was not always easy to convince people of the need to adopt integrated control. Cultural Revolution–era documents from Big Sand speak of capitalists and other "class enemies" sabotaging the work, and of the age-old problem of "superstition"—the notion that pests were sent by heaven prevented people from putting faith in insect control technologies of any sort. Party officials reportedly countered these forces with political education and class struggle. A document filed in the early 1980s, after the end of the Cultural Revolution, no longer cast the problems in those terms, but local resistance nonetheless remained a recognized problem: now people were said to have resisted because the technologies were too cumbersome (麻烦) and not as fast or effective as chemical control.[12] Yet, if the archives are to be believed, the scientific experiment movement in Big Sand was successful enough that cadres and the masses alike celebrated its results with a clever rhyme: "Integrated control is best: it uses insecticides less, and production it protects" (综合防治好，农药用得少，生产有保障).[13] The road from resistance to acceptance, and from failure to success, involved much political maneuvering, social engineering, and rhetorical finessing. This was the job of Mai Baoxiang and his colleagues on the ground in Big Sand.

Between Top-Down and Bottom-Up

When he traveled to China with the US Plant Studies Delegation in 1974, Philip Kuhn noticed the ideological significance of the term "experiment," but as he further noted, "Much of the experimental work at this [grassroots] level actually consists of demonstrations to show peasants the greater yield of improved seed or more advantageous planting densities, and thereby overcome conservative prejudices."[14] The difference between demonstration and experiment was not just a semantic question. What Kuhn noticed was a classic tension between bottom-up and top-down that lay at the heart of both Chinese revolutionary politics and agricultural research and extension around the modernizing world. (And indeed, this tension is a fundamental problem for people devoted to change anywhere.) "Experiment" suggested faithfulness to local needs and provided room for the agency of local actors. "Demonstration," on the other hand, facilitated the spread of new practices approved by people with authority. The temptation, of course, was to use demonstration to push through desired changes while cloaking it in the politically more satisfying language of experiment—and it is clear that this was often the rule. However, the state also actively sought to make even demonstration a revolutionary force. At the center of all these tensions lay the local state agents—in particular local cadres and agricultural technicians.

Sometimes cadres and technicians emerged from the ranks of local villagers, and sometimes they were sent from nearby cities or even from other regions entirely to "squat" at an experimental point. Like the point-to-plane system more generally, the practice of "point-squatting" (dundian, 蹲点) linked agricultural extension with political organizing.[15] Many of the experiences of sent-down agricultural technicians would be instantly understandable to agricultural extension agents around the world: their core responsibility is traveling from research universities to rural outposts to facilitate the flow of new agricultural knowledge and practices. However, in China the red revolutionary context was as significant as the green, and agricultural technicians saw themselves as part of the larger system of sent-down cadres. Even before 1949, the Communist Party had routinely sent political cadres and technical specialists to provide expertise for a time at the village level. The specific term "point-squatting" emerged in the mid-1950s, became widespread in 1960, and took off with the socialist education and scientific experiment movements of the mid-1960s.[16]

In 1965, Jiangsu provincial party secretary Xu Jiatun applauded the 70 percent of agricultural science and technology personnel in Jiangsu who had gone to point-squat, undergoing both the Socialist Education Movement and the practice of agricultural demonstration (样板). He further explained how advancing agricultural extension and deepening the socialist revolution were meant to work together: point-squatting synthesized the green and red revolutions. "Only by going to point-squat at demonstration points and realizing three-in-one integration with scientific and technology personnel and the peasant masses will leading cadres be able to personally participate in the practice of the three great revolutionary movements, strengthen survey research, study the leadership skills of class struggle and the struggle for production, and study agricultural science and technology. Only then will they be able to transform their own subjective world while also transforming the objective world and work hard to become both red and expert. Only then will they be able to teach by both word and example, serve as a model, and revolutionize the thought of science and technology personnel. Only then will they be able to discover problems in time, resolve them, use the points to guide the whole area, and revolutionize science and technology."[17]

The same year, Jin Shanbao—a wheat scientist who headed the Chinese Academy of Agricultural Sciences from 1965 to 1982—spoke of the role of agricultural science workers. Before Liberation, he said, if agricultural science workers went down to the countryside, the "reactionary party" suspected them of seeking to stoke revolution; reactionaries wanted them to stick with research and not inquire into production. And so, said Jin, quoting the Confucian *Analects*, the agricultural science workers "did not exert their four limbs and could not distinguish the five grains." When the Communist Party called on them to go to the countryside, they were scared of the hardship, but given how divorced they had become from reality and production, they were still more scared that peasants would ask questions they could not answer. However, recently science workers had taken up the party's call to undergo "long-term point-squatting, eating, living, and working with peasants, putting down their arrogance and building up friendships." Focusing on demonstration fields "made concrete the policy of science serving production" and "made agricultural science revolutionary." To provide a sense of what it meant for science to be revolutionary, Jin added: "What counts as *shuiping* [水平, literally, 'level,' here meaning 'attaining a scientific level']? Some think that if it makes it into the scholarly journals, it counts as having *shuiping*; some think that if experts nod their heads, it counts as having *shuiping*. To me, if it solves production problems and explains scientific principles, it counts as *shuiping*. If it undergoes practical testing and peasants nod their heads, it counts as having *shuiping*."[18]

Because of the effect they were meant to have on cadres and technicians, demonstration fields were understood to be revolutionary and not merely technocratic. As Xu Jiatun pointed out, "The majority of science and technology personnel are intellectuals, and are thus susceptible to the corrosive influence of capitalist ideology [and its emphasis on] individual fame and profit, and to the 'three departures' [from the masses] in research ideology, research style, and research methods."[19] Point-squatting at demonstration fields provided an opportunity to revolutionize their thought. In Jin Shanbao's words, "Demonstration fields are the battle lines of scientific experiment, and are also the front line of class struggle and the struggle for production. When agricultural science and technology workers personally participate in the three great revolutionary movements and practice the 'four togethers' [eating, living, playing, and laboring with peasants], through real-life struggle they can quickly transform their old worldview and create a revolutionary worldview armed with perspectives based on class, dialectical materialism, production, and the masses."[20]

Our skepticism of propaganda—sharpened by its grinding repetitiveness—may prevent us from recognizing some truths in such political framings of agricultural science work. In forging solidarity with local people and solidifying their commitment to ensuring good harvests, local state agents often did cultivate what we might think of as a "revolutionary worldview." However, as this chapter will show, bottom-up thinking was perhaps less likely to challenge their own sense of superiority or entitlement and more likely to assist them in challenging the mandates of the central state itself.

"Cultivating People"

Figure 19 depicts a famous propaganda poster that eloquently captures the political ideals local cadres were meant to embody. The book and pencil represent the party secretary's *wenhua* (文化, culture/education); the tools represent his willingness to get his hands dirty alongside the people below him on the political ladder; that he is reviewing his notes while lighting his pipe suggests that he keeps his mind on production even during his brief moments of leisure. And it is not coincidental that the "old party secretary" was male. Although gender equality was a political priority, and so the participation of girls and young women in scientific experiment was a common theme, the leaders were almost always men. Even agricultural technicians were rarely female, and when they were, their gender could limit their opportunities: during my interviews in Guangxi, the one woman I encountered who had served as a technician in the Mao era complained that because she had two small children, she was assigned office work, "like a secretary—that's how it was then."[21]

Cultural Revolution–era propaganda materials were carefully selected for dissemination by the state to enforce and reinforce specific political values. Oral testimonies by agricultural technicians reminiscing today about their Mao-era experiences often underscore those same values, but for somewhat

FIGURE 19. Listed as Liu Zhide, "Lao shuji" (Old party secretary) (Shanghai: Shanghai renmin chu-banshe, 1973) on http://chineseposters.net/posters/e27-321.php. This is one of the most famous of the paintings produced by the celebrated Cultural Revolution–era peasant-artists of Huxian. Reproduced from Fine Arts Collection Section of the Cultural Group under the State Council of the People's Republic of China, *Peasant Paintings from Huhsien County* (Peking: Foreign Languages Press, 1974), 14.

different reasons: nostalgia for the past helps to highlight disappointments about the present. For example, I heard from several former agricultural technicians about a Cultural Revolution–era regional party secretary named Yan Qingsheng. Yan was beloved for his close attention to agriculture. He frequently made personal visits to the Department of Agriculture to learn new technologies alongside the technicians. He was even known to accompany technicians in the middle of the night to check on the seedlings in a greenhouse. And when a peasant in Pubei County bred a new variety of rice, the party secretary immediately brought him up to work in the agricultural science institute.[22] In both content and tone, these stories sound remarkably like the tales of party support that run through the Yuan Longping biographies explored in chapter 3, and they resonate as well with Cultural Revolution–era propaganda about wise and considerate leadership. Despite some important differences, all these sources are informed by a political culture that has produced powerful narrative forms: they "agree" on fundamental points, since both the upper-level state officials and the peasants themselves needed local state agents who could bridge the gap.

Local leaders were admired when they did well by their communities. At the lowest level of the administrative hierarchy, in the words of one agricultural technician I interviewed, "everything depended on the production team leader."[23] Commune-level officials like Big Sand's Mai Baoxiang were admired for similar reasons. In historical documents and interview narratives alike, the success of local cadres is usually portrayed as depending on the care they showed people, their intellectual resources (*wenhua*), their willingness to make sacrifices, and their initiative in actively pursuing opportunities wherever they might arise.

One of the chief marks of a good cadre was his ability to cultivate talent among the people in his jurisdiction.[24] The connection between "cultivating" plants and cultivating people was not accidental; it is an ancient analogy in China that has survived in the present day. Cao Xingsui explained to me and a group of agricultural technicians that in addition to this shared use of the term "cultivate" (栽培), the term "select the best" (选拔) so commonly employed in reference to identifying and promoting human talent originated in the ancient practice of selecting seeds to improve crop varieties.[25] Cultural Revolution–era materials also highlighted this relationship. A 1971 article from Xin County, Shanxi, cited the "broad masses of poor and lower-middle peasants," who reportedly said, "To cultivate sprouts you must first cultivate seeds, and to cultivate seeds you must first cultivate people." Thus revolutionary leaders had to compose the scientific experiment groups by selecting "sprouts" from among the peasants and educated youth and encouraging them to engage in the three great revolutionary movements.[26]

Stories of exemplary local cadres also highlight their personal character, especially their generosity. One former sent-down youth told me how grateful the old people in her village were that the government sent a wonderful cadre to point-squat and serve as their production team leader. At first people resented him for assigning arduous work in soil improvement. But then he sold his Shanghai-brand watch to buy new seed varieties and fertilizer. The watch was worth more than 120 yuan at a time when 10 yuan could feed a person for a month, including meat. The cadre served as a model for the community and produced much positive feeling for scientific farming.[27] Another former sent-down youth spoke emotionally of the production team leader who recognized his potential and not only promoted him to be an agricultural technician, but also assigned him two pieces of the best land on which to plant a new variety of seed he had obtained. The leader's generosity inspired the young man to work very diligently, reading books to decide just how to plant these precious plots.[28]

I had the chance to meet one of these Mao-era local leaders in the person of Cao Xingsui himself. Wherever we traveled, his charisma, energy, and resourcefulness inspired old friends to greet him warmly and strangers to listen respectfully. (This was especially noticeable on the train, where we shared crowded quarters with several other people who were quickly captivated by Cao's knowledge and charm.) Observing him today, I was not surprised to learn that when he was a sent-down youth in northwestern Guangxi, in spite of his terrible class background (his father fought for the Guomindang), he was assigned the position of production team leader. During his tenure, Cao especially gained the appreciation of women on his team for his consideration of their hardships and his initiative in addressing them. During the day, women had to participate in production team work; in the evening they had to take care of many children, closely spaced, and also had to grind corn for the next day. So Cao invested in an electric corn grinding plant. The women loved it, and people also came from other production teams to grind, leaving behind the chaff, which Cao's team could feed to their livestock. Another burden on women's time was sewing—whenever rain kept them from the fields, they spent the whole day sewing clothes and making shoes for their families. Cao used money earned by selling sand dredged from the river to purchase sewing machines. He first went to learn how to use the machines to make simple clothes, then returned and taught ten women to do the same.[29]

Because they were typically stationed at the commune's agricultural extension station, and so had to spread their energies throughout all the production teams in the commune, agricultural technicians did not have as intimate a relationship with peasants as the local cadres did. Nonetheless, the success of a technician was grounded in his or her ability to cultivate relationships with people at the grassroots, along with people in "brother" communes and officials higher up the chain. Today, when people talk about why agricultural extension was so successful even during the Cultural Revolution, a common refrain is that agricultural technicians were "quality" (素质好) people and a "bargain" for the state because of their willingness to labor long and hard for little pay.[30] As one former technician told me, they had a very deep sympathy for the peasants in their communes and worked alongside them "through sweet and bitter." He recounted with feeling one episode when he went down to a village and the peasants gave him some water spinach to take home with him. "They didn't have any money, but they still gave me vegetables—oh, what feeling they had."[31] Another said, "Grassroots agricultural technology extension workers could really eat bitterness [endure suffering in the service of a cause]. We were truly devoted. We went out with the peasants through wind and rain, working arduously, laboring with the peasants, verifying with them, doing experiments with them. . . . We were integrated with the peasants and had a very deep sympathy with them."[32] And speaking from his experiences as a production team leader, Cao Xingsui echoes this assessment: "They came from the big cities into the villages where conditions were so much worse, but they never uttered one word of complaint. Their families were also there, living with us, with never one word of complaint. Nowadays you'd never have that."[33] These nostalgic depictions are clearly at least somewhat hyperbolic and speak as much to people's negative impressions of social relationships in the current era as to actual relations in socialist times. Nonetheless, they testify to the importance of the social aspect in the work of Mao-era agricultural technicians.

The relationships that technicians developed with local cadres and members of the scientific experiment groups were especially important, and here the production teams geographically close to the commune center were especially fortunate. Cao Xingsui speaks of the close friendships he and other members of his scientific experiment group made with the technicians at their extension station, which was not too far away. "Our production team had a big fish pond. We would catch fish and invite the technicians to come eat fish, or we would send them fish to eat. Or we sent them wonderful vegetables that we could grow and couldn't be bought in the city. So the technicians were very dedicated in providing guidance to us." They were especially close with the veterinary technician, since he came to the village in all seasons and veterinary knowledge was particularly difficult for laypeople to master. "When the vet came, we gave him lots of good food and drink. When he came, we would first invite him to go eat, just as we went out to eat yesterday. We killed a chicken to give him to eat and only then went to see the sick animals." (Here Cao sought to impress upon me the importance of hospitality by referencing one of the many banquets we were given as we stopped to visit agricultural extension offices around Guangxi. I preserve this detail to underscore once again the connections between the context of the interviews and the reminiscences of an earlier time they produced.)

Relationships were not always that close. One technician I interviewed expressed frustration with the head of the scientific experiment group who was also the production team leader where the technician was attempting to extend a variety of improved rice. The team leader doubted the new variety and raised such objections that the technician felt obliged to promise (without any real ability to make good) that the team would be compensated in case of a poor harvest. The team leader's doubts must have irritated the technician, because the technician reported triumphantly that the success of the demonstration field left the leader with "nothing to say."[34] More often, technicians report genuine affection for and from the local people at their "points." Unlike many tales from propaganda materials, however, they attribute this affection not to the celebration of peasant expertise, but rather to peasant appreciation for the expertise and care provided by visiting technicians.

That these statements occur within a larger nostalgic discourse that critiques today's materialism and lack of public spirit does not negate the significance of these personal relationships during the Cultural Revolution. They were deeply meaningful to the people who experienced them. They were also instrumental in accomplishing the tremendous technological and sociopolitical transformations of the Mao era.

Self-Reliance and Local Responsibility

The central state could not avoid depending on local agents to realize both its technological and its political goals. It could, however, make virtues of this necessity. The policy of self-reliance was one such virtue: every local cadre who increased a production team's self-reliance saved the higher levels from having to supply precious resources. And so materials relating to the scientific experiment movement quoted Dazhai's famous leader, Chen Yonggui, as saying that "self-reliance is a magic weapon" (自力更生是法宝).[35] It was indeed powerful. And it could cut many ways. Whether it served the state's cost-reduction priorities or the interests of local people lay in the hands of the local state agents who wielded it.

Foreign observers of all political stripes saw self-reliance as one of the most important principles of the socialist Chinese approach to science, though what they understood self-reliance to mean varied greatly.[36] Within China, self-reliance was also multivalent. Born of the revolutionary era in Yan'an, self-reliance continued during the Mao era—especially after the Sino-Soviet split—to represent national pride in China's being able to go it alone in the hard world of the Cold War. At this level, the Chinese state further sought to become a model for Third World countries who shared China's struggles with the effects of imperialism and the uncertainties of the bipolar system the United States and USSR represented. At the same time, however, local communities in China were also being asked to embrace self-reliance as a principle for their own economic development. The widely recognized standard was for localities to move from a grain-importing to a grain-sufficient or even grain-exporting status. State-produced documents and interviews alike relate stories of advances in scientific farming that allowed communities to "take off the hat" (i.e., remove the stigma) of eating grain supplied by the state.[37]

While the policy of self-reliance had many merits, it was also a way for the central state to justify its failure to supply communities with what they needed to fulfill the very policy mandates the central state was handing down.[38] And so memos from central authorities to provincial authorities, and from provincial to local authorities, document the transfer of money to support the mass scientific experiment movement along with entreaties to practice the "spirit of self-reliance, diligence, and thrift."[39] At times the stinginess produced blowback: In 1970, the Guangdong provincial science and technology leading group allegedly first failed to provide promised funding to the grassroots scientific experiment movement, and then devised a plan where they would provide seven times as much to science and technology stations serving industry and transportation as to those serving the poor and lower-middle peasants at the forefront of the agricultural scientific experiment movement. Radicals delivered a memo that charged: "Can we see what flag [they] are flying, what road they are traveling, and what line they are implementing?"[40] But the state's encouragement of self-reliance in this as in other matters continued. For example, a 1975 report on the scientific experiment movement in Nanzhao County, Henan, reported that the county's agricultural science station relied on funding from enterprises at the production brigade level and from extra enterprises that the station itself could engage in to make the most of local assets.[41] Here much depended on the entrepreneurial initiative and innovation of state agents.

Self-reliance for local communities thus first and foremost meant making do with less. In a section titled "Be Self-Reliant and Practice Science in a Hard-Working and Frugal Way" (自力更生，勤俭办科学), the proceedings of a 1965 Beijing conference emphasized mobilizing the masses to figure out solutions, making do with simple materials, and replacing the "foreign" with the "native" (以土代洋).[42] Other examples throughout the Cultural Revolution abound; they are vividly represented by the 1975 poster from Huarong "Self-Reliance; Practice Scientific Research with Diligence and Frugality" (see figure 5). The poster's caption praises Huarong for "persistently drawing on local resources, using local methods, and improvising equipment, such that they met the needs of agricultural scientific research and drove forward mass-based scientific farming activities." The left picture tells the story of an attempt to build a model greenhouse; because it was too expensive to build, they could not popularize it. Only after the substitution of cheaper local materials (e.g., mud bricks and wood for red bricks and reinforced concrete) did the new "native [tu] greenhouse" become widely accepted. The right picture similarly emphasizes the need to mobilize the masses "to select methods that are crude and simple, substituting the native for the foreign, and in this way resolve the equipment needs of scientific experiment." One of the examples it cites of using local resources should be familiar from Yuan Longping's story: ordinary clay bowls once again served as replacements for specialized seedling trays. The central state's interest is still more clearly articulated in a 1976 article from the popular magazine *Rural Scientific Experiment*: "There are two attitudes: One involves depending on the masses, self-reliance, proceeding with native methods, and learning by doing; the other involves reaching for handouts from above, waiting/depending/demanding, and hankering after the grandiose and foreign."[43]

To avoid having communities "reach up" for assistance, the state further encouraged grassroots production of seed, fertilizer, and insecticide. The approach to seed production was called "four selfs and one supplement" (四自一辅)—that is, the locality should take on the responsibility of selecting seed, propagating it, storing it, and using it, while supplementing their own stocks with appropriate transfers from other localities. The slogan originated in the National Conference on Seed Work held in 1958 and continued as policy throughout the socialist era.[44] When the scientific experiment groups came into being, they were obvious candidates for organizing this work.[45] A 1975 collection on the four-level networks celebrated the policy for allowing communities to leave behind the old system of requesting from above when short of seeds and relying on outside places for improved varieties, replacing it with a system in which the community itself became responsible for selecting seed every year and thereby standardizing and maintaining improved varieties.[46]

The notion that individual communities ought to be responsible for seed propagation should interest critics of the green revolution and agribusiness more generally. One of the most problematic aspects of new seed technologies—including both the production of hybrid seed and the use of GMOs—is the way it strips farmers of the ability, or even the right, to preserve and select their own seed for future plantings.[47] The introduction of hybrid seed propagation in Mao-era China was highly unusual in its emphasis on building local expertise and self-reliance, even to the extent of sending young peasants from local communities in rice-growing regions throughout China to Hainan Island for training. It is hard to know how this system would have fared if political conditions had continued to favor mass science and local self-reliance. However, by 1976 a county in Guangxi was recommending the centralization of seed production at the commune or brigade levels, rather than the production team level; and by 1978 a district in Hunan was opposing seed propagation at the brigade or team levels, advocating centralization at the county or commune level to achieve a state of being "big and public" (一大二公).[48]

In the 1970s, many communities in China were building factories to supply their fertilizer needs, not only chemical fertilizer plants, but also facilities for producing bacterial fertilizers such as 5406 and for using high-temperature techniques to compost farm and household waste. But whatever the type of fertilizer, the key was that it was produced locally, allowing the community to become self-reliant. The same was true of local production of insecticides—whether chemical, plant-based, or bacterial. Production of the bacterial insecticide *Beauvaria* depended on a network similar to Huarong's celebrated system: county-level factories supplied high-quality starter culture to the communes, each commune managed a second-tier factory to produce more starter to pass down the line, and every production team managed a third-tier factory to produce the *Beauvaria* needed in the team's fields.[49] Peasants I interviewed in Guangxi Province reported making their own insecticides out of tobacco leaves and also mining lime to control insect pests. One man especially praised tobacco leaves for their effectiveness against both insects and bacteria. He considered them to be as effective as the chemicals, but now they are not used much because the higher authorities do not promote them. Back then, he said, "the government promoted doing things for ourselves, self-reliance."[50]

All this emphasis on self-reliance made local state agents still more important to the state. But if the state saw self-reliance as a "magic weapon" that could sever dependence on overtaxed purse strings, it could also become a magic shield to protect localities against potentially damaging directives from above.

Self-Reliance as Resistance

One of the favorite topics of conversation among local state agents reminiscing about their Mao-era experiences is the irrationality their higher-ups exhibited. It seems everyone has a story of this type, and they all appear to stem from the highly interventionist attitude characteristic of both modernizing (green revolutionary) and socialist (red revolutionary) ideologies. As one technician neatly expressed it, "We were not allowed to *do* agriculture. We could only *transform* it, we couldn't *do* it."[51] In other words, they could not just get on with their jobs but were constantly being called upon to reinvent things.

More specifically, technicians and local cadres sometimes encountered problems of bad direction from above by officials who did not understand agriculture and did not listen to the people who fathomed conditions on the ground. One technician told me that the system of "administrative decrees" had both positive and negative consequences. On the plus side, it made possible rapid implementation of good new technologies, but "there were also mistakes." For example, during the Great Leap Forward "somebody got all hotheaded [头脑发热] to extend short-grained rice, and so very quickly there was a decree to plant." However, it soon became clear that it did not suit the southern climate and it was also very hard to thresh manually. Still, he added, "there were not many of these failures, and they were due to hot-headedness."[52] He also noted that the pressure to introduce multiple harvests of rice in one year was a mistake, and this was elaborated during an interview I conducted with another technician. In some areas of Guangxi, the season was not long enough to accommodate two harvests. But higher-ups had "elementary school educations" and moreover refused to listen to the technicians, who knew better. They handed the technicians a timeline simply requiring the harvests to be earlier by a certain number of days every year.[53] In another conversation, a party secretary at the county level, who had been transferred from his home in the North, came in for special scorn. He had decided that what Guangxi really needed was to start growing apples, so he invited a few peasant experts from Shaanxi Province to plant the trees. Of course, the significant climate differences between North and South made the project a dismal failure and seems to have soured the technician on the whole idea of "peasant experts." On the other hand, technicians who had encountered peasant experts from the more carefully cultivated model of Chaoshan were more positive or at least neutral in their assessments.[54]

Local cadres and agricultural technicians did not just gripe about bad direction from above; they often actively resisted it, and one of the most effective means came out of the state's own toolkit: self-reliance. One of the causes of failure most commonly identified in post-Mao critiques is the inappropriate application of models: celebrated, stereotyped practices were forced on local communities in defiance of on-the-ground realities. The classic oral-history study *Chen Village* relates just such a scenario: compelled to emulate Dazhai by leveling hills to plant trees, the village invested vast amounts of labor and resources and ended up producing acres of land with insufficient topsoil to support the new crop.[55] Similarly, in the Shaanxi village of one former sent-down youth I interviewed, villagers were mobilized to make Dazhai-style terraces in the hills, but since the soil contained no rocks, it was impossible to complete the terraces.[56] Debacles like these—and they were undoubtedly common—are certainly worthy of criticism. But it is important to recognize the attention devoted to such problems during the Mao era by state agents: the critique of top-down technology transfer was already well established long before post-Mao critics picked up their brushes.

Throughout the 1960s and 1970s, we find rural experiment groups dedicated to testing new seeds and methods to determine local suitability and producing new varieties and techniques on-site that better matched local conditions (因地制宜). The second principle cited at the 1965 National Conference on Rural Youth in Scientific Experiment proclaimed the need for tailoring experiments to "suit local conditions" and "serve production in the here and now."[57] That same year, the proceedings of the Conference of Activists in Beijing Municipal Rural Scientific Experiment Groups specifically tackled the relationship between "studying the advanced technological experience" of models like Dazhai and then "acting according to local conditions." For example, one nearby brigade had achieved high yields with a new millet variety and sought to encourage wide extension of the new seeds. "The masses" reportedly complained that the results could not be trusted, since the experiment group had used the very best land for its experiment field. The second year the experiment group used more average land, but because they had used more fertilizer than would ordinarily be available, people still complained. The third year, the group was finally able to demonstrate that the new millet variety produced high yields under the prevailing local economic conditions.[58]

The concern about the potentially damaging consequences of importing inappropriate technologies continued throughout the Cultural Revolution and into the post-Mao era. In 1971, a county in Shanxi Province documented big agricultural losses attributed to "blind" extension of imported varieties. The authors bemoaned their failure to learn about the characteristics of imported corn varieties before planting them on a large scale.[59] This orientation was also in play when research institutes throughout China were mobilized to work on hybrid rice. In Liaoning Province, for example, researchers noted the need to compare the hybrid varieties specifically with the improved varieties that had already been extended locally, so as to be careful that they did not just accept a new technology from the outside without making sure it measured up to local standards.[60] Interestingly, Dazhai itself became a model for local cadres in resisting pressures to adopt new technologies. As a 1974 document explains, "The experience of Dazhai tells us that improved varieties have conditions. Areas are different; soil and climate are different; planting times and needs for fertilizer and water are also different." Thus any given "superior" variety would not be "superior in all places." The authors thus criticized the idea that breeds from outside were necessarily "improved" breeds.[61]

Mai Baoxiang of Big Sand Commune remembers with frustration being compelled by higher authorities to direct Big Sand peasants to plant a Mexican variety of rice. He thinks that because the new variety performed well in northern China, people assumed that it should be extended throughout the country. However, when the crop failed the first year, Mai was able to use the failure as evidence that the new variety was not suited to Big Sand, and he was not required to promote it the second year. Local leaders were not always able to use the "local conditions" argument to ward off outside pressures, but they did have some latitude, especially when they could blame failures on the widely recognized problem of "blindly" adopting imported seeds and practices.[62]

The policy of self-reliance was a mixed bag for local communities. On one hand, it did promote the development of skills and capabilities at the grass-roots, in contrast with the pattern in other places and in post-Mao China, where, for example, local communities became dependent on large seed and chemical companies to supply inputs. On the other hand, it gave the state a convenient excuse not to provide materials and expertise necessary to accomplish some of the tasks it was demanding of local communities. The key here was the initiative of state agents: in the hands of the right people, self-reliance became useful rhetoric for advocating for local needs and resisting pressures to conform to inappropriate outside models.

Managing Peasant Resistance

In addition to mitigating bad directions from above, state agents also had to manage peasant intransigence. A 1965 account of attempts to introduce chemical fertilizer told of strenuous resistance owing to four different types of "fear": (1) the fear that spending money on fertilizer would reduce payments to commune members; (2) the fear that applying chemical fertilizer would be a waste of effort; (3) the fear among cadres that the "lower levels" (下楼) would "raise opinions" (i.e., complain); and (4) the fear among commune members that chemical fertilizer would require more watering and thus be more trouble.[63]

Resistance was a profound threat to the socialist Chinese state: it was real enough and multifaceted enough to demand sophisticated responses from state agents. Materials produced throughout the Mao era document peasant resistance to newly introduced agricultural technologies. However, political constraints often led to a framing of the issues that does not illuminate the real tensions very well. To take just one of many examples, a 1976 report by a scientific experiment group in Guangxi on the introduction of hybrid rice complained of resistance but chalked it up to a "two-line struggle" with people on the wrong side worrying about how much fertilizer they were using and other concerns that supposedly reflected their "slavish conservatism" (因循守旧) and "conventionalism" (墨守成规). Indeed, the experiment group found itself using lots of high-temperature compost and manure accumulated over years in pit toilets.[64] The relevance of a "two-line" ideological struggle is dubious, but such documents are highly informative about the common concern of local peasants that precious fertilizer be used wisely, and perhaps also about the ways in which the introduction of new technologies sometimes led to waste of resources. Resistance and the responses appear most clearly in documents from 1965 and 1966, when the rural scientific experiment movement had begun in earnest, but before the Cultural Revolution had heightened the stakes of rhetoric. These early documents reveal the power of peasants to resist and the need for the state to find ways far beyond simple orders or threats to bring peasants on board with new technologies.

Narratives of overcoming resistance typically followed a very familiar arc: a new technology was introduced, people resisted, they were presented with various forms of persuasion, and in the end they accepted the new technology. The narratives served political purposes—from offering strategies for cadres in overcoming resistance, to reinforcing, again and again, that resistance was temporary and acceptance the inevitable end result. Sometimes stories of overcoming resistance related to the specific political campaign of the day. For example, the Socialist Education Movement of 1964–1965 generated many tales of local wrongheadedness brought into line through political tutelage, and stories of the introduction of agricultural technologies served this purpose as well as any other kind. Sometimes the narrative testified to the more long-lasting lesson "Scientific experiment is best" (还是科学实验好). Sometimes peasant resistance in the narratives even proved at least partially correct; this too was politically useful, since it reinforced the rightness of mass-science approaches.

But the narratives were not spun just from the imaginations of talented masters of propaganda. They do require critical reading: neither the characterization of opponents as "class enemies" nor the unfailingly happy endings ring true to what we know from other sources. However, there is every reason to believe that the conflicts described in the narratives reflected real tensions at the local level.

Resistance from the peasants took a number of forms in such narratives, many of which have been made famous in James Scott's *Weapons of the Weak*. In resistance narratives, "the masses" (otherwise known as "commune members," "the lower levels," "old peasants," etc.) often appeared in the form of a chorus in a Greek play. At their best, they "raised opinions" (意见); at their worst, they "talked all over the place" (议论纷纷). In either case, they presented a headache for cadres caught between the mandates of higher authorities and the resistance of the grassroots. Chapter 4 recounted how boar keepers threatened by a new state-endorsed pig-breeding station spread their own propaganda to convince sow owners that they should stick with the tried and true, and how local people reportedly used clever analogies and rhymes to mock attempts to promote scientific farming: "Girls doing *scientific farming*?! That's like frogs at the bottom of a well trying to grow feathers and fly away!" Earthy epithets also bolstered resistance, as when women in one place refused to use ammonium bicarbonate fertilizer, or what they called "stinky fertilizer" (气儿肥), because it stung the eyes and nose and smelled like urine: "I'm not doing it. The cadres can do it!" they allegedly said.[65]

Peasants also resisted by speaking directly from their knowledge and experience. Here we may compare peasants' claims to experience with state agents' claims to self-reliance in resisting unfavorable new technologies: both types of claims drew power from state ideology. As one technician told me, it was usually the older peasants who balked at new technologies, saying, "I've farmed for decades and never done this. It's no good."[66] Sometimes they could be very powerful. When peasants criticized poorly conceived experiments with the argument "We'll have to sell our wives and children if we farm like that," they evoked at once the stark truths of what happens to rural people when harvests fail and also the frequent refrains heard in "speak bitterness" narratives that the state encouraged to help people remember just such traumas from the old, "pre-Liberation" days.[67] Peasants also drew on their knowledge of local agricultural needs to offer specific objections to new technologies. For example, a peasant in Guangdong in 1964 voiced his opinion that a newly introduced dwarf variety of rice would "stuff-to-death the ducks and starve the oxen"—that is, the grain would all fall on the ground and be eaten by ducks, and there would not be enough straw for the oxen. This would seem to demonstrate his deep knowledge of the relationship between grain production and livestock management in his local farming system, though he was later "proven" wrong with respect to this specific introduced variety.[68] Meanwhile, in Ningxia Province, commune members proclaimed that "each place has its own kind of water and soil"; they were concerned that a newly introduced variety of rice might work for areas with clear water, but not for areas like theirs where the water was turbid.[69] Even more than the previous example, this objection suggests an agricultural approach rooted in long-lasting ideas about the special environmental characteristics of a given place, often described as the specific *qi* of one locale that makes it different from another.

Official reports and propaganda accounts presented a variety of responses to resistance. In many cases, resistance was associated with "superstition"—for example, in 1965 peasants in Guangdong were said to believe that planting in a certain field would bring death, and in 1966 peasants in Fujian were said to believe that peas attracted ghosts, so they were unwilling to plant them.[70] Particularly during the more politically volatile periods, resistance was often framed as attempts at sabotage by "class enemies." The solution in both types of cases typically included a refresher course in Mao Zedong Thought.

Beyond these predictable and easily criticized responses, however, were many strategies that suggest a far more dynamic approach to state-society relations. In Big Sand Commune, Mai Baoxiang discovered that peasants were willing to accept new technologies that were "simple and cheap"—hence, their adoption of chemical herbicides instead of the "intensive cultivation" that peasants from land-poor Chaoshan promoted.[71] Documents from other places similarly highlight the importance of basing research directions on local production needs and engaging with local experience in designing scientific experiments. This is epitomized in a 1966 report from Ningxia, which cites "starting from the practical needs of production" as "the crucial point determining whether scientific experiment will achieve the masses' support and whether science and technology will take root, germinate, bloom, and bear fruit in the countryside." It goes on to insist that scientific experiment at the local level must combine the expertise of high-level scientific research authorities with local traditional experience (传统经验): "Through experiment, we will quickly localize [本地化] advanced technologies from outside places."[72] Peasants thus could not be expected simply to accept any and all introduced technologies; the technologies and the methods through which they were introduced had to be carefully selected to suit the local context and not unjustifiably burden local people.

Involving "the masses" was a conscious strategy to gain local acceptance of new technologies. The proceedings of the 1965 Beijing conference spelled this out clearly. First, it said, scientific experiments had to be based on the masses' "opinions." Second, if the scientific experiment group encountered problems, it should solicit the "masses' collective wisdom" and work with them to devise a plan to address the problem. After the completion of the experiment, it should invite the masses to a meeting to evaluate the results and collaborate on extending the new technology if it was successful. If the experiment failed, the experiment group should explain the situation clearly to the masses, analyze the reason together with the masses, and find a way to correct it.[73]

If peasants in these accounts were sometimes "conservative" or "backward," they were also typically open to convincing based on hard empirical evidence. In a 1959 article, Pu Zhelong reported that peasants had objected to the use of parasitic wasps, thinking that little wasps would be ineffective, would sting people, or would simply be too troublesome. "They had never in their whole lives heard of using bugs to eat bugs, so without seeing results how could they believe it?" One effective solution was to split open the insect pest's eggs and use a magnifying glass to show peasants how they had been parasitized.[74] As another 1966 report emphasized, "With respect to the masses' conservative thinking and farming habits, the experiment group paid attention to the process of experiment through which facts could be used to satisfy the masses."[75] In the case of the "pea ghosts," officials reported that when encountering some of the masses whose thinking was "blocked" (思想不通), the best method was for cadres to take the lead and let the model "speak" to the masses and educate them. When offered the freshly harvested peas, "as soon as the peasants' lips started moving, their thought cleared up."[76]

Technicians remember the slow, laborious process of convincing peasants of the value of the new technologies at each "point" in the "point-to-plane" system.[77] One recalled the introduction of dwarf varieties of rice: "Extension was difficult. At first the peasants didn't believe in it. Aiya! They saw how short the plants were and thought the yield would be low, so they said, 'No good!'" So the technicians did comparative experiments and let the peasants verify the results (验收), after which the peasants very quickly came to believe in the dwarf rice.[78] Here again, it is striking that what would be called "demonstration" in most extension contexts was referred to by Chinese technicians as "experiment." A technician who began as a peasant and the leader of a scientific experiment group in Big Sand offered me his own perspective on what worked in extension, a perspective clearly informed by the theory of knowledge found in Mao's "On Practice": "You need to get peasant acceptance of each new technology. First you need to do demonstrations [样板] and make experiment models [做试验示范]. In the process of studying the technology process, [peasants] gain rational knowledge. When experiment models achieve expected results, peasants obtain perceptual knowledge. Only then will the technology be easily accepted and used by the peasants."[79]

The extension strategies described in these documents reveal an understanding of not only the role empirical evidence plays in persuasion but also the importance of authority and trust. What people believe about science depends not only on their own interpretation of "the facts" but also on who delivers those facts.[80] And so in the case of the pea ghost dilemma, the scientific experiment group recruited an "old peasant" from a neighboring community who had good experience with peas to stay on as an adviser through the harvest to help convince the old peasants there. And in the case of the "stinky fertilizer," a woman member of the scientific experiment group helped explain the benefits of ammonium bicarbonate to the women who had complained of the smell and swore they would not use it again.

If detractors could not be convinced by their peers, the best method might be to bring them on board themselves. Such was the case with a peasant named Zhang Jiatao, who doubted the value of a youth's experiments (see chapter 6): he was reportedly convinced when he was invited to participate in a friendly competition. This seems to have been a fairly standard story: it appears also in an account from Beijing, where a seventy-nine-year-old peasant named Zhao Guang compared close planting with "beating people up." Zhao Guang was allowed to decide how to plant one field, and the scientific experiment group used close planting on the other; Zhao Guang reportedly ended up having to acknowledge the superiority of the new method.[81] The same document recounts the story of an old peasant named Wang Chenglin who objected that adding water and chemical fertilizer to a plot of seedlings would cause them to grow too fast such that they would fall over before they produced well. But when the experiment turned out well, he began to attend the lectures the experiment group organized, and he subsequently became a "backbone" member of the group.[82] Highlighting such cases in propaganda materials was a way not only of celebrating peasant participation in "mass science," but also of emphasizing the important role that local state agents could play in cultivating compliance among resistant peasants.

Resistance and the Meaning of Scientific Experiment

These stories reveal just how desperately the Chinese state needed the cooperation of people at the grassroots. They show it in the celebration of "wise" local cadres able to carry out state initiatives while keeping their communities afloat, and in the efforts to bring on board "old peasants" who might otherwise have been sticks in the mud when it came to modernizing agriculture. Extending this theme, the next two chapters will chart the state's attempts to woo rural youth back to the farm with the promise of meaningful work as agricultural technicians and even leaders of "scientific experiment." Understanding all of this is necessary to grapple with the question of what "scientific experiments" meant: they were often as much a form of negotiation between state and rural society as they were programs to advance agricultural knowledge—as much about the social world as the natural world. And the chief negotiators were the local cadres and agricultural technicians on the front lines of the green revolution.

The previous chapter examined some of the cultural customs and social relationships that, for better or worse, the new agricultural technologies threatened. State officials had to choose their battles carefully. For example, while they invested enormous energy in promoting new varieties of grain, they often left the choice of livestock breeds to local people, who tended to prefer (and still do) the flavor of local breeds. Some conflicts were useful in furthering the ideological priorities of the Chinese Communist Party. When young women practicing "scientific farming" competed with the men who ordinarily provided boars for mating services, this served a double function of promoting a new breed of swine and challenging existing gender relations, and so it was prime material for propaganda. But officials could not afford to turn every effort at agricultural transformation into an opportunity for social revolution. In many other cases, state agents worked within the system: for example, the extension station would employ an old man without relatives to raise an improved breed of boar and bring it around the village every few days to see what sows might need his services.[83] The "boar man" was an important fixture in many Chinese communities, and the system helped people with no other means of support.[84] Even by choosing its battles the state was by no means guaranteed total victory. Peasants were happy to accept new varieties when they tasted better than the varieties they replaced.[85] When they did not, local cadres had to find ways to accommodate the continued cultivation of traditional varieties—even if that meant finding hidden fields in the hills to grow them out of sight of inspectors from higher levels of the state.[86]

These stories of overcoming—or in many cases, accommodating—resistance provides another way of thinking about what the Chinese state called "experiment" and what appeared to Philip Kuhn and other observers to be nothing but "demonstration." By the 1960s and 1970s, the significance of the "experiment" concept was so deeply associated with Chinese revolutionary ideology that it actually fed back into agricultural science (whence it had emerged),[87] such that the insistence that what was going on was in fact "experiment" owed at least as much to the pressures of political culture as to the pressures of scientific culture. That is, claiming to be conducting agricultural "experiments" was at least as much a claim to revolutionary authenticity as it was to scientific rigor. Recognizing that the terms "science" and "experiment" held political significance in Mao-era China, we can see more clearly that the rural scientific experiment movement was in fact testing a great deal.

New technologies had to be "tried out" in the locale: in order for them to "work" they had to suit not only the local environmental conditions but also the people who were to implement them.[88]

Scientific experiment groups working on plant breeding, fertilizers, hormones, and other projects were not producing new scientific theories, but they *were* testing the limits of what people in the countryside with rudimentary equipment could do to produce and apply new agricultural inputs. And they were also testing the local suitability of such new technologies: Did the new technologies work in those specific rural places? Very often, they did not, and then the question of whether to bemoan the failed experiment as a waste of time and resources, or accept it as a bend in the oft-referenced "winding road" of science, depends on many factors, the most important of which are again social and political: How much did higher-ups listen to people at the grassroots? And how much trust did people have in their leaders' intelligence and good faith? Although stories repeat themselves through the written and interview sources, when it comes to questions like these, there would seem to be as many answers as there were villages in socialist China.

The Lei Feng Paradox

A young man who grew up in the capital of Liaoning Province, Shen Dian-zhong was "sent down" to a state farm during the massive rustication move-ment that between 1968 and 1975 relocated an estimated twelve million urban youth to the countryside. Before leaving home in September 1968, Shen had already developed a passion for science and had thought deep, if adolescent, thoughts about the history of quantum physics and the philosophy of science. Just days before his departure, Shen Dianzhong pondered in his diary: "Why do we do scientific experiment? Is it to 'make a name' for oneself as an indi-vidual, or do we do scientific experiment for the revolution? Does one close the door and do it by oneself, or do we go into the wide fields and unite with the masses?"[1] A few years later while on the farm, Shen had to decide whether to join a scientific experiment group. What might seem a simple decision given his profound interest in science in fact required considerable soul-searching. He wrote, "The process of doing scientific experiment is in actual-ity a process of struggle. This kind of struggle is a struggle I love, a struggle I want to go for, a struggle I welcome, a struggle I support." As if preparing in advance for an argument, he went on to list the three main reasons in favor of his participation: for three years he had been enthusiastic about the scientific experiment activities others were pursuing on the farm; he wanted to make a contribution to the "battle to transform Liaoning's agriculture"; and it would be a learning experience and a way to transform himself.[2]

Behind Shen's deliberations lay a central tension defining the experiences of many young people during the 1960s and 1970s: the drive to achieve some-thing important as individuals, to be heroes; and the deep understanding of the impropriety—and even danger—associated with individual achieve-ment. Shen greatly desired to immerse himself in scientific experiment, but he feared that his desire would be seen as—or perhaps even that it truly stemmed from—a bourgeois impulse to "make a name" for himself rather than a purely revolutionary commitment.

This was the Lei Feng paradox. Lei Feng was a young soldier whose post-humously published diary inspired legions of young people around China to write diaries of their own, and burdened them with a near-impossible mandate to "emulate Lei Feng" in his self-effacing worship of Chairman Mao. Lei Feng's claim to fame was his humble life studying Mao's thought, which ended in a suitably humble death under the weight of a fallen utility pole. He achieved recognition and glory, ironically, for being utterly common-place and unprepossessing. The campaign to emulate Lei Feng crystallized from a much broader struggle that Cultural Revolution–era youth faced, as they wrestled with the conflicting calls to be revolutionary heroes and mere "bolts" in the revolutionary machine.[3]

The contradiction was implicit in the very name "educated youth" (知识青年), which included both urban youth "sent down" (下乡) to the country-side and rural youth who "returned" (回乡) to their villages after graduating from urban secondary schools. Memoirs by urban people who came of age during the Cultural Revolution dominate Western perspectives on the Mao era, such that our mental picture of young people in rural China collapses onto an image of the urban, sent-down educated youth like Shen Dianzhong. For the state, the sent-down youth program functioned as a safety valve to reduce the pressure of urban unemployment, a means of easing the disruptions caused by Red Guards in the early years of the Cultural Revolution, an opportunity to deepen the revolutionary values of a generation who had been born after the founding of the socialist state, and a way of bringing needed knowledge to the countryside. But long before 1968, rural-to-urban migration of young people had already presented the twin problems of urban unemployment and rural brain drain.[4] And so rural youth who traveled to their county seats to receive secondary education were sent back to their villages to rejoin agricultural labor. These returned educated youth were the original targets of Mao's declaration (a Cultural Revolution stock phrase, but dating to 1955): "All such educated young people who can go and work in the countryside should be glad to do so. The countryside is a big world where much can be accomplished."[5]

The contradiction faced by educated youth, both urban and rural, was closely related to the tensions between the *yang* (professional, transnational, elite) and *tu* (earthy, native, mass-based). On one hand, educated youth were celebrated for having knowledge and "culture" (*wenhua*, 文化); on the other hand, the toppling of educational elitism ranked highest on the Cultural Revolution agenda, and propaganda carried repeated warnings against trusting to ivory-tower values. Youth were encouraged to see themselves as possessing, by virtue of their modern schooling, the *wenhua* necessary for the revolution—and simultaneously warned of the bourgeois or even counter-revolutionary tendencies inherent in their intellectualism. It was a fine line to walk: on one side lay the risk of being stuck in manual labor for the rest of their lives, and on the other the danger of being labeled a "stinking ninth" or other epithet associated with intellectuals. But youth could not shed their "educated" label even if they wanted to, since official rhetoric frequently spoke of "the intellectuals and student youth" as a single category. The seminal "Little Red Book" (also known as *Quotations from Chairman Mao Zedong*) even conflated the two by including in the chapter entitled "Youth" a quotation that dealt solely (and critically) with "intellectuals" and mentioned "youth" not at all.[6]

The equating of youth and intellectuals may have been a legacy of the May Fourth era (c. 1915–1925), when young minds both literally and symbolically represented the "new thought tide." In a more concrete way, it was an outcome of the emphasis on "educated youth" as a force for cultural change in the new society. Echoing Mao's declaration that youth were the "least conservative" social force, propaganda materials frequently trumpeted the ability of youth to "accept new things." This of course is the key ingredient in transforming agricultural practices, the "conservatism" of farmers being the bane of agricultural reform efforts around the world. And it became a core part of the identity of youth themselves: in survey responses and interviews, former educated youth attribute their participation in the scientific experiment movement to the fact that "young people can accept new things."[7]

Aspiring to be a scientist could be risky for youth if in so doing they acquired too much of a "bourgeois" air. Like their counterparts in the humanities, scientists faced persecution in Mao-era China because of their status as "intellectuals," "authorities," and "experts." But scientists had several things going for them that substantially reduced their risk. First, science was seen as both essential to socialist construction and capable of disproving traditional ways of thinking. Second, science often involved physical work—sometimes dangerous physical work—and thus had some claim to the privileged category of *labor*.[8] Thus it was never science itself that came under attack, but rather elite, bourgeois authority in science. Thinking back to earlier chapters, science could be *tu*: it could call for getting dirty, and it could even entail bodily danger and sacrifice in the pursuit of revolutionary ideals. This in turn shaped the way propaganda presented youth participation. As an energetic, courageous, antiestablishment force, youth could make great contributions to revolutionary science. But when viewed as intellectuals, their role was suspect. How they navigated this contradiction is the central theme of this chapter.

The Politics of Book Learning

The notion that educated youth possessed superior knowledge by virtue of their schooling conflicted with one of the most fundamental tenets of the Cultural Revolution. And so state-produced materials repeatedly emphasized the perils of relying on book learning and the need to learn from peasant experience and root one's own knowledge in practice. A 1972 *People's Daily* article told of a group of Nanjing youth sent down to Jiangsu who were said to have discovered the hard way that what they read in books about seed germination would not necessarily hold in practice. They achieved better results once they followed the advice of old peasants in their experiment group and changed their watering regimens.[9] In another story, when an educated youth first arrived in his new village, he was excited to improve crop production through application of cobalt chloride, which he had learned about in school. But the team leader gently admonished him that the coming heavy rains would make such efforts useless. At first the youth assumed that the team leader was merely ignorant of science, but when not long after a tremendous rain flooded the fields, he reportedly realized the truth of Mao's words: "If intellectuals do not unite with the masses of workers and peasants, then all they do will come to naught."[10]

This is the familiar anti-intellectualism of the Mao era, especially the Cultural Revolution. Nevertheless, propaganda of the 1960s and 1970s frequently referred positively to books and book learning. The most important books, of course, were those containing Mao's own words. But books on science and technology also played an important role. And indeed, it is hard to imagine how the state could have expected youth to fulfill the mandates of agricultural transformation without such books. As one former sent-down youth explains, the knowledge that youth encountered in school was "very formalistic" and without many "concrete details." At the same time, "it was not easy to transform oneself to have a peasant's consciousness." Since "no one provided guidance on the concrete things," the only way for sent-down youth to do what they were charged with doing was to rely on books to supply the needed information. So, for example, he read in a book about how to select seeds for the next planting by identifying middle-height stalks and saving the seeds from those plants. "I didn't invent this. I read it in a book and followed the rules."[11]

Once we recognize the practical necessity, it becomes less surprising that state-produced materials celebrating the contributions of educated youth to the scientific experiment movement regularly highlighted the importance of knowledge obtained from books. One account from a 1965 collected volume asserted that a rural youth's failure in breeding successful strains arose not only from his lack of production experience but also from his lack of scientific knowledge. The solution in the latter case was to "chew on some books." This was difficult for a rural youth with limited education, but going character by character, he reportedly managed to read twenty-two volumes of agricultural theory and thereby achieved better results in the field.[12] A story from a collection published in 1974 handled the practice/book learning dilemma in an innovative manner. Wang Chunling, a rural "returned" youth, was said to have experimented by taking weak piglets and placing them on the front nipples to help them grow more quickly. When she presented these findings at a "conference to exchange experience on livestock work," an "old comrade" explained that this had already been described long ago in books: she had taken a "circuitous path." Criticizing Liu Shaoqi, Lin Biao, and the "theory of innate genius," she concluded that books are the "synthesis of earlier people's experience." Thus, in addition to pursuing one's own practice, it was important to read books so as to minimize circuitous paths.[13] Here books became a way of avoiding the classic bourgeois fault of operating too much as an individual.

Books, magazines, and pamphlets came to educated youth in a variety of ways. Cao Xingsui remembers reading "lots of books" while stationed in the countryside. Some were produced by the extension station and related specifically to agricultural technologies. Others were produced by provincial and county agencies governing sent-down youth. These included not just agricultural materials but all kinds of books: "At that time, they needed educated youth to study, so they provided [books on] archaeology, anthropology, philosophy, Marxism, biology, taxonomy—everything." And of course they included a series of books on agriculture, covering breeding, fertilization, pest control, storage, and other key topics.[14] Another sent-down youth remembers that in the absence of teachers youth "learned from books and pamphlets" provided free by the commune's party committee. He especially recalls the *Plant Protection Handbook* (植保手册) published by the Guangxi Academy of Agricultural Sciences. It was intended to be a multivolume set, but the Cultural Revolution interrupted the work so he had only the one volume on diseases and pests in rice. "I was very interested. I flipped through it all day long. It was my first specialist book." He also recalls many pamphlets on green manure, which was being heavily promoted and was highly effective, and the capture of methane for energy production, though this was only discussed and not promoted in his area until later.[15]

For some youth, the books they received for free from the government were not sufficient to satisfy their need to read. Interviews, diaries, and propaganda accounts all testify to young people's efforts to acquire books in the countryside.[16] One man I interviewed remembers going to all the bookstores in the provincial capital and buying every book he could find on scientific rice farming; Shen Dianzhong's diary records a similar trip to the provincial capital to purchase books.[17] An interviewee who belonged to an experiment group in northwestern Guangxi went to the book store in the county seat to buy books on agriculture. He was interested in the subject, but his main motivation was his lack of formal training and his need for knowledge to fulfill his obligations to the experiment group. After reading about how to make fertilizer, he found he was able to put it into practice.[18] A story in a 1965 volume on youth in the scientific experiment movement used the phenomenon of book buying to emphasize the ideals educated youth were meant to embody. Deng Yantang was an urban youth who had already spent thirteen years in the countryside. Abstaining from smoking and tea houses, Deng saved his money to buy books and magazines. He would go to town intending to buy food, enter bookstores to flip through a few books, and end by buying the books and returning home with an empty stomach. He began by reading pamphlets, then, later, books on agricultural theory. If he needed to understand the meaning of a term, he would look it up in one of his books or write to an expert at the provincial agricultural institute. He reportedly concluded, "In [acquiring] knowledge of breeding, practice is the foundation and books are the path."[19]

Youth sometimes acquired reading material in more unorthodox ways. The Cultural Revolutionary countryside presented youth with a strange combination of restriction and freedom: on one hand, much of the world they sought to explore—from sex to philosophy—was officially off limits; on the other hand, the ceaseless political campaigns on top of the ordinary hard work of rural life often prevented adults from paying too much attention to their activities. Under such conditions, many youth were not above breaking the rules to obtain desired reading material. Of course, we cannot hope to find such stories in propaganda, but memoirs and works of semi-autobiographical fiction emphasize the importance of illicit literature to Cultural Revolution–era youth. Perhaps the most vivid account appears in Dai Sijie's *Balzac and the Little Chinese Seamstress*, in which the forbidden books stashed in one sent-down youth's suitcase provided inspiration not only for his sent-down comrades but also for a young peasant woman they befriended.[20] Still more striking was a story I heard from a former sent-down youth whose friend had a relative employed in a library. They managed to steal the key and absconded with all sorts of books, including some on literature and history, which they then read voraciously. He credits the learning he acquired in that way for his later success on the college entrance examination.[21]

Educated youth went beyond just hoarding illicit books: they actually wrote their own stories and circulated them among their friends.[22] Especially successful stories spread throughout the country, as youth copied the stories by hand, often adding to the narratives along the way. The most widely read of such hand-copied literature was an unpublished novel by Zhang Yang titled *The Second Handshake* (第二次握手), which featured a romance between two scientists, one of whom had traveled to the United States.[23] This was a dangerous business. When they discovered it in 1975, officials objected to the descriptions of American cities, the prominence in the narrative of Premier Zhou Enlai (then out of favor with the radicals), and the risqué love scenes, which no doubt became more elaborate as copiers added their own details. Official critics blasted Zhang Yang for suggesting that science, rather than Marxism, would "save China," a charge Zhang did not deny.[24] Possessing the book brought the risk of imprisonment or worse, and Zhang Yang himself endured four years of prison. No small price to pay for literature, but many youth in the Cultural Revolution proved remarkably brave on this front.

"I Had Knowledge"

Though the pleasure of reading warranted some risk, this was not the only significance books held for youth in the countryside. Reading was essential to their identity as educated people possessing *wenhua*. Today Mao-era educated youth easily recall the sense that their knowledge set them apart from the ordinary peasants in the villages to which they were assigned. As one sent-down youth told me, "The way we thought about it at that time was that we knew that we were knowledgeable, and so we should tell" peasants how to improve agriculture.[25]

The knowledge that educated youth were perceived to possess extended beyond what they would have learned in the classroom. Their *wenhua* equated to modernity and gave them authority over any technology deemed "advanced." Former educated youths I interviewed often dismissed the notion that old peasants could engage in advanced technologies such as spreading chemical fertilizer or insecticide (figure 20). One former educated youth recalled his experiment group's first forays with chemical insecticides. They

FIGURE 20. This photograph from 1964 shows the typical handling of insecticides (in this case, DDT). The young men in the foreground are filling a backpack tank, while those in the background are preparing to spray. The image originates in the same set of small propaganda posters depicted in figure 3, all of which relate to the ways in which industry can serve agricultural production. The caption for this poster identifies the men as workers from the Chongqing Insecticide Factory, who are obtaining "first-hand" experience with their product by testing it in a cotton field, thereby helping peasants kill bollworms while also "systematically conducting survey research." Reproduced from Xinhua tongxun she, ed., *Wei nongye shengchan fuwu* (Serve agricultural production) (Beijing: Xinhua tongxun she, 1964), 10.

had attempted using lights at night to observe the populations, but they could not see anything and gave up. So they put in a requisition for funding and traveled to the county seat to buy BHC, DDT, and two backpack sprayers. "At that time, these were very advanced—no one had them in [the whole commune]. When we used them, the peasants all stood in a circle to watch and thought it very curious and very advanced. After a few hours bugs started dying quickly. . . . The peasants thought it was great."[26] The image of a group of ordinary peasants watching the educated youth in appreciation or even amazement as they demonstrated a new technology captures well the widely accepted notion that educated youth embodied modern culture.

Feeling they possessed education and culture invested youth with enormous confidence. When I asked another sent-down youth whether he was ever concerned about the toxic effects of the pesticides he used, he replied, "Because we were educated youth, we had some *wenhua*, so we used masks and went with the wind. For example, if the wind came from this direction, I would spray like this [pointing an imaginary nozzle in the opposite direction]. We were not like the peasants. Peasants maybe didn't understand this. . . . After all we were educated youth, we had knowledge. I was never poisoned. . . . I used very little . . . like this [miming cautious spraying] . . . not like people on the production team, who sprayed it all over the place whether or not there were any insects—very wasteful. After all we were educated youth, so I thought I was upholding the honor of educated youth. Because I had knowledge. We heard the peasants say, 'I've spent decades farming but am not as good as you who have spent three years farming.' That's not what I said, it's what they said."[27]

Not all peasants esteemed the knowledge that educated youth brought to the countryside. I spoke with one peasant who was middle-aged during the 1970s. He said that peasants had experience and so taught educated youth how to do things such as killing insects with plant-based insecticides. "Their knowledge was still lacking. They did not have innate [knowledge]."[28] Here the man echoed Cultural Revolution–era efforts to debunk "theories of innate genius." And for their part, educated youth sometimes truly disparaged the knowledge of "old peasants." One former educated youth shared with me an unpublished essay he had written on his experiences in which he referred repeatedly to "uncultured old peasants" (没文化的老农) who were still using practices of water management that "had not changed in a thousand years," who failed to understand the "scientific" practice of sun drying the fields, and who were reluctant to learn from him on the subject. "Fortunately," the production team leader backed him up so the team was able to benefit from what he had learned in books.[29]

There are good reasons to critique the scope of the knowledge possessed by educated youth. As agricultural historian and former sent-down youth Cao Xingsui points out, reports of youth actually breeding new crop varieties through hybridization were undoubtedly greatly exaggerated if not downright false—it was simply too complex an undertaking for people with secondary school educations working under the celebrated "crude" conditions of the Chinese countryside.[30] But Huang Shaoxiong—another former sent-down youth who now holds an official position—urges further consideration: "When educated youth brought the knowledge they got in school to bear on agricultural production, of course it was crude and superficial . . . but you can't judge that time by today's standards. In those days, scientific experiment technology was backward; it was then that it started [developing]."[31] These two former sent-down youth are caught in a historical debate with high emotional stakes for their entire generation: on one side is the desire to celebrate the role of young people who sacrificed so much for revolution, red and green, and whose contributions were unquestionably significant; on the other is the queasiness that comes when those celebrations start sounding too much like the propaganda of the Mao era, and so ring worse than hollow. Despite these differences of perspective, educated youth shared a sense of pride in their books and their *wenhua.*

Revolutionary versus Bourgeois Science: The View from Propaganda

Although the state could not avoid encouraging educated youth to rely on book learning in the scientific experiment movement, it made up for this with incessant alarm bells warning of the dangers that bourgeois values posed for intellectuals. The very first principle embraced at the 1965 National Conference on Rural Youth in Scientific Experiment was that scientific experiment served the revolution and was not for the purpose of gaining fame or private profit.[32] A *People's Daily* article on the conference hammered this point home: where youth embraced revolutionary ideals, they succeeded; and where they pursued science for personal fame or profit, they failed.[33]

An account of a brigade-led scientific research team included in a collected volume on the scientific experiment movement published in 1971 in Shanxi Province took a square shot at technocracy and elitism in science. All the members of the team came from poor and lower-middle peasant backgrounds and were said to be members of the "young people's militia on the front line of the three great revolutions" of class struggle, the struggle for production, and scientific experiment. However, in the past they had allegedly suffered from the "poisonous influence" of Liu Shaoqi's "counterrevolutionary science and technology line." Some of them had become "technocratic" (技术挂帅): their "two ears did not hear the world around them," while their minds "burrowed deep into technical books" filled with theory divorced from the struggle for production. The poor and lower-middle peasants criticized them, saying, "Speak of theory and heaven's flowers fall all around, but talk about practice and they're all sloppy mud and runny eggs."[34] ("Heaven's flowers fall all around" is a Buddhist expression indicating the artifice and emptiness of clever speech. "Sloppy mud and runny eggs" presumably implied they lacked not only coherence but also even the pretense of prettiness.) The peasants reportedly further complained, "Their clothing is lavish and their speech foreign; they've lost the appearance of the poor and lower-middle peasants and soaked up the stench of the capitalists." Worse yet, the account accused the Liu Shaoqi–influenced youth of having "taken the technologies that the poor and lower-middle peasants slaved to teach them and secreted them away in their own sacks instead of sharing them with others."[35]

Similar examples abound. In 1972, *People's Daily* introduced a group of youth sent down from Nanjing to a production brigade in rural Jiangsu: "At one time because some of the youth had been influenced by capitalist-class ideas about fame and profit, their experiment topics departed from the practical needs" of the production brigade. They were "seeking overnight fame" and kept "holding out their hand for chemical fertilizer" so they could achieve a high yield. The party secretary organized them to engage in revolutionary criticism so they would realize that "scientific experiment is not about making individuals famous, but about transforming the face of the countryside."[36] In another example, a group of sent-down youth in Hebei Province set up an experiment station with the support of the local party office. Instead of allowing themselves to be guided by the "old peasants," however, they allegedly pursued impractical ideas in an attempt to "startle" people with their innovation, such as by hybridizing cotton and paulownia to create a perennial "cotton tree." When local party officials became aware of the problem, they reportedly educated the youth about the importance of uniting with the masses, such that the youth became very successful in designing new forms of pest control and fertilizer that served real needs rather than merely winning attention.[37]

In an effort to distinguish "bourgeois science" from revolutionary science, propaganda often strove to portray youth efforts in the scientific experiment movement as gritty and daring. Youth were said to possess the energy and courage necessary to try new ideas and to withstand physical hardships. During the Great Leap Forward, the media celebrated the discovery of large deposits of minerals by a "young girl" and geologist "hero" named Liu Jinmei. Liu "traversed some 7,000 kilometers in the towering ridges of the rugged Changpai Mountains in Northeast China, the haunt of tigers and bears."[38] Agricultural science offered few opportunities to scale "towering ridges," so propagandists highlighted instead the dizzying effects of agricultural chemicals and other hazards. In 1973 Lang Yuping, a returned youth in Miyun County (near Beijing), sought to control a wheat virus with a highly toxic chemical: "One time I really was poisoned, dizzy, nauseated, sweating. . . . I was scared to death that I . . . would lose the fall crop. I went to the clinic, got an injection, and continued work. The party secretary told me to go to the hospital to get a checkup and rest a few days, but I didn't go."[39] Nominally self-critical, the account in fact emphasized the youth's courage and willing self-sacrifice. Other examples abound of young people celebrated for braving cold, rain, mosquitoes, and sweat, and going so far as to refuse medical treatment when ill, all because of their dedication to science and to production.[40]

And so, even as propaganda warned youth against sliding into the character of ivory-tower, work-shirking, fame-seeking, capitalist children, it also tempted them to imagine themselves as revolutionary heroes. Encouraging youth to aspire to heroic deeds, propaganda materials provided plenty of room for celebrating individual efforts and achievements (figure 21). The books that came out of the youth conferences on scientific experiment highlighted the experiences of notable individuals to offer models and inspiration. Consistent with a distinctively Maoist form of knowledge production based on "personal subjectivity [as] the basis for the pursuit of objective reality" and inspired by Mao's investigations of rural conditions during the

FIGURE 21. From a collection of small propaganda posters designed for display in rural areas. Here we see Hebei sent-down youth Cheng Youzhi (程有志) "with his comrade in arms, researching the use of sterility in wheat." (Plants with male sterility are of use in producing hybrid seed—see chapter 4.) The photograph's caption, which calls Cheng a "*tu* expert," further highlights his accomplishments working with "the poor and lower-middle peasants" to research effective pruning of fruit trees and his successful breeding of more than seventy improved varieties of crops, which beyond Hebei had reportedly been extended in fifteen other provinces and municipalities. Reproduced from Xinhua tongxun she, *Zhishi qingnian zai nongcun* (Educated youth in the countryside) (Beijing: Renmin meishu chubanshe, 1974), 9.

revolutionary period, young people who related their stories of scientific experiment spoke of "my own personal practice" (自己的亲身实践) as their source of knowledge.[41] And third-person narratives portrayed youth in highly individualized, even romantic ways. Deng Yantang "had a tanned face and short hair, wore a blue shirt with bare feet, and was dirty from head to toe." This exquisitely humble young man pursued new strains of rice "like a brave explorer finding the path"; his "influence over youth throughout the county grew day by day," and he achieved a following of other young technicians who continually sought him out to ask questions and learn from his experience.[42] A 1974 magazine celebrated an urban youth named Xin Wen, recently graduated from junior high school, who volunteered to be sent down to Yunnan to plant cinchona (the tree from which quinine is made). Elected unanimously as the leader of her experiment group, Xin quickly began demonstrating strong leadership qualities. According to the article, she sacrificed her siestas to experiment with different ways of addressing evaporation and the cinchona trees' weakness in pushing through thick soil. Soon, the entire group began using her methods. When the weather turned cool and rainy, she determined that the group should heat soil in pots to keep the young trees warm. She worked the longest hours, shouldered the heaviest responsibilities, and made all the big discoveries. And she was honored for it.[43]

Propaganda thus offered youth very mixed messages about their participation in science: they were simultaneously encouraged to imagine themselves as heroic individuals and warned about the political dangers of "seeking fame." This profound ambiguity carried over as well into the positions that youth were expected to take with respect to local peasants. Propaganda narratives and images depicting youth in science almost invariably emphasized not only the guidance of trusted cadres (figure 22) but more importantly their cooperation with, or even subservience to, peasants (figure 23). On the other hand, propaganda also sometimes characterized peasants as repositories for backward, conservative, or even reactionary thinking, and accused peasants of viewing scientific experiment as a frivolous activity engaged in to avoid real work in the fields. According to propaganda accounts, youth had to endure snide comments referring to their scientific experiment work as "a new plaything."[44] (A returned youth I interviewed confirmed with irritation that some people thought scientific experiment was "just playing," when in truth "it was not playing—it was hard"; a sent-down youth similarly told me that a few people in his production team "made sarcastic remarks about youth liking to go play.")[45] The contradiction inherent in official attitudes toward peasants meant that propaganda presented far from clear guidance on when to bow to peasants and when to stand up against them.

FIGURE 22. Wang Junliang, "Jingxin peiyu" (Cultivate meticulously), 1972. Youth receive guidance in their scientific experiments (probably producing either plant hormone or microbial fertilizer) from a cadre, identifiable by his hat and jacket. Stefan R. Landsberger Collection, International Institute of Social History, Netherlands, http://chineseposters.net.

Chapter 5 examined narratives of class struggle in state-produced materials for clues about peasant resistance. These narratives were also plentiful in materials specifically on educated youth, where they served to illustrate the political stakes involved in the adoption of new agricultural technologies. A story written in late 1965 recounted the repeated efforts of a youth named Zhang Yankao to convince local peasants of the worth of his experiments. In 1957, some "opinionated" members of the collective complained that he had taken a perfectly good field and planted it unevenly, with some patches growing long and some shorter such that the field "looked like a spotted leopard." Yankao patiently explained why the patches had to be different. When he left for meetings at the county seat, one of these naysayers went so far as to cut down the rice plants and replace them with corn, so Yankao explained again

FIGURE 23. Here a sent-down youth, Lin Chao (林超), credited for having bred the widely used improved variety "white breast king soybean," receives guidance in "soybean management technology" from an "old peasant." Reproduced from Xinhua tongxun she, *Zhishi qingnian zai nongcun* (Educated youth in the countryside) (Beijing: Renmin meishu chubanshe, 1974), 9. For an account of the breeding of this specific variety, see Heilongjiang sheng, "Bai ying dadou."

the need for experiment fields. In 1962–1963, a "big struggle" between forces for and against scientific experiment came to a head. Yankao established a scientific experiment group with four specialized research teams for seeds, fertilizer, cultivation, and plant protection. The group received widespread support from poor and lower-middle-class peasants, but the vice-team-leader—an upper-middle-class peasant named Zhang Jiatao—allegedly considered scientific experiment too "cumbersome" and worried that such waste of resources would lead to food shortages. The dispute was reportedly settled through friendly competition, with each side permitted to pursue its own methods; when Yankao's group achieved better results, Jiatao conceded, admitting the reasons for Yankao's success.[46]

In many stories, opposition from reactionary forces was balanced by the enthusiastic support of the "poor and lower-middle peasants." But sometimes poor peasants themselves were said to have opposed the research. Deng Yantang reportedly succeeded in hybridizing rice strains but encountered criticism from old peasants, who said that the hybrids looked good but tasted bad, and that "high lanterns see far, but not close." The party secretary affirmed the peasants' criticism and reminded Yantang that scientific experiment must serve production. Yantang reread Mao's "Serve the People" and his thoughts clarified: "Yes! Cultivating superior breeds appears to be a question of technology, but first is the political-orientation question of breeding for whom and serving whom. If the orientation is wrong, the experiment will go off track."[47]

In the more heated climate of the Cultural Revolution, the class politics of such accounts sharpened considerably. In one story, published in 1974 about an incident in spring of 1969, county leaders called on returned educated youth in a three-in-one experiment group to hybridize sorghum. This was a "brand-new thing" for them: no one had known before that sorghum plants were male and female or that they had "sex." Many people were doubtful, and the youth wondered if they should go forward, but they determined that "science is about making a new road with new methods." When the initial experiments failed, "class enemies" took the opportunity to attack science. Rich peasants said, "If you're going to do that, we'll have to prop our mouths open [把牙支起来, i.e., resign ourselves to starving]." But the production brigade party secretary organized the youth with the commune masses to struggle against the class enemies and study Mao, with the result that the youth learned to sex the sorghum plants more accurately and ultimately achieved success.[48]

Peasants could offer their criticisms with remarkably clever rhythm and rhyme, and this was well captured in accounts of youth experiment activities, where conservative factions or class enemies were said to have delivered taunts like "Experiment, experiment; grain, but no gain" or "Little children conducting scientific research, that's like a rabbit trying to pull a horse cart."[49] Hearing such attacks reportedly made the science team members realize that their research was not just about increasing production but also about fighting a political battle.[50] And it is no surprise that young women in these accounts were expected to confront especially charged abuse: stories of their ordeals highlighted not just class struggle but also struggles against patriarchy. A report from the 1965 National Conference on Rural Youth in Scientific Experiment explained that in the early days of their work, some people scolded young women engaged in livestock breeding, saying, "You spend all day mating donkeys and horses" (成天配驴配马).[51] An article from 1974 recounted a female educated youth who set out to learn swine veterinary skills and encountered the scorn of "conservative-minded people," who said, "Female comrades can't do this kind of work; imagine girls castrating pigs and having no shame!"[52] Practicing scientific agriculture was thus said to be a way of overcoming sexism and conservative thinking, waging the kind of cultural revolution that youth had been told to lead.

Revolutionary versus Bourgeois Science: Youth Perspectives

So much for the propaganda: How did youth themselves feel about science? Diaries, memoirs, literature, and interviews speak to this question. They demonstrate that some youth took to heart the cautionary tales of bourgeois temptation even as they aspired to scientific heroism. The sources further indicate that many youth embraced the state's portrayal of science as revolutionary; at the same time, it is clear that youth often found other elements of the state propaganda less appealing. And today we find former educated youth with very diverse perspectives on the meaning of their experiences, including on the degree to which their spirit and ideals mattered to history.

Between the frequent warnings in propaganda and the skepticism of some peasants, it is not surprising that youth worried about the impressions they made. Sent-down youth Shen Dianzhong was sensitive to the perception that people who pursued scientific research were doing so for the wrong reasons. Before beginning his involvement in the scientific experiment movement, he penned a diary entry in which he quoted Marx on the relationship between studying science and serving the people: "Science must not be a selfish pleasure. Those who have the good fortune to be able to devote themselves to scientific pursuits must be the first to place their knowledge at the service of humanity."[53] Nonetheless, he later invited criticism by talking too openly about wanting to write a book: others were quick to chastise him for seeking fame. Shen complained that his critics were "focusing on the motivation question"; he defended himself by saying that his "experience could offer profound warnings to people"—he emphasized that this, and not self-aggrandizement, was his true ambition.[54]

Perhaps the most compelling evidence of youth commitment to revolutionary science lies in the phenomenon of the hand-copied novel *The Second Handshake*. The story revolves around two patriotic young scientists who fall in love in the early decades of the twentieth century. Fate separates them when the woman, Ding Jieqiong, goes to the United States to study and then to work on the atom bomb project, which she eventually exposes as a weapon for killing civilians. In 1959 she is reunited with her old flame in the fatherland—but too late to pick up where they had left off.

The author of *The Second Handshake*, Zhang Yang, has recently published a book-length account of his writing of the novel in which he traces the childhood origins of his passion for science and his deep affection for scientists. (The memoir's tremendous resonance with the values of the post-Mao era— the adulation for Zhou Enlai, the bitterness about the treatment of intellectu- als, the faith in science over politics—reminds us to use this helpful resource with caution.) Zhang recalls following the political travails of intellectuals through the 1950s and 1960s, suffering vicariously when the Anti-rightist Movement crushed the hopes of the mid-1950s "March toward Science" promoted by Zhou Enlai, and worrying about the fate of "regular scientists" when the Sixteen Articles on the Cultural Revolution called for "greater pro- tection for scientists making contributions." As he explains his decision to write the story of *The Second Handshake*, "Since I could not become a scien- tist myself, I used my pen to portray scientists, to represent them, to eulogize them so that my readers could understand them, respect them, and love them just as I did!" The scientists he chose to present were of his uncle's generation (the same generation as insect scientist Pu Zhelong), educated in the "old society" but ready and enthusiastic to participate in revolutionary work— for example, by researching methods to combat the germ warfare allegedly waged by US forces during the Korean War. Notable also is his decision to place a woman in the most prestigious scientific role: Ding Jieqiong inspired untold numbers of young women with scientific aspirations—in later years an official with a similar name was driven to distraction by fan letters asking her advice and encouragement.[55]

The book's presentation of scientists as courageous, patriotic, and roman- tic clearly resonated with young readers. It also strikingly mirrored many revolutionary themes found in state propaganda. In 1979, after the end of the Cultural Revolution and the birth of a new regime, China Youth Press printed 3.3 million copies of a cleaned-up version of the novel. By the late 1980s, it was "the most widely circulated story of any kind in the history of the People's Re- public."[56] That youth in the Cultural Revolution would choose such a novel in defiance of political authorities speaks volumes about their internalization of the value of science. And that the scientists it described were so imbued with heroism testifies to the degree to which the state-sponsored vision of science had become their own.

Elements of that vision continue to come across vividly in the memories of people who participated in scientific farming as youth, evidenced in the written responses to a survey I circulated among participants in the insect control work at Big Sand. Chen Haidong was a returned educated youth with a story that could have been torn from the pages of a Cultural Revolution– era magazine. He hailed from Dongguan—an area of Guangdong then famous for lychees (now famous for export manufacturing and migrant worker riots)—and so had early on heard of Pu's work with *Anastatus* wasps on lychee pests.[57] As he describes his introduction to scientific experiment at eighteen years old: "Young people can accept new things, and so I participated in the brigade's scientific research group." The group employed "native methods" (土法)—for example, using "crude and simple" (简陋) equipment to produce *Beauvaria* bacteria to control insect pests. He recalls that the work required "not only abundant scientific knowledge but also the spirit and persistence to do scientific research."

Although youth embraced the revolutionary, *tu* vision of science promoted in state propaganda, they did not embrace all aspects of that vision equally. Heroism, patriotism, endurance, scientific spirit, and self-sacrifice were popular themes. As in propaganda accounts, a common theme in interviews with former educated youth involves close encounters with toxic chemicals—for example, the routine of mixing the insecticide BHC with their bare hands.[58] Taking risks and engaging in back-breaking labor to "serve the people" and improve production were sources of pride. The valorization of peasant knowledge and class struggle, however, left many of them cold.

When I asked one former sent-down youth whether he saw his experiences in the scientific experiment movement in terms of class struggle, he replied decisively, "Actually, class struggle was just talk. We never did anything with it. People in the production team just wanted to resolve questions of food. Peasants didn't want to talk about class struggle; only the government did, only Mao Zedong." But he did not mean that youth lacked passion for other revolutionary ideals. He continued, "Serving the people was different. Everyone believed in that. . . . In those days, speaking from our hearts, we truly were serving the people. . . . We were truly willing to offer our abilities and our knowledge to the local peasants, truly serving the local people, we were very happy in our hearts to be serving the peasants."[59] Another sent-down youth I interviewed struggled to put her feelings into words, but in the end she said something very similar: "Educated youth being sent down, taking root in the countryside . . . I think lots of people nowadays would not be willing to say this, but at that time we listened very much to Chairman Mao. We very much wanted to make revolution, and yet it was not really revolution, but we very much wanted to do something for our fellow villagers. . . . It was like that then, that was your state of mind to want to do something for your fellow villagers." Both of these former educated youth had very ambivalent feelings about Mao's calls to engage in "class struggle" and "make revolution," but they nonetheless look back on their time in the countryside as one of passionate commitment to the revolutionary ideal of serving the people.

These ideals were, and remain, so important to educated youth that memories of transgressions carry a sharp taste of shame. A conversation I had with three former educated youth, two men and a woman, demonstrates this. It was a lively discussion, and one of the men had to try several times before he could put a word in edgewise to get his story off his chest. It turned out that he and his comrades in a scientific experiment group had "played a joke": charged with fertilizing a large orchard, they opted instead to bury all the fertilizer around just a few of the nearby trees. When the trees died soon afterward, their action became a "political incident." The woman attempted to put a gentle spin on the misdeed, characterizing it as a "result of not understanding science," but the second man declared it "irresponsibility." The culprit himself called it "laziness." But he added, "If we had guessed what would happen, we would have died rather than do such a thing."[60]

Former educated youth today struggle to understand the historical significance of the revolutionary spirit they were expected to display—and that many of them "truly" (as one sent-down youth quoted above repeatedly emphasized) felt. A conversation among three other youth shows how important, and how complicated, the issue remains for them. Huang Shao-xiong is now vice-chair of the Hezhou Municipal Chinese People's Political Consultative Conference (CPPCC). In 2010, he edited a collection of essays by former educated youth of Guangxi Province, and in 2012, he organized a research symposium on the history of educated youth under the auspices of the Guangxi CPPCC. In keeping with other such events organized in recent years, his main theme was the tremendous energy, enthusiasm, and spirit that the youth brought to their work in the countryside. Pan Yiwei is a former sent-down youth who participated in the conference and whose essay appears in the collected volume. As the next chapter will show, he is one of the many sent-down youth for whom life in the countryside did not lead to future opportunities; those years were the highpoint of his contribution to society, and it is perhaps not surprising that he agrees with Huang's take on the significance of sent-down youth. Chen Yongning is a former sent-down youth who studied plant protection science in college and parlayed this knowledge into a series of successful business ventures, most recently in an endeavor that combines organic farming and ecotourism.

PAN: At that time people's thoughts were more revolutionary than they are today. . . . No one would say, "I'm not going [to participate in the experiment group]." They would compete to go. It didn't matter if it was returned youth or educated youth or what. Everyone did it. I had lots of skills, and I read lots of books. There was no television, no Internet or whatever.

HUANG: Educated youth had a spirit of delving into study. . . . And when they participated in production, they feared neither hardship nor death, they would get muddy from head to toe and temper their whole red heart. . . .

CHEN: I have a different view. . . . Actually at that time science and technology extension and research were highly emphasized [by state officials], otherwise why would they have established scientific experiment groups? Did you yourself set it up? No, right? First it went to the commune revolutionary committee, then to the production team, and then on down. There was already a system and a kind of administrative pattern. Even though the level wasn't that high, overall there was this administrative awareness. There were calls from administrative departments, including our experiment groups. Why was I able to become the group leader? Not because I made the call, not because I took the lead. It wasn't like that. We were organized and assigned to jobs. It was the production team leader who did the assigning, [saying] "You, be the group leader!" Then whoever became the members of the experiment group, everyone who did it with you, this was also organized administratively. . . .

HUANG: To be an educated youth, you needed enthusiasm to participate in the activity. Administration was one thing, but in the end educated youth needed to have enthusiasm, otherwise you wouldn't have put your name forward.

PAN: That's how it worked where I was. . . . It wasn't a mandate, it was us ourselves. . . . The upper levels handed down administrative mandates, but they didn't carry them out at the production team level.

CHEN: First of all, officials, leaders needed this awareness in order for it to work. So at that time they had to consider how to use scientific farming as a factor in driving production.

PAN: During the Cultural Revolution, I was in the rebel faction. Why am I saying this? The motivation for those of us in the rebel faction was that we had received unfair treatment, so people like us really wanted to do something and had something to prove. I wasn't that kind of person—like what Chen said, stuff being promoted level by level through the administration, through whatever commune, brigade. . . . If our production team leader hadn't emphasized improving varieties, I would have had no way of doing it.

CHEN: Right. He gave you a platform.

PAN: He knew how poverty tasted. As soon as he saw that you had some *wenhua*, [he would say], "I'll give it to you to do and I'll see if you can do it, and if you can do it, then I'll benefit too."

The conversation among these former sent-down youth parallels a larger debate over the historical significance of their generation—to some extent, this also shaped the discussion between Huang and Cao above. Like many other officials and former sent-down youth, Huang and Pan are actively promoting a view on the Cultural Revolution that emphasizes the agency of educated youth. Chen, on the other hand, follows a different school of thought that recognizes instead the significance of top-down policy and administration. As a successful agricultural technician and businessman, Chen has considerably less at stake than Pan in particular when it comes to the historical verdict on their generation's role in the Cultural Revolution. Pan's star rose and set during his time in the countryside; he is one of many former sent-down youth for whom the post-Mao era did not bring opportunity but rather disappointment. If the revolutionary spirit of sent-down youth was not the driving force it seemed, the disappointment is still greater.

Bourgeois Science? The Post-1978 Transformation of Youth and Science

Reading the Cultural Revolution–era criticisms of bourgeois values in science, the tendency may be to dismiss them as overly ideological and ultimately ridiculous. More importantly, it is undeniable that the criticisms contributed significantly to the climate of harassment that resulted in physical and mental injury for vast numbers of scientists and young people. The use of the specter of "bourgeois science" to persecute people was inexcusably immoral and undeniably destructive. However, when we compare the vision of youth participation in science during the Cultural Revolution with that promoted just a few years into the reform period, it becomes clear that bourgeois science was not just a silly fantasy concocted by propaganda spinners. There are real decisions to be made about how to conduct scientific research, what questions to pursue, and the relationship between scientists and other members of society—and all these decisions influence the way people imagine the role of youth in science and the role of science in young people's lives. A brief look at what happened to the state's vision of science and youth after 1978 will help put bourgeois science into perspective.

After the death of Mao and arrest of the "Gang of Four," urban sent-down youth and their supporters began increasingly to resist the rustication policy. In 1978 and 1979, coinciding with the wave of prodemocracy actions throughout the country, great numbers participated in protests calling for a return of sent-down youth to their homes.[61] The leadership had no ready solution to the problem of the rusticated youth. On one hand, Hua Guofeng and Deng Xiaoping had both condemned the travesties wrought by the Gang of Four.

On the other hand, bringing fourteen million young people back to the cities and finding jobs for them was no easy matter. The publication in 1979 by the Agricultural Press of a collection of stories about urban, sent-down youth entitled *Young People Bravely Scaling the Heights: The Scientific Experiment Achievements of Educated Youth on State Farms* reflected these political tensions.

In some ways, the stories in this 1979 collection were very much in keeping with earlier accounts. They celebrated the positive scientific contributions youth were making in the countryside. They continued the familiar narrative in which youth faced opposition from politically suspect characters but flourished under the guidance and with the enthusiastic support of the peasantry and the party (now represented by Chairman Hua). And they still emphasized the courage of youth in overcoming adversity and boldly trying new ideas. Whether skipping meals and sleep or braving bad weather to observe conditions in the fields, youth were participating in the same kind of heroic scientific endeavor that had been celebrated throughout the 1960s and 1970s.[62]

Nonetheless, the 1979 stories differed from those of recent years. Not only did they identify the Gang of Four as the principal obstacle to scientific progress, but they emphasized book learning in a far more unambiguously approving way. In several of the stories, the youths in question were positive bookworms. In one story, a youth was celebrated for having resisted the "anarchism" of the Cultural Revolution and the Gang of Four's political labeling by continuing on his own to study math, physics, mechanics, and combine harvester theory and design.[63] The title story of the collection introduced a youth who since middle school had loved to read science fiction stories and science magazines. When assigned to be a "plant protector," he bought books from the bookstore on preventing plant diseases and pests. Once, he sacrificed a bus ticket to buy books, even though it meant traveling back to the farm by boat. He read during every possible moment—while eating, instead of sleeping, and even while walking down the street. One day, lost in thought about the books he was studying, he walked right into a utility pole before coming to his senses.[64] This story resembles the 1965 account of Deng Yantang, who abstained from smoking and even went without lunch to buy agricultural books. But in the 1970s, an anecdote presenting an urban youth with such pronounced bookishness would undoubtedly have signified the need for intellectuals to get their heads out of the clouds and forge solid relationships with peasants who had real experience in agriculture. Now, such bookish leanings were unambiguously a point of pride, indicating a keen mind not distracted by such mundane things as buses or utility poles.

Young People Bravely Scaling the Heights was among the last gasps of support for the rustication movement, a policy that could not be saved in the new political climate. The tide was turning: intellectuals would no longer be called to be reeducated in the countryside. In keeping with this trend, in 1979 the popular science magazine *Scientific Experiment* suddenly and decisively accomplished a shift in orientation. Where in previous years agriculture along with other obviously "mass" sciences had filled the pages, it now disappeared as a category in the index. Instead, the magazine began publishing more articles of general interest to urban youth—for example, on breakthroughs in computer technology or new knowledge about Mars from NASA's Viking space probes. We can imagine them curiously poring over the stories in their city homes, without thought of applying the knowledge in the here and now. Not only were these urban youth returned to urban settings, but rural youth had disappeared almost entirely from the stage.

The pages of a brand-new science magazine, *Science for Children*, reveal still more clearly the changes in values that the post-Mao era would bring. In the inaugural issue, published in January 1979, we find an article titled "The Future Countryside" (figure 24). What is "bourgeois" about this vision is not the technology. The high degree of mechanization featured in the accompanying illustration, including an airplane dusting the crops, represents a continuation of a dream prominent since the 1950s (see figure 3); even the notion of remote-controlled field machinery had a precursor in Great Leap– era visions of the future.[65] What is startling, however, is how quickly the perspective has shifted away from that of peasants and urban people tempering themselves through immersion in the countryside. Instead, depicted here are a young urban girl and her grandfather zooming by the fields in a private convertible. There is certainly an implied reduction in labor (not to mention employment) for peasants, but the real promise seems to be the replacement of the "sent-down" program with "just-visiting" tourism of rural areas for urban people.[66]

Beyond catering to urban fantasies, the magazine also offered a more inspiring vision, though one that still very much reflected what in the Mao era would have been labeled "bourgeois science." The inaugural issue began with the reprinting of a letter Mao wrote to his sons in 1941 urging them to "take advantage of your youth to study more natural science and talk a little less about politics."[67] This was followed by a poem by the famous science popularizer Gao Shiqi entitled "Spring," which brought to children a theme China's top leaders had embraced for the beginning of the new era: "Springtime for science" (科学之春).[68] The idea was that China would now move away from the political struggles that had dominated recent years and instead

FIGURE 24. A cartoon accompanying an article from the inaugural issue of *Science for Children*. Compare this vision of the future of agriculture with that in figure 3. Both imagine the use of airplanes for crop protection, but the protagonists have shifted from peasants to urbanites. Reproduced from Shan Ren, "Xiangcun de weilai" (The future countryside), *Shaonian kexue* 1979.1: 38–44.

focus on modernizing the country by investing in science and technology—which along with agriculture, industry, and national defense constituted the "Four Modernizations" enshrined as cornerstones of Deng Xiaoping's platform. "Where is spring?" Gao asked, and then answered, "Spring is you; You are the ancestral country's spring." He further elaborated, "Today you study science culture; tomorrow you'll turn around and realize the great responsibility of the Four Modernizations."[69]

The use of spring as a metaphor for science and youth was certainly not a dramatic break from the past. Rather, it strongly recalled the May Fourth legacy that had also infused the 1960s and 1970s, when young people were called to give their "youth" (青春, literally, their "young spring") to their country, and when scientific experiment was contrasted with the old, dead knowledge to be found in the ivory tower. Nonetheless, there is a subtle but important way in which this spring was unlike that evoked during the Mao era. Instead of being called to transform the present with their energy and courage, young people after 1978 were portrayed as China's future. In issue after issue of *Science for Children*, distinguished scientists wrote articles encouraging children to study hard because "the future of science rests on your shoulders."[70] This is far more similar to the conceptualization of the value of youth and children found in capitalist countries: they are to be invested in for future dividends. It is no coincidence, then, that this shift occurred as the post-Mao leadership began to steer China in a direction dominated by the logic of a market economy rather than revolutionary politics. It was part and parcel of the more sober sense of planning that characterized Deng's economic program, which celebrated not the glory of struggle today but the possibility of a better tomorrow.

Opportunity and Failure

When I first met Pan Yiwei, it was obvious he was bursting with desire to tell his story. He began by crediting Huang Shaoxiong, who had brought him to the meeting, for organizing a recent conference on educated youth and allowing him to participate. "If it weren't for him recommending [推荐] me, no one would know who I am," he said. "Why? I'm a low-profile kind of person without any influence in society. But in farming at that time, I was a complete expert." Pan's self-deprecating words and the high-energy enthusiasm with which he delivered them sparked a ripple of sympathetic laughter through his audience—three other former sent-down youth and myself. It was a perfect preface for the story he proceeded to tell, in which success and disappointment were twisted tightly together.

During his years as a sent-down youth, Pan's abilities had been recognized by his production team leader. He had helped introduce a new variety of rice and for this had been granted the title of agricultural technician. But his accomplishments in the countryside did not translate into any further opportunities. Rather, he failed to gain entrance to college, and he was "too knuckleheaded" to become a party member when it was offered, a lasting regret. His greatest claims to fame in the decades since his return from the countryside were his participation in Huang's conference and a few essays he submitted to newspaper contests: In 2004, he was first runner-up for an essay contest on a "dream trip" that described his visit to the village of his sent-down years and his hope that he might help the inhabitants by raising money for a new irrigation system; the story of his sent-down experiences, return to the village, and winning of the commendation also received newspaper coverage (figure 25). And in 2007, he wrote an essay entitled "1977: The Lost Dream of College" for a contest marking the thirtieth anniversary of the re-

FIGURE 25. A compelling visual illustration of the layering of historical memory. I clipped this from a typed-up copy of an article former sent-down youth Pan Yiwei provided me during my interview with him in 2012. He clipped the photograph from a newspaper article based on an interview with him in 2006. The photograph shows the journalist holding a Cultural Revolution–era photograph Pan had shared during that encounter, with Pan seated in the background. The original photograph, which Pan treasures to this day, is dated 8 October 1972 and depicts Pan and a fellow sent-down youth performing seed selection in a rice field; it commemorates Pan's contributions to scientific farming and the improvement of production in his village. Cai Limei, "Lao chaqing de 'yinshui meng'" (The "irrigation dream" of an old sent-down youth), *Nanning ribao*, 10 January 2006, back page.

sumption of college admissions examinations after the Cultural Revolution. For this effort, he won second prize—worth 500 yuan (less than $100). The arc of failure defines his life story, but so too does his lasting sense that he did something important, and he is understandably proud of having taught himself about new agricultural technologies and having made them work for his team.[1]

Pan's story, like many other accounts of youth in the countryside during the Cultural Revolution, tends to vacillate between two starkly differing narrative arcs: one defined by "opportunity" and the other by "failure." Propaganda accounts during the Cultural Revolution hewed closely to a framework inspired by Mao's 1955 declaration, "The countryside is a big world where much can be accomplished." Sent-down youth were encouraged to see themselves as heroic for their efforts to build socialism under difficult conditions, and were urged to "temper" themselves through hardship and struggle, the way steel is tempered in a furnace. The countryside was said to offer opportunities to do great deeds and transform oneself in the process, and for many sent-downers joining the scientific experiment movement was

PLATE 1. "Pigs Are 'Fertilizer Factories' as well as 'Treasure Bowls.'" Reproduced from "Zhu shi 'huafei-chang' you shi 'jubaopen'" (Shanghai renmin meishu chubanshe, December 1959). Stefan R. Landsberger Collection, International Institute of Social History, Netherlands, http://chineseposters.net.

自力更生 勤俭办科研

华容县四级农科网认真贯彻执行"自力更生,艰苦奋斗",勤俭办一切事业的方针,把大搞农业科学实验放在依靠群众自己力量的基点上。他们领导群众开沟挖渠,平整土地,建立起科研基地,坚持就地取材,土法上马,自制设备,解决了农业科研的需要,推进了群众性科学种田活动。

一九七一年春,为了推广温室育秧,华容县曾建立了一座"示范"温室。由于造价贵,推广不了。新河公社新建大队用土砖木材代替红砖和钢筋水泥,用薄膜代替玻璃,用户苗做秧板,仅花了十多元。这种"土温室"受到群众的欢迎,很快在全县得到了推广(左图)。各级农科组织,发动群众采取因陋就简,以土代洋的办法,解决科学实验需要的设备。他们用筷子代替天平,用土铁代替发芽箱,用灶头加温代替温箱。这是景港公社知识青年用土铁进行科学实验(上图)。

PLATE 2. "Fertilizing the Cotton Fields," by Zhang Fangxia. Reproduced from Fine Arts Collection Section of the Cultural Group under the State Council of the People's Republic of China, *Peasant Paintings from Huhsien County* (Peking: Foreign Languages Press, 1974), 34.

PLATE 3. "Self-Reliance; Practice Scientific Research with Diligence and Frugality." Reproduced from Xinhua tongxun she, ed. *Dagao kexue zhongtian, jiasu nongye fazhan* (Greatly undertake scientific farming, accelerate agricultural development) (Beijing: Renmin meishu chubanshe, 1975).

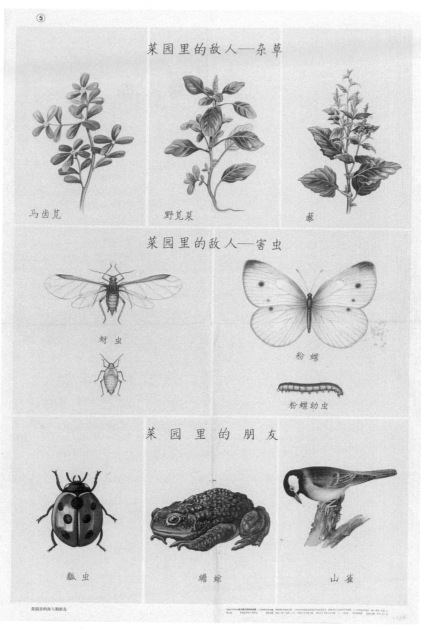

菜园里的敌人——杂草

马齿苋　　　　　　　野苋菜　　　　　　　　藜

菜园里的敌人——害虫

蚜虫　　　　　　　　粉蝶

　　　　　　　　　　粉蝶幼虫

菜园里的朋友

瓢虫　　　　　　　蟾蜍　　　　　　　山雀

PLATE 4. "Enemies and Friends in the Vegetable Garden." Reproduced from Renmin jiaoyu chubanshe, ed., "Caiyuanli de diren he pengyou" (Beijing: Jiaoyu tupian chubanshe, 1956).

PLATE 5. "Eight-Character Charter." Reproduced from Xu Jiping. "Nongye bazi xianfa" (Shanghai: Shanghai renmin meishu chubanshe, December 1959). Stefan R. Landsberger Collection, International Institute of Social History, Netherlands, http://chineseposters.net.

大学办到咱山村

PLATE 7. "The University Has Moved to Our Mountain Village." Reproduced from Hong Tao, "Daxue ban dao zan shancun" (Beijing: Renmin chubanshe, November 1976). Stefan R. Landsberger Collection, International Institute of Social History, Netherlands, http://chineseposters.net.

PLATE 8. "Cultivate meticulously." Reproduced from Wang Junliang, "Jingxin peiyu," 1972. Stefan R. Landsberger Collection, International Institute of Social History, Netherlands, http://chineseposters.net.

PLATE 9. "Thanks to Wufeng, I Learned Scientific Breeding." Photograph taken by the author in Huarun Baise Hope Town.

PLATE 10. "Local" chickens for villagers' consumption wander through idyllic courtyards in Huarun Baise Hope Town. Photograph taken by the author.

the most promising of those opportunities — certainly it was one that fit their intellectual aspirations as educated youth. The far greater number of rural youth who returned to their villages after graduating from urban secondary schools were arguably even more important targets of the Mao-era "opportunity" narrative.[2] It took much convincing for rural youth to accept that a return to the farm would constitute a great opportunity for accomplishment (大有作为) rather than a waste of their talent (取材). Hence the significance of rural programs to encourage youth to engage in scientific experiment.

Overshadowed in the historical record by the splashier optimism found in propaganda posters, the failure narrative too has roots in the Mao era. Looking specifically at materials written on the scientific experiment movement uncovers much mention of failure. And for good reason: failure was a common experience and youth needed encouragement to view it in positive terms. Hence the frequent repetition of the old axiom "Failure is the mother of success" and the characterization of science as a "winding road" filled with obstacles and false leads on the way to new discoveries.

In recent times, the narrative of opportunity has found new life in nostalgic literature and exhibitions on the sent-down program under the slogan "No regrets for a lost youth."[3] However, the failure narrative is far more familiar to Western audiences, thanks to the large and compelling genre of "scar literature" — including many memoirs and works of semi-autobiographical fiction written by former sent-down youth specifically for audiences in North America and Europe. It lands the sent-down program a sorry historical verdict: futile, wasteful, irrational. What we gain from such accounts is an understanding of the coercion youth faced, the violence they encountered, and perhaps most importantly the deep sense of frustration they felt when called to exert themselves in endeavors that were poorly conceived and ultimately fruitless. The bitterness with which many look back on their "wasted" youth and missed educational opportunities, combined with the succeeding regime's interest in discrediting what it calls the "ten years of turmoil" (i.e., the Cultural Revolution), has led to a privileging of such stories. Although the theme of the "lost generation" has been assumed to refer specifically to the tragedy of the urban sent-down youth, many rural youth also had dreams that went unrealized; such youth remember the Cultural Revolution with a similar sense of waste and loss.

Thus, narratives of both opportunity and failure existed already in the Mao era, where they applied equally to the experiences of urban educated youth sent down to the countryside and rural educated youth returned to their home villages. In postsocialist times, these narratives continue to exert a powerful shaping force on memories of what it meant to be an educated youth during the Cultural Revolution—though the significance, and even the very existence, of the rural educated youth is often forgotten in public memory. Whether urban or rural, recognized publicly or privately, many of the stories are like Pan's, caught in a kind of magnetic field between the two poles of opportunity and failure. This chapter explores those stories and their larger significance in the history of China's red and green revolutions.

"A Great Opportunity"

The copious published accounts of educated youth finding fulfillment in the countryside were undoubtedly meant to assuage uncertainty and pessimism about the prospect of "taking root for the rest of one's life in the countryside" (扎根农村一辈子), as the slogan urged. The many stories that highlighted agricultural science had the more specific goal of convincing youth to see living in the countryside as an opportunity to fulfill their educational dreams, which the state's own dreams of agricultural modernization demanded.

For the urban sent-down youth, agricultural labor meant unprecedented physical challenges, and rural life unprecedented physical hardships; moreover, many urban youth landed in remote places where they had no family to help them adjust to their new lives. A quintessential account that spoke to their concerns can be found in a 1972 *People's Daily* article on a sent-down youth from Shanghai named Mei Minquan. The story went that young Mei traveled to the "Great Northern Wilderness" in response to Mao's call to "achieve great things" in the countryside. Finding himself assigned to a forested region with low productive value, Mei reportedly hatched a plan to import from the Shanghai area the fungus delicacy known as "silver ears" (or white tree ears) and make the forest "blossom with silver flowers." When he told others of his idea, he met with opposition. Some objected that the climates were too different; others questioned whether a middle school graduate had the necessary microbiological knowledge. But the poor peasants were reportedly very supportive, saying, "For years it's been obvious that the timber has not been worth cutting. You youth have ambition and culture: go for the gusto!" Mei felt encouraged by the peasants' support, and he also appreciated the words of caution from the others. After some additional training at the agricultural institute in Shanghai, Mei succeeded in growing beautiful silver flowers in the Great Northern Wilderness. The moral of the story, as the newspaper account put it, was that "silver ears and educated youth alike could settle down in the wilderness, and alike they needed to undergo struggle" to make that transition.[4] The story thus at once justified the rustication movement, celebrated struggle, and offered inspirational testimony of the possibilities for improving agriculture through science.

Rural returned youth faced a different, but no less difficult, set of problems. While they were not raised in intellectual homes, rural youth perhaps had a more clearly defined sense of personal ambition. They had intimate knowledge of life in the countryside, but this in itself could be a problem, since they knew all too well the limited opportunities that faced them. Attending secondary school in the city offered the hope of urban employment; returning to the countryside was a bitter disappointment.[5] Stories about returned youth acknowledged such feelings but vehemently asserted the tremendous ultimate worth of an educated rural life; the chance to participate in scientific experiment offered a key incentive.

One case that received special coverage was Chunwan Commune in Yangchun County, Guangdong. Chunwan faced special challenges in holding onto its educated youth, since most of its residents were recent migrants from different urban areas and many had little background in farming.[6] According to the Communist Youth League and the Yangchun Science and Technology Association, the solution was to tempt Chunwan's educated youth to stay on the farm by offering opportunities to participate in agricultural science. In the beginning, youth from Chunwan reportedly thought that participating in agricultural labor would be a waste of their talent (大材小用).[7] Not content to stay in the countryside, they hankered to move to Guangzhou where they imagined a better future lay. But following the establishment of a scientific experiment group, they decided they would "try it and see."[8] Their experiences participating in agricultural science reportedly showed them that in the countryside there was much they could learn and much they could contribute: gradually they stopped talking about food and clothing all the time and instead began talking about science and studying technology.[9]

A 1974 collection of stories about urban and rural youth involved in the scientific experiment movement in the Beijing area returned again and again to the notion that the countryside offered unparalleled opportunities not only to apply knowledge gained in school but to acquire new scientific knowledge. (Though these accounts are written in the first person, they were packaged as propaganda and cannot be considered evidence of youths' actual perspectives.) One youth wrote, "I recalled that when we graduated, some of my classmates were worried that in participating in agricultural production they would lose the knowledge they had learned—how incredibly funny! Now I think not only was the knowledge I studied in the past not losable, but it was really deficient. Take forecasting as an example. If you want to forecast

whether a certain insect pest is going to appear, you need to use entomological knowledge to research the insect pest's life habits and how climate,
geographical environment, and other conditions affect it, and you need to
do lots of survey statistics; when you control the pest using chemical pesticides, if you don't have the necessary chemical knowledge then you'll have
safety and usage problems. In sum, the physics, chemistry, and mathematics
you study in secondary school are all needed." Another wrote, "For educated
youth, going to the countryside isn't the end of the study mission but rather
the beginning of an important educational stage. The countryside is another
big school, the poor and lower-middle peasants are our excellent teachers,
and scientific experiment is one of the courses in this big school." Sending
middle school graduates back to the countryside was not a case of "using a
talented person in an insignificant position," or of "water buffaloes jumping
into a well." It was not a "waste of talent" but a "great opportunity."[10]

Propaganda further presented scientific experiment as a choice, thus emphasizing the nobility of individual commitment. In a 1966 Women's Federa-
tion document, a model youth named Huang Chunlai reported that others
had encouraged her to take a job as a worker at the state farm because the pay
was good and the workday only eight hours long. But she remembered that
the party had sent her to an agricultural high school to prepare her for building the new countryside, and so she continued to devote herself to the struggle for production and scientific experiment.[11] In 1972 Miyun County opened
a chemical fertilizer plant. People were excited because the plant represented
"modernization" and there was a real future in becoming a technician there.
A returned youth named Lang Yuping debated whether to head to the new
factory or take the opportunity presented by the brigade party secretary to
study to become an insect pest forecaster: "That night I was restless thinking
it over. *Which one should I choose?* Then before my eyes floated an image of
the party secretary carrying an insecticide sprayer and directing us to exterminate armyworm. He said that we can't let armyworms in just like we can't
let class enemies wreak havoc in the fields. . . . I realized that the countryside
needed me and decided to stay and do pest forecasting."[12] In choosing the
rural scientific experiment movement, youth were said to be making the best
possible use of their precious education—for themselves, but more importantly for the countryside, the party, and the country as a whole.

There were clear political goals driving propagandists to portray the
youth as making conscious decisions to participate in scientific experiment.
However, diaries and interview sources provide evidence that at least some
educated youth did in fact have some degree of choice. A young woman from
Beijing remembers that when the call came for someone to go receive train-

ing in scientific experiment, local peasants thought it would be a waste of time; it was "her own motivation" that led her to pursue this opportunity.[13] A young Nanjing man sent down to Inner Mongolia wrote in his diary in 1971 that recently many youth were responding to the call for industrial workers by entering factories. He had decided not to join them because his work producing the plant hormone gibberellin (in Chinese, "920") would suffer. The production brigade had invested time and money in scientific farming; to abandon the scientific experiments at this early stage would be a big blow to the brigade. "So I have decided to subordinate my individual interests to the interests of the revolution; I will stay here and not go."[14] Shen Dianzhong reported no such alternative option, but he still savored his decision, which for him amounted to a deeply philosophical consideration. On 26 November 1971, he wrote, "Of course, I still haven't made up my mind, because the conditions are not yet ripe." On 12 and 16 December, with passionate language, he committed himself to the project of producing 920 in the laboratory: "Whether I live or die, I'll do this work well."[15]

For these educated youth, the chance to participate in the scientific experiment movement offered a rare opportunity to choose their own path—the choice itself gave a sense of liberation, and the work involved more autonomy than regular agricultural labor. Chen Yongning remembers the meeting in the production team in which they discussed the need to improve scientific farming technology by creating an agricultural science group. He volunteered and was made the leader of the group. The members of the group still spent most of their time working in the fields, but when they had experiment responsibilities, they told the production team leader and went off to do the experiment work by themselves. When he entered college in 1978, he decided to study plant protection at the Guangxi Institute for Agriculture. This was his own choice, an interest formed when he was a sent-down youth.[16] At a time when people in all areas of society have reported being buffeted this way and that by the political winds, Chen experienced the road to science as remarkably full of choice and even some amount of independence.[17]

Rural youth Chen Haidong similarly experienced science as a choice and an opportunity. In response to the survey I circulated among participants in the Big Sand research, he wrote, "When I began participating in the scientific research group in 1969, I started to have the determination to participate in scientific research work. So even if I couldn't go to college, I could still continue to do scientific research." But in the end, Chen did get to go to college. He was selected for the honor after, as production team leader, he "brought out the activism of the great people's masses, used the scientific knowledge I had studied to guide production, and greatly increased rice production, such that the average yield per *mu* reached 650 pounds, thereby resolving the peasants' problem of grain self-sufficiency." When he arrived at Sun Yat-sen University, he opted to study insect control. "Since I came from the country-side . . . I had a deep knowledge of the uses and influence of scientific knowl-edge on agricultural production, and knew that we needed to rely on science if we wanted to transform my hometown's poverty and backwardness. So I chose to study entomology in order to serve agricultural production even better in the future." His participation in Pu Zhelong's research at Big Sand led to a long and successful career in insect control science.

Young peasants from Big Sand also had the opportunity to participate in the research through their local scientific experiment groups (figure 26). Most of the more than two hundred workers in the "integrated control leadership groups" were returned educated youth from Big Sand.[18] For one survey re-spondent, Luo Zhongbi, this early involvement led to decades of meaningful work in agricultural science despite his lack of college education. Luo recalls

FIGURE 26. Pu Zhelong with urban and rural youth discussing the rearing of parasitic wasps at Big Sand Commune. Reproduced from Gu Dexiang, ed., *Pu Zhelong jinian yingji* (Pu Zhelong memorial album), 2002, 23.

nostalgically that when Pu came to Big Sand he fostered a program of truly "open-door education" (开门办学, a Cultural Revolution phrase signifying mass participation in education and academics).[19] He wrote that to engage in agricultural science, one "must raise one's level of science and technology, must understand science and technology, and integrate study and practice." And so when he left school and began work in agricultural production, he "set his mind" to studying agricultural technology and "personally subscribed to *Guangdong Science and Technology News*." Luo's hard work paid off. He became a plant protection specialist at the agricultural technology station, and eventually came to direct the station.[20]

Of course, looking at the country as a whole, we see that such success stories were rare, especially for rural youth. Why then did youth choose to participate in scientific experiment activities when opportunities arose? For some, the realities of poverty in the countryside provided an immediate, very material incentive. Cao Xingsui recalls, "We were extremely hungry back then, there was never enough grain to eat."[21] And as Pan Yiwei explained, "I actually didn't have any lofty aspirations. Why did I do it? Because I was hungry and there was nothing to eat." He said this not just to me but also to the villagers themselves when he returned in 2004 for a visit. They said he had "put them on the road to wealth" and called him a "shake-money tree" (i.e., a magical tree that drops money when shaken), but he said modestly, "I was not thinking about that, just having enough to eat."[22] Cao and Pan were experiencing, and now participate in the narration of, what Gang Yue has called the Chinese communists' "hungry revolution": from Yan'an-era depictions of want in the old society through postsocialist remembering of sent-down deprivations, Chinese literature has "both reflected and nurtured the peasants' revolutionary hunger," and Chinese people more broadly have learned (through bodily experience and cultural inculcation) to speak a language in which hunger serves as "a historical master trope for social reality" and thus as the most convincing possible reason justifying any course of social action.[23]

In their groundbreaking interview-based study of rural China, *Chen Village*, Chan, Madsen, and Unger suggest another motivation for youth who joined the scientific experiment movement: "the amateur researchers enjoyed their adventures" because, "however futile the results, they liked the opportunity to use their initiative."[24] Participating in scientific experiment was indeed an enjoyable venture and a welcome break from the dull routine of farm labor. The experiment fields themselves broke the monotonous landscape in a pleasing way. One young woman sent down to rural Shaanxi recorded in her diary on 8 April 1977: "Spring has come. . . . A bright afternoon, carrying a hoe

I came to the rapeseed fields. This is a big experimental plot where wheat and rape are sown in tandem. Between two rows of lustrous green, evenly brushed wheat stretches a row of dazzlingly beautiful rapeseed plants bursting with yellow blossoms. In the flat landscape, the rape mixed among the wheat looks like an emerald-green rectangular carpet embroidered with gold thread, truly beautiful." The pleasures of working in the experiment fields gave way that day to a still greater pleasure. She had just started hoeing when the old peasant who led the experiment station approached her and said, "We're going fishing!" "So that was that," she concluded. "No work this afternoon."[25]

But participation in rural scientific experiment was for many youth far more than just a way of filling their bellies, an amusing escapade, or an escape from boredom. The evidence presented on *The Second Handshake* in the previous chapter, together with evidence from interviews, diaries, and memoirs, supports the conclusion that youth were strongly attracted to science in the Mao era and that participation in scientific experiment was deeply meaningful to them. Even as intellectuals of their parents' generation were suffering humiliation, imprisonment, and physical abuse, a surprising number of youth dreamed of becoming scientists—and indeed of making serious contributions to science while they were still young. In his study of the Cultural Revolution–era earthquake prediction program, Fa-ti Fan interviewed about ten people who participated as youth and who testified that they "felt that they were doing something valuable, important, exciting." Fan emphasizes that they were not "doing kid's science or learning to do science simply as part of science education." Rather they "actually took part in an official national science project. In other words, they were already doing science and contributing to the national effort against natural disasters."[26]

We have already seen the passion that Shen Dianzhong expressed for science and the eagerness with which he joined his scientific experiment group. Former rural youth Zhao Yuezhi remembers a feeling of pride that her uncle was the leader of the science and technology group, a sense of the significance of scientific and technological innovations, and a desire to achieve something of such importance.[27] One woman student who participated in the Big Sand research wrote in her survey response: "Twenty of us classmates shouldered our backpacks, and in April 1975 went to Big Sand. . . . In 1975, from April to November, every day we were on the front lines of production, or in the fields doing surveys, or in the lab raising natural enemies, or surveying the ebb and flow patterns of spider species, insect pests, or rove beetles (a natural enemy of insect pests). The first time I observed eggs laid by rove beetles and the larvae hatching out, my heart was overjoyed."[28]

Similarly, several contributors to a recent volume of memoir essays by

Chinese women who grew up during the Cultural Revolution specifically recall positive memories of their engagement in science and aspirations to careers in scientific fields. One remembers how much she loved the children's encyclopedia collection *One Hundred Thousand Whys* (十万个为什么): "I had learned from it why there are little holes in bread, why a zebra has stripes on its body, why hens lay more eggs in summer, and why I would have a different weight on Mars. I wanted to be a scientist or an astronaut so that I could ask more whys and publish the answers in books."[29] Another contributor remembered hoping to become a biologist.[30] A former Red Guard relates, "I believe many little girls and boys of my generation dreamed of being a geological prospector. . . . Propaganda for recruiting young people to work in this area was very effective. When my neighbor's daughter was accepted by the geology department of a prestigious university, we all envied her for her future prospects of an adventurous life."[31] (She makes clear in her account that she considers such propaganda to have been generally a positive influence, or at least more positive than the propaganda that leads American girls to aspire to be cheerleaders.)

For many youth, participating in the scientific experiment movement was a desirable opportunity. To some extent, their feelings about science reflected the success of state propaganda designed to produce just such sentiments among China's educated youth, who were desperately needed to fill the gaps caused by China's low numbers of scientists and technicians. However, youth also poured their own dreams into the propagandists' vessels. In their eyes, science was intellectually stimulating, adventurous, promising in career terms, and, yes, an escape from the drudgery of a lifetime of ordinary agricultural labor.

The Scope and Limits of Opportunity

Of course, "opportunity" and "choice" are relative terms. Volunteering to work in the scientific experiment group was a choice made when other choices were out of the question. Cao Xingsui remembers feeling torn between wanting to use his knowledge to help the very poor local people in the village to which he had been assigned and thinking about his own life ambitions. He told me, "We felt if we had the opportunity we would want to leave the village. But while we were still there and had no way out, we were very happy to have the chance to help the local peasants."[32] In other periods, urban children—especially the children of elites—would have expected far bigger opportunities. If we take these ordinary expectations of urban youth as the point of comparison, the "opportunity" to participate in the rural scientific experiment movement seems pretty weak. However, compared with the options facing vast numbers of rural youth, it looks very different. The organization of night schools, short-term training courses at the commune or county level, and scientific experiment groups in the production teams all provided genuine educational opportunities for millions of young people around the country.

And for some, participation in scientific farming opened doors to truly exciting paths. The most successful youth had the chance to tour other communes to share new technologies and travel to regional or even national conferences where they presented their accomplishments (figure 27). Aside from those like Deng Yantang and Xin Wen, whose stories appeared in nationally circulated collected volumes cited often in this book, many others—including my collaborator Cao Xingsui—had their turn in the spotlight as they attended conferences and received local media attention.

A few such mini-celebrities were even selected to receive foreign visitors to showcase Maoist science at work. One example is Sun Zhongchen, the research group leader for a brigade called Daxiyu in Jilin Province. When Norman Borlaug and his colleague Haldore Hanson traveled to China with an international wheat studies delegation, they were deeply impressed by Sun. Hanson wrote in his diary, "Informal training, combined with a bright mind, diligent work habits, appealing personality, and leadership capability, has produced the kind of rural talent which Chairman Mao must have visualized when he urged that peasants should be incorporated into collaborative research with research institutes and academies."[33] Sun's work was also highlighted in celebratory magazine articles on scientific farming. One article that credited Sun as author specifically referenced the international attention his project had received: "There was an insect investigation group that came from a capitalist country and saw peasants cultivating *Beauvaria* to control corn borer, with over 90 percent effectiveness. They couldn't but acknowledge that in biological control of corn borers China is already a frontrunner in the world." The moral of the story was Mao's dictum that "the lowliest are the smartest and the most elite are the dumbest"—a turning of tables that favored people like Sun, rural-born youth trained as agricultural technicians.[34] Sun was elsewhere trumpeted for having "undergone several years of training and practice" that allowed him "not only to grasp fertilizer, plant protection, cultivation, and the fundamental theory behind crossing two varieties, but also to resolve some key technological problems related to production," such that he was reportedly seen by the peasants as "a 'red and expert' expert."[35]

But not everyone had the chance to participate in the scientific experiment movement. Many people who wanted to participate in experiment groups

FIGURE 27. Two of a set of photographs preserved from a 1971 conference on agricultural scientific experiment in Xin County, Shanxi. The vast majority of the participants were clearly youth. Young men appear to have outnumbered young women considerably (the relatively few young women in the top image are identifiable by their braids), but the women are featured prominently in the photographs, suggesting once again the ideological significance of their participation. Xinxian diqu dan wei keji xiaozu, *Shanxi sheng nongye kexue shiyan xianchang jingyan jiaoliu hui zhaopian xuanji* (Collected photographs from the Shanxi Provincial Conference for On-Site Exchange of Experience in Agricultural Scientific Experiment), 1971.

must have been disappointed. Chen Yongning remembers that his site had a dozen sent-down youth and between forty and fifty rural youth, among whom the leaders picked just twelve.[36] Cao Xingsui recalls that "everyone, whether they were returned or sent-down educated youth, wanted to participate in the groups because of the opportunity to go to the commune or county level to receive training." However, before assigning them to the group, the production team leader had to evaluate their sense of responsibility, their understanding of agriculture, and their responsiveness. Surprisingly, Cao recalls that family background did not play much of a role in the decision—and indeed he himself was chosen to be production team leader despite the fact that his father had fought for the Guomindang. This undoubtedly differed depending on the locale, but his recollection was that peasants in the production team were not interested in discussing family background and so it did not factor into the decision. In any case, the existence of any vetting process is a clear indication that not everyone who aspired to participate was allowed the opportunity.

In addition to those who were unable to participate at all, for many of those who did, the work was grueling and ultimately dissolved without opening any further opportunities. Like Pan, some sent-down youth engaged in agricultural science and technology in the villages only to find themselves later shut out of the far more significant opportunity of attending college. Although Pan passed the college entrance exam in 1987, his eventual success could not erase the pain of his original failure. Indeed, he says that the "belated fulfillment of my college dream and the dream I dreamed in 1977 cannot be discussed in the same sentence"; he bitterly regrets the ten years he lost as a result of that initial failure.[37]

For rural returned youth, the lack of opportunity to attend college was a far more common experience, and just as disappointing. The story told by one man I interviewed, Wang Xiaodong, exemplifies this. After he had graduated from secondary school in 1974, he returned to his village, participated in the scientific experiment group, and in 1977 became the third of three plant protection specialists to serve his production team. When I asked him how he studied the material, he said it was not really "studying." He had some books he consulted, and every day he recorded what insects he saw and how many. When I asked whether as a student he had been particularly interested in science, he replied that because it was the Cultural Revolution, there was no point in being particularly interested in anything since there was "no opportunity": by the time the college entrance exams started again in 1977, it was too late for him to apply.[38] His response was similar to comments I heard in various forms from several people who felt the need to correct what appeared to be my misunderstanding of a fundamental aspect of the Cultural Revolution: they emphasized that personal interests were irrelevant because all work was "assigned" (分配)—there were few opportunities and no choice in any case. My sense is that for Wang and many others, the bitterness of not having had the chance to take the college entrance exams colors their perspectives on everything else about their education and participation in agricultural science and technology. Successful people can look back on their road as one filled with choice despite the high degree of state intervention. Unsuccessful people may, like Pan, highlight their own agency during the Cultural Revolution as a way of celebrating their one moment in the sun, or they may, like Wang, dismiss any notion that life ever afforded them opportunity at all.

Women faced an especially complicated set of opportunities and obstacles in agricultural science. We have seen that the state made explicit efforts to link the rural scientific experiment movement with the revolutionary transformation of gender relationships in the countryside. Propaganda frequently highlighted the agricultural science accomplishments of both female educated youth (rural and urban) and "old ladies" as a way of demonstrating the revolutionary character of the scientific experiment movement. Millions of women throughout China had the opportunity to engage in agricultural science and technology work—and even if what they were doing was merely using night soil to fertilize fields and other activities not noticeably different from "traditional" practice, the fact that it was called "agricultural science" meant something important in the new society.

However, it was striking that of the twenty-nine former participants in the scientific experiment movement I interviewed during my trip to Guangxi in 2012, only two were women. Most of my interview opportunities had been arranged by my hosts, and when I noted the gender disparity, they all acknowledged the great numbers of young women who participated as educated youth in the movement. Why, then, were they not called to be interviewed? Most had not gone on to careers in agricultural technology—perhaps, as two people suggested to me, because their parents objected to their daughters enduring physical hardships, or perhaps because of discrimination that all the red revolutionary attempts at social transformation had not been able to eliminate. My hosts had their closest contacts among people with jobs in agricultural bureaus; moreover, they had strongly defined understandings of what constitutes an "expert"—people with degrees and official positions—and expected that I would be best served by interviewing experts. So there are reasonable, practical, mundane reasons I interviewed so few women. Yet, taking a step back, we can see all these reasons as part of a strong set of cultural understandings about agriculture, science, and society that differs in profound ways from the set that Mao-era radicals attempted to implement. Young women had many opportunities to participate as "educated youth" in the scientific experiment movement, and their participation was used by propagandists to further the explicit state goal of conquering (or at least claiming to conquer) patriarchal attitudes. However, these women's efforts rarely translated into long-term career opportunities.

Initiative and Connections

Youth displayed remarkable resourcefulness in pursuing agricultural science. Former sent-down youth Ye Wa remembers that the state provided test tubes and books, and that the village carpenter used locally available materials to make her a box for microbial fertilizer production according to specifications provided at her training. Supplies were limited, however, and soon she found herself buying more test tubes with her own money.[39] Here as elsewhere, it is striking how much interview narratives resonate with state propaganda from the Mao era, which applauded youth who traveled to the city to buy technical manuals and materials for scientific experiments with their own money.

Personal and family connections offered important resources for educated youth—in particular, urban sent-down youth—seeking to make things happen in the countryside. Ye Wa recalls asking her mother to mail her agar to use as medium for growing microbial fertilizer. Chen Yongning similarly attempted to tap personal connections. When Chen's classmate traveled to his ancestral village in the North, he wrote letters to Chen talking about the various corn varieties they were introducing. Chen replied asking his friend to mail some corn seeds. He received the seeds, but they did not work in the southern climate.[40] Pan Yiwei was more fortunate in this regard. The key contribution that brought Pan Yiwei to his production team leader's attention was his introduction of improved varieties of rice; he too had obtained the seeds by making good use of personal connections. He had fortuitously bumped into his teacher's son at an event celebrating National Day and arranged to visit him where he worked at the agricultural academy. His teacher's son took him to the agricultural science institute where they had a number of improved varieties that had not been widely extended because of the political disruptions of the Cultural Revolution. Pan used 7.5 yuan of his own meager funds to buy more than ten different varieties, which he then took back to the production team to test. While in Nanning, he also took the opportunity to purchase every book he could find on planting rice, and he spent his evenings back in the village poring over them under a kerosene lantern. The new varieties achieved far higher yields than what they had been using, and production tripled in just three years.[41]

Many sent-down youth have similar stories of taking initiative to make the most of personal connections for the benefit of their rural communities. Cao Xingsui was chosen in 1974 to be the leader of his production team. At that time, his team was still planting tall varieties of rice, and the harvests were only about 300 *jin* per *mu* (about 1,800 pounds per acre). There were some dwarf varieties available that had been bred in the 1950s, but these did not perform well so few people planted them. However, as Cao says, "At that time I already knew that there were varieties in the outside world that would yield 700–800 *jin* mer *mu*. Because I was an educated youth, I knew this, and a technician also told me this." When he explained to his father that the production team had chosen him to be their leader and that he very much wanted to help them, his father told him of a distant relative, whom he could call "Uncle," who had graduated in 1939 from Jiangxi Agricultural Academy and was then employed as an agricultural scientist in Guangxi. So Cao went to Nanning by horse cart, paying the driver out of his pocket 3.6 yuan per day for the three days it took to get there.[42] At night, he and the driver slept on the grass under the cart.

Cao's efforts were well rewarded. His "uncle" and others at the Guangxi Academy of Agricultural Sciences had bred a new variety called "one-shear even" (一剪平) because of the way the rice grew to an even height—"very even, very magnificent, you could tell just by looking at it that it would be high-yielding." It had already undergone testing throughout Guangxi Province and had been deemed very "reliable," but in 1966 the Cultural Revolution threw the academy into disarray and Cao's uncle underwent criticism, after which people had stopped paying attention to that variety. Cao's uncle informed him that there was still some seed in storage that would then be nearing the end of its shelf life and would soon be discarded in any case, so Cao should feel free to take it. This lack of oversight was the flip side of the Cultural Revolution's chaos, and it made many otherwise impossible things possible. Cao was able to bring 300 *jin* of the seed back to his production team without anyone noticing or caring.

The following spring Cao skipped the stage of planting in experiment fields and instead planted all 300 *jin* in an area of about 100 *mu*. "I was very bold at that time," he says. He simply trusted what his uncle had said "because he was a scientist and he said it was good so I believed him," though looking back and considering the risk involved, "it makes me a bit scared." Even before the harvest, the fields looked so good that the commune leadership called an "on-the-spot meeting" (现场会) to let the "masses" witness the success and invited every production team leader along with old peasants, technicians, and even people from other communes. "And so all of a sudden I became very famous in that place." As a result, Cao was invited to present at conferences for educated youth, where he spoke also about insect control, preventing epidemics, developing aquaculture, and raising ducks. "But the reason I became famous was the rice." Given the political situation at the time, he could not discuss the origins of the new variety—after all, his uncle was still considered a counterrevolutionary. Instead, he used the accepted vocabulary of "relying on the masses."[43]

The frequency with which sent-down youth appealed to friends and relatives in the cities for help supplying everything from simple agar to precious new varieties of seed flew a bit in the face of the policy of relying on locally available resources. In a very real way, however, educated youth were one of the most important local resources a production team had. And when those youth had both initiative and connections, their value increased tremendously.[44]

Failure

One of the most striking themes in the literature on youth and scientific experiment is that of failure. The frequent return to this subject makes clear that experiments commonly failed and that the state faced a major challenge to convince people that failure was acceptable, and even revolutionary. So soon after the massive famine that followed the experiments of the Great Leap Forward, rural people needed a lot of convincing if they were to overcome fears that new experiments would end equally badly, leading to loss of valuable land and consequent lack of sufficient food.[45] Thus the 1965 National Conference on Rural Youth in Scientific Experiment embraced as a core principle "When experiments fail, we must diligently analyze the causes and explain it clearly to the masses."[46] And in 1969, when Huarong County (Hunan) established a new model network for scientific experiment, the plan highlighted the need to "help people develop a correct understanding of the relationship between success and failure."[47] Most of the inspirational propaganda stories on youth and science included some degree of failure before the eventual success of the experiments: this provided opportunities for kindly party secretaries and poor peasants to offer encouraging words like "Failure is the mother of success" and reminders of Mao's wisdom—especially the "winding road" that led to the production of new things and the need to emulate the Foolish Old Man Who Moved the Mountains and who was not afraid of failure.[48] In such stories, initial failures only sweetened the feelings accompanying success. And youth were expected to grow as a result. As a story published in 1974 explained, "The failure was a loss for the collective's production, but for the science team, especially for us youths, it was a great education: it made us deeply experience the process of integrating theory and practice and the process of receiving reeducation from the peasants and changing our worldview."[49]

The actual experiences of youth involved in failed experiment projects were often less sanguine. Sent-down youth Ye Wa enthusiastically volunteered to join the scientific experiment group when offered the chance to get out of the village for a little while and go to the county seat. She learned to create bacterial fertilizers and acquired an improved variety of hybrid sorghum to introduce to the village. Upon her return, she was given a piece of the most fertile land to plant the new variety. As the plants matured, she suspected the harvest would not be as good as expected—or even as good as an ordinary crop. High on the plateau, the village had a cooler climate than the areas where the hybrid sorghum had been successful, and that year the autumn rainy season came early. Ashamed to have wasted the land that the peasants gave up to her, she strategically arranged to leave the village when harvest time approached. She recalls, "Everyone saw the failure. I was ashamed and scared, and had no courage to face the reality. I felt that I had betrayed the villagers. There are still no words to describe how I felt at the time. The thing was, when I finally returned to the village in the winter, hanging on the wall inside the storage house . . . someone had placed several large ripe sorghum spikes to leave for the next year's planting. You can imagine how touched and overwhelmed I was when I saw the spikes. However, we did not plant them next year."[50]

Because she is a friend, my interview with Ye Wa traversed an unusual amount of personal territory and so uncovered an especially rich set of experiences that speaks to the significance of this failure. Like that of millions of other sent-down youth, Ye Wa's time in the countryside opened her eyes to the bitter poverty that peasants in rural China continued to face after more than twenty years in the "new society." Her status as a sent-down youth only partially protected her from the hunger that the villagers experienced after the terrible harvest of 1970. By the time spring came around, there was nothing left to eat in the village. In that tiny village of about one hundred people, three children died. Two were twins whose mother did not have enough milk to keep them alive; the third was a five-year-old girl with the unlucky name "Flying Bird" who died of dysentery—her soul "flew away" as if destined to do so. A few months later, Ye Wa was called upon in her role as leader of the women's team to organize the village women in the May work of thinning seedlings. As they were walking toward the fields, Ye Wa and Flying Bird's mother were taking up the rear, when suddenly a woman at the front of the group froze, turned around with her face ashen, and said, "We're not doing it. We're not doing it after all." Ye Wa said, "We have to work on this piece of land today. How can we not do it?" The woman just repeated, "Not doing it, not doing it, not doing it." Looking back, Ye Wa thinks Flying Bird's mother understood the reason before Ye Wa did: the women in front had come upon Flying Bird's clothes, a flowery red shirt of padded cotton cloth that Ye Wa remembered Flying Bird wearing, but now frayed and rotten. Too weak from hunger to dispose of her body properly, the old man charged with burying her had dropped Flying Bird's body in a nearby ditch. Today, Ye Wa remembers the episode with painful vividness; at the time, this and other experiences with extreme poverty undoubtedly did much to provide the framework within which she viewed any failure on her part, any waste of good land, any poor harvest.[51]

Although the context was not as tragic, Shen Dianzhong's diary offers further poignant evidence of how well youth learned to anticipate failure, and how devastating that failure could nevertheless be. Shen's early entries on his decision to embark on 920 production are filled with words of self-warning and self-encouragement. "If I really do it, I may encounter setbacks and losses, and I may have to travel a winding road. I must really make failure into the mother of success." He went on, "I deeply understand that the road before me will have many difficulties, including some difficulties I cannot even imagine and some that would give people thoughts of faltering."[52]

On 22 January 1972, he recorded his sadness upon the event of his first failed experiment. More than a week later he was still preoccupied with it. "Everything I have prepared for has come to pass. Now, only now, do I understand something; now, only now, have I been put to this real, solemn, profound, merciless test." Then on 12 February, "Another failure has come before my eyes. This kind of blow is really too severe, it just makes it hard for me to breathe, just makes me fall over. But I cannot, absolutely cannot, I must straighten up and move on, must stand and be steady, must coldly and tenaciously persevere, working without stop. If I fall I must crawl, if I fail I must do it anew."[53]

Spring Festival came and went with no mention—just a short entry on enthusiasm and the need to persevere. The next day he sought out the production brigade leader, who gave him a ray of hope about the future of the work but warned it would get harder, not easier. This turned out to be the case, and Shen plunged into increasingly despairing discussions of failure. In June he wrote a lengthy "summary" of the 920 work in his diary, and in July he composed a "report" to submit to the district party committee youth group, which subsequently discontinued the experiments and apparently offered Shen little in the way of consolation. Shen wrote, "Recent events have stripped me of any right to 'work.' . . . Who will work with me in the future, and whom will I be able to work with? Maybe nobody! . . . I had best face death calmly. Of course, this is not a death of the living flesh, but a death of my political life. Although I'll never be able to accept this death, I will ultimately find some significance in it and survive. . . . I will not be pessimistic or timid, but will pledge my life to upholding the truth."[54] It is tempting to smile at the adolescent passions Shen's diary exhibits and how large the small failures of a teenager can loom. However, in the context of 1970s rural China, when the political stakes were so high and the consequences for people's lives so significant, Shen's fears were by no means unreasonable.

Rethinking the "Failure Narrative"

The theme of failure provides an arc of continuity across Mao-era and post-Mao sources, personal accounts and formal histories. However, it takes on very different kinds of significance in different narratives. In the years after the death of Mao Zedong and repudiation of the Cultural Revolution, the new leadership came to define the entire radical approach to science, youth, economics, and everything else as a "colossal failure" (巨大的失败). Where it was once obligatory to blame failings on Liu Shaoqi, Lin Biao, and unnamed counterrevolutionaries, it has for decades now been almost as obligatory to blame the "Gang of Four" or leftism generally. A recent Chinese historical survey of youth and the Communist Youth League estimates that "because of stereotyping and not proceeding from reality," "excess formality and flash-in-the-pan" approaches, among other reasons, only 20–30 percent of pre–Cultural Revolution youth-based scientific experiment groups achieved success. The authors emphasize "leftism" as the root cause of these problems, since everything came down to applying Mao's works.[55] A book from the same series on agriculture acknowledges the usefulness of the "four-level agricultural scientific experiment networks" in the countryside during the Cultural Revolution but nonetheless concludes that it involved enormous wasted effort because of "leftist influences."[56]

The experiences of many sent-down youth in scientific experiment would uphold the verdict that many experiments were "failures" and that much good effort was wasted because of excessive emphasis on ideological correctness. Even so, the post-Mao "failure narrative" does not do justice to the full significance of youth participation in the scientific experiment movement. To begin with, it obscures the direct engagement with questions of failure in Mao-era propaganda. Far from blanket optimism, propaganda emphasized the difficulties associated with agricultural experiment and recognized failure as a common—almost universal—experience.

Moreover, the failure narrative does not recognize the real and complex issues at the center of Mao-era agricultural work. Many of these issues are found also in other parts of the world: agricultural experiments often do fail, and not always for political reasons. Moreover, there is no consensus on many of the most basic questions of how to balance economic development, environmental protection, labor justice, human health, and cultural values. If state-led efforts to promote a new variety of rice led to increased production but required expensive, polluting fertilizers and displaced varieties considered more nutritious and delicious by local people, is that "success" or "failure"?

We should also question whether we are using the same analytical toolbox to analyze socialist China as we use for other times and places. One of the most important insights afforded by recent decades of work in the sociology of science and technology is that it makes little sense to explain successful or otherwise "good" science in "purely" scientific terms, while reserving social or political explanations solely for scientific work that fails or otherwise deserves censure.[57] Criticisms of socialist China leveled in the postsocialist era carry an inevitable, though typically implicit, comparative message: Mao-era socialism failed, thus the superiority of market capitalism is upheld.[58] With a failure narrative already in place, it becomes easy to discredit anything associated with the radical line—the more characteristically Maoist, the more obviously flawed it appears in retrospect. In particular, we may find ourselves laying the blame at the feet of "politics" and "ideology" in ways seldom applied outside the socialist world, though politics and ideology exist everywhere and everywhen.

As an exercise in historical imagination, consider the possibility that the US economic and political system collapses and is replaced by something fundamentally different. How would people in the new era evaluate the agricultural youth program 4-H? Would former participants testify that many of their projects were not in fact very successful? Would they dismiss the program as a waste of time and resources? Back in 1973 Jim Hightower's provocative *Hard Tomatoes, Hard Times* already connected some of these dots, charging the state extension service system with serving chiefly agribusiness and calling 4-H a "frivolous diversion of 72 million tax dollars."[59] As an arm of the US Department of Agriculture, 4-H has undeniably had its share of influence from pesticide corporations, and its youth activities reflect and reinforce capitalist agricultural economic relations. As Gabriel Rosenberg puts it, "4-H developed as an integral part of [the] broader push towards mechanized, industry-backed agriculture and the politics of progressive agricultural reform that eventually rendered rural America safe for agribusinesses." And as he further shows, during the Cold War 4-H developed an international arm—including a chapter in Vietnam run by the US military—that sought to combat communism and advance US-style capitalist agriculture in the Global South.[60]

Would historians in the new era point to these political linkages and ideological biases as reasons for the ultimate failure of 4-H? Maybe, and maybe with some justification. But such an analysis would be insufficient to capture the complexity of the issues that 4-H program officials, local leaders, and youth participants sought to address, the variety of their experiences, and what 4-H meant to them. Turning to a real historical example, we see that the verdict on Republican-era Chinese attempts to reform agriculture has often been that they failed because of a lack of attention to politics: none of the reforms, however technically good, could work without attention to larger social relationships.[61] Again, failure: but it is not unusual for observers to call for greater emphasis on politics in a nonsocialist context, while politics is held to blame for the problems of socialist-era China.

Part of the difficulty lies in the murkiness of the term "politics," which fails to distinguish between politically aware scientific practice on one hand and narrow application of specific political ideas to criticize and demean honest efforts on the other. Treating science as a political struggle or "revolutionary movement" can be helpful or even inspiring if, to take just one example, it opens avenues for girls to work in nontraditional roles—as it undeniably did in Mao-era China. But what about the experience of Shen Dianzhong? He was made to feel himself a failure and deprived of hope for his future, despite his sincere and passionate devotion to the revolution and to mass science. The challenge of conducting rural scientific experiments in crude conditions and in compliance with highly complex political standards placed a terribly heavy emotional burden on young people of 1960s–1970s China.

On the other hand, looking at Shen Dianzhong today, "failure" is not what comes to mind: he is now director of the Institute of Sociology at the Liaoning Provincial Academy of Social Sciences. Shen's experience is certainly not generalizable, and we know that many sent-down youth did not go on to successful careers. Yet neither is he unique. Several of the other sent-down youth discussed in this book have also done very well for themselves, and in some cases their experiences in the scientific experiment movement had a direct impact on their future careers. It was certainly rarer for rural, returned youth to gain opportunities in the reform era. The case of Wang Xiaodong recounted above exemplifies the disappointment that many experienced when the promise of the scientific experiment movement evaporated. Another interviewee who grew up in rural China testified that her cousin, whose participation in the scientific experiment movement she found deeply inspiring and enviable, returned to standard farm labor during the reform era—work that had nothing to do with the scientific knowledge she had cultivated.[62] But then there are the rural youth who participated in Pu Zhelong's research and went on to satisfying careers in agricultural science.

And there is Sun Zhongchen. In a 1988 article titled "Privately Owned Businesses Raise Peasants' Commercial Consciousness," *People's Daily* applauded the fast pace of rural China's economic transformation. As a case in point it raised the example of Elm Village in Nanwanzi township, Shaanxi Province. Land in Elm Village averaged 1.5 *mu* (about a quarter of an acre) per person; in the past, all the villagers tilled the land and were "frightfully poor." But after 1978, many industries sprang up that grew to employ 80 percent of the village's labor force. In the decade since, eighty-seven households had taken the initiative to rent out their land to "farming experts" who could then farm on a bigger scale and thus reap bigger profits. One of these experts was none other than "a peasant named Sun Zhongchen"—the same Sun Zhongchen who had so impressed foreign visitors in his role as spokesman for socialist China's unique brand of mass science in agriculture and whose name had graced an article trumpeting that "the lowliest are the smartest." Sun had amassed a whopping 220 *mu* of land. Through mechanized planting and harvesting, and the application of chemical herbicides, he had raised his income to 150,000 yuan. This representative to the People's Congress was quoted as saying, "The reason farmers' incomes are so low is that they do not have enough land to manage, which wastes their labor power. Running private businesses to create an outlet for surplus labor power is a good way of doing things!"[63]

Sun's story captures many contradictory truths about China's transition to a postsocialist economy beyond the celebration of the market economy that was the focus of the *People's Daily* article. One is that the scientific experiment movement, and agricultural research and extension more generally, provided a strong foundation for the economic boom of the Deng Xiaoping era: Sun Zhongchen was an "expert" because he had been afforded extraordinary opportunities as a rural educated youth to engage in *tu*-style agricultural science. Another is that the chief beneficiaries of the new economic system were people in good political positions with the influence necessary to secure access to resources when the communes were decollectivized and opportunities were created for entrepreneurs. Sun's 220 *mu* of land represented 146 times the per capita holding in his village; moreover, tractors were by no means ubiquitous. How he came to be in a position to acquire all these resources is not discussed in the *People's Daily* article, but we can guess that it had something to do with his political connections.

A third implication of Sun's post-Mao fate is that for all the blood and sweat expended through the decades of political campaigns, once Mao died it took very little time for the Chinese nation to effect a massive transformation from the radical politics of class struggle to the technocratic politics of economic development. China's red revolution was apparently eclipsed remarkably quickly by its green revolution. Apparently. To explore the post-Mao fate of the red and green revolutions more fully, the epilogue will consider some of the recent efforts to transform Chinese agriculture once again.

In 1978, Deng Xiaoping rose to leadership of the Chinese Communist Party, and the Chinese countryside began the process of decollectivization. At the same time, agricultural production increased rapidly. The coincidence of these two phenomena has led many to conclude that decollectivization was the most important factor driving the increase in agricultural production, and in turn that the collectivist rural economy was the crucial impediment to agricultural robustness.[1] Looking at the dismantling of the communes and the triumph of technocratic solutions to rural poverty similarly suggests that the green revolution won over the red, in keeping perhaps with Francis Fukuyama's famous contention that the 1990s witnessed the "end of history" as capitalism finally won the struggle with communism. However, the "three great differences" between the cities and the countryside, between mental and manual labor, and between urban and rural people that Mao targeted for elimination have by no means disappeared. Moreover, current state policies and efforts by various social actors to address these lasting inequities suggest strong continuities with Mao-era agricultural extension and the scientific experiment movement—though translated into the new context of global capitalism.

These continuities are not well recognized today. Popular understandings of contemporary China suffer from an illusion of newness. Just as the Mao-era fashioners of China's red and green revolutions emphasized rupture with the ancient regime and elided the tremendous influence of US agricultural science on Chinese agriculture and on the Chinese Communist Party's policy process itself, so post-Mao agricultural reformers have often sought to avoid plowing up the politically uncomfortable past, portraying their activities as breaking new ground. Rescuing the diverse and meaningful Mao-era experiences of scientific farming from historical erasure has been the chief goal of this book; the epilogue will demonstrate their lasting significance in the present day.

Red and Green Revolutions in Postsocialist China

The Mao-era agricultural extension system and scientific experiment movement were instrumental in the post-Mao era's so-called agricultural miracle. Indeed, the miracle doesn't look quite so confined to the post-Mao era when we overlay the chronology of the green revolution on that of the red revolution.[2] Agricultural extension work and scientific experiment movement activities were necessary both to convince peasants of the worth of new technologies and to teach them the skills associated with their adoption.[3] The new technologies resulted in tremendous increases in food production, and the consequent widespread availability of cheap food in turn had an important role in the economic growth of the post-Mao era.[4] The effect of these changes on the quality of people's lives cannot be overestimated, which helps explain why the reforms have been so popular, despite rising inequality, environmental destruction, and other significant repercussions. There is something deeply compelling about eating well, and Chinese people today eat far more meat and a wider variety of fruits and vegetables than they had dreamed possible in the old days.

The economic transformation has produced dramatic social changes. The movement of young villagers to urban factories represents the largest internal migration in world history. Just as in the Mao era, rural youth flee to urban areas whenever they can, and even those who stay often have little motivation to participate in agriculture. Education and culture (*wenhua*) continue to be seen as the antithesis of peasant identity; people who have education do not typically aspire to farm. If we take all the propaganda of the Mao era proclaiming that peasants are knowledgeable and that farming is not a waste of one's education as evidence that the opposite perspectives were in fact rampant, we find a formidable continuity in popular attitudes across the socialist and postsocialist eras. The difference is that post-Mao leaders do not try as hard as their predecessors to keep rural youth in the countryside, though they continue to recognize the problem.

Even when the only jobs to be had are in factories under grueling and lonely conditions, the desire to escape rural poverty drives young rural people from the land. In a village in Guangxi where I lived for two weeks in 2012, it was clearly the older generations who were farming and who felt compelled to do the arduous (辛苦) farm work, a phenomenon widely recognized in

空心化 农村之痛

为了摆脱贫苦，许多农民工不得不把家人留在农村，自己单枪匹马到城市闯荡。我国农村"男耕女织"的传统生存方式在许多地方已不复存在。由此，农村便形成了一个以妇女、儿童和老人为主体的留守群体，他们被戏称为"386199部队"。调查显示，目前全国有8700万农村留守人口，其中包括2000万留守儿童、2000万留守老人和4700万留守妇女。

FIGURE 28. The first of a series of thirteen photographs titled "Guangjiao jing: Kongxinhua nongcun zhi tong" (Wide-angle lens: Emptying out, the suffering of the countryside), *Xinlang tupian*, 2 September 2011. The caption identifies the couple as sixty-seven and sixty-five years old, from a village in Shaanxi Province. The introductory text (included here) reads: "In order to cast off poverty, many peasant laborers must leave their family members in the countryside while they go alone to the cities. The traditional Chinese way of life known as 'men tilling and women weaving' no longer exists in many places. For this reason, the countryside has become a caretaker community of women, children, and elderly people. They have come to be known as 'the 386199 army.' Surveys show that now in China the caretaker population has reached 87 million, among whom 20 million are children, 20 million are elderly people, and 47 million are women." The romantic evocation of a utopian past in the phrase "men tilling and women weaving" is striking here, since of course that system is long gone in most of the country. The number "386199" refers to 8 March, 1 June, and 9 September—holidays honoring women, children, and the elderly.

the mainstream Chinese media (figure 28). When I remarked to my friend how impressive it was that her mother, now in her sixties, continued to carry heavy buckets of fertilizer across her back to fertilize the family's fruit trees in a distant field, my friend replied that she often pleaded with her mother to ease up, saying, "It's not the collective era anymore. You don't have to work that hard!" But in her family, the trees not tended by her parents must simply be abandoned, and some of her neighbors rent out their fields for nonagricultural purposes. The families in this village are eating better and enjoying more material comforts than in the past, but the future of agriculture under such conditions is very uncertain. Moreover, the vast wealth gap has produced not just a widespread feeling of injustice but also its own share of poverty: hunger is by no means a thing of the past in many parts of China.[5]

The market economy has also led to new kinds of vulnerability for farmers who are now faced with a much wider "choice" of seed and chemical vendors, but little confidence in determining which choices can be trusted. Although the Chinese state has been notable in its efforts to prevent monopolization of the seed business by Monsanto or any other single corporation, it effectively no longer participates in the production or supply of seeds. Significantly, farmers demonstrate a strong preference for buying seed from what they perceive as state institutions: in a poorly regulated market, "stealth seeds" (including unapproved transgenic varieties and pirated first-generation transgenic seed) and "fake" seeds that lack the advertised productivity or pest resistance have proliferated, and farmers put their trust in the state rather than private companies. However, the county seed companies and agricultural extension stations themselves have been privatized, and the state in fact has much less control over the seed market than farmers typically imagine.[6] The problems faced by Chinese farmers are part and parcel of the far more widely reported problems faced by consumers in China and around the world because of the state's inability to control product safety and quality in the expanding economy.

Even before the introduction of market reforms, the green revolution in China had created the conditions for much environmental destruction. One of the key narrative tropes on this issue—which, like all powerful narrative tropes, is materially manifest in people's daily lives—involves the death of the irrigation ditches. Travel across China and you will hear people everywhere speaking of the animals that used to inhabit the irrigation ditches that run between fields, and their disappearance with the increase in agrochemicals. The fun that youth had catching fish and crustaceans, as well as the pleasure of cooking and eating them, explains the focus on this particular aspect of what is a more general recognition of environmental transformation.[7] As one of my interviewees explained when describing the early enthusiasm for pesticides, "Of course now we don't think [pesticides] are such a great technology because of the environmental effects. At that time the environment was still very good—the pollution problems hadn't yet emerged. The trees were full of birds, the fields had lots of frogs, and leeches would chase you [in the rice paddies]."[8] (Leeches may be unpopular when they attach, but they are pretty when they wriggle through the water, and in any case their absence is not a good sign for the ecology.)

Awareness of China's environmental problems is of course not confined to China itself but has become a global preoccupation. In 2008, amid widespread concerns about declining honeybee populations, *Newsweek* shared an alarming report from China:

> For 3,000 years, farmers in China's Sichuan province pollinated their fruit trees the old-fashioned way: they let the bees do it. . . . When China rapidly expanded its pear orchards in the 1980s, it stepped up its use of pesticides, and this age-old system of pollination began to unravel. Today, during the spring, the snow-white pear blossoms blanket the hills, but there are no bees to carry the pollen. Instead, thousands of villagers climb through the trees, hand-pollinating them by dipping "pollination sticks"—brushes made of chicken feathers and cigarette filters—into plastic bottles of pollen and then touching them to each of the billions of blossoms.[9]

Here indeed is an odd echo of the mass mobilization efforts in Mao-era agriculture. In the 1970s, observers around the world celebrated the tapping of the collective force of China's rural masses for making possible labor-intensive but ecologically sensitive solutions to insect pests. Decades later, *Newsweek* cites a very similar type of mass mobilization to paint a far more negative picture: those same rural masses must now replace the beneficial insects lost to the chemical excesses that, despite the best efforts of people like Pu Zhelong, swept the country.

Half a century after Pu Zhelong and other scientists in China and around the world began sounding the alarm over the consequences of chemical insecticides, a growing sense of environmental loss has led to another kind of "green revolution"—the popularization of organic agriculture. In fact, people in China tend to associate the term "green revolution" with the organic movement rather than the mid-twentieth-century adoption of dwarf varieties and agrochemicals, which is more often called the "seed revolution" or more generally "scientific farming" in Chinese. Organic agriculture has captured the interest not just of middle-class people but of peasants as well. Some peasants are refashioning their crops for the luxury market, and some just value the organic methods and traditional plant varieties and animal breeds they think produce tastier grains, vegetables, eggs, and meat.[10] The Chinese state sees potential in this, as do private interests and nongovernmental organizations.

Alongside the worries about social dislocations and the increasingly criti cal perspective on agrochemicals, some leftist scholars have challenged the dominant perspective that credits decollectivization for the rise in agricultural production. They point to signs that production had already begun increasing in the early 1970s, to policy changes during those years that were stimulating growth, and to the underappreciated role of green revolution technologies, whose effects would be expected to produce exactly the pattern of rapid growth followed by leveling off that in fact occurred during the 1970s and 1980s.[11] Viewed within the intersecting global histories of the "red" and "green" revolutions, such leftist critiques face something of a dilemma. Attributing the economic triumphs of the post-Mao era to the green revolution rather than to decollectivization does help salvage the reputation of the collective economy; but in the process it risks validating not only technologies widely condemned by leftists in other parts of the world for their environmental and social consequences, but also a technocratic approach to rural transformation embedded in Cold War–era geopolitical strategies that facilitated the development of global capitalism.

Critics of the post-Mao reforms go further in their challenge to free-market triumphalism, not only questioning the cause of the economic growth, but suggesting that the reforms were in many places imposed from the top down and not the result of clamoring from below, and arguing that capitalism is fundamentally inconsistent with the healthy development of rural China, such that the Chinese countryside now faces a three-dimensional rural crisis (*sannong*, 三农): peasants, rural society, and agriculture (农民、农村、农业). Proponents of this perspective promote reorganizing rural China along cooperative lines. Calling themselves the New Rural Reconstruction movement (新乡村建设运动), they take inspiration especially from the work of James Yen, Liang Shuming, and others who established "experimental" rural projects in the 1930s, and who were in turn influenced by American agricultural extension practices.[12] As Li Changping, one of the leaders of the movement, argues, NRR is the third major rural cooperative movement in modern Chinese history, the first occurring in the 1930s and the second in the 1950s.[13]

The New Rural Reconstruction movement is increasingly influential on state policy, as is evident in the state's 2006 launch of yet another effort to "build a new socialist countryside" (建设社会主义新农村). This is an important example of what Elizabeth Perry has called a "managed campaign," though squeamish adherence to Deng Xiaoping's call to "rely on the masses, but do not launch campaigns" prevents the use of the word in official discourse. (The Chinese term *yundong* [运动] means both "campaign" and "movement," a conflation of official and grassroots political activity characteristic of the self-styled revolutionary state.) In managed campaigns, the state takes an "engineering" approach: it makes use of many of the same tactics as in earlier decades—for example, the sending down of officials, experts, and youth to carry out initiatives—while engaging in a more "pragmatic" acceptance of a variety of methods and styles (including even Confucian or Christian practices, where locally relevant), all under the umbrella of "scientific development."[14]

NRR advocates have in turn sought to associate with the state's initiative. Alexander Day has explained that NRR proponents differ in important ways from state policy in their perspectives on global capitalism: while NRR advocates seek to provide a space shielded from capitalist forces for the Chinese countryside to develop along alternative paths, the effort to "build a new socialist countryside" typically encourages ever-stronger links between villages and globalized markets.[15] However, the lines sometimes become blurry as alliances form. Moreover, there are other projects underway that share the goal of improving life for China's peasants but position themselves in still different ways with respect to both the market and the Mao-era legacy. The next two sections will look first at examples of state-directed, market-oriented efforts to "build a new socialist countryside" (which draw to various degrees on NRR for inspiration and legitimacy), and then at a transnational project influenced by Western social science and committed to a more bottom-up approach to agricultural transformation.

Model Villages of Today

During my trip to Guangxi in 2012, two of the institutions I visited to interview Mao-era agricultural experts arranged tours of nearby model villages. The first took me to one of a number of "Hope Towns" established jointly by local governments and the Huarun Charitable Fund. Huarun (also known as China Resources) is a major state-owned conglomerate that includes an agricultural products company, a beverage company, a chain of supermarkets, and a cement manufacturer, among other enterprises. Construction of the Huarun Hope Town at Baise (华润百色希望小镇社区) began in 2008 and was completed in 2010.[16]

The Huarun Hope Towns reflect the engagement between intellectuals committed to New Rural Reconstruction and the state's policy of "building a new socialist countryside." One of the most important leaders of NRR and the person most associated with calling attention to the "three-dimensional rural crisis," Professor Wen Tiejun of People's University, is the chief consultant for the project; at the same time, a web article associated with the project, "A Village to Dream Of: Introducing Huarun Baise Hope Town," positions it within the broader context of the state's rural development policies. Consistent with the perspectives of Wen Tiejun and others associated with NRR, "A Village to Dream Of" describes current efforts as the most recent stage in a longer history of movements to cooperativize the Chinese countryside, beginning with 1930s-era rural reconstruction in the style of Yan Yangchu and Liang Shuming, and then proceeding also through Mao-era collectivization.[17] Like Wen and others, the website's authors also seek to distance the project from much-criticized Maoist policies. Wen associates his own work specifically with that of Yan Yangchu and Liang Shuming, and considers Mao-era collectivization as an anomaly in the history of rural co-operative movements because of its top-down character, its failure to tune itself to local conditions, and its treatment of the countryside as a repository of resources to be extracted for industrial development.[18] The website similarly acknowledges that the Mao-era collectives were too "rash" (冒进) and so "ended in failure," but maintains that they still deserve to be considered, along with the 1930s efforts, as "experiments in village reconstruction." To what extent Wen is satisfied with the direction of the Huarun Hope Towns is unclear; however, the collaboration is clearly of political use to both Wen and the Huarun Corporation.

Despite criticism of Mao-era collectivism, what comes through very strongly in the promotional literature, and came through also in my visit to the site, is the project's deep resonance with the 1960s and 1970s experiences that are the subject of this book.[19] According to "A Village to Dream Of," in the early days of the town's construction, Huarun transferred more than ten "backbone" youths from its various companies to Baise Hope Town to "form a project group" that would serve "on the front lines." Echoing Mao-era discourse on sent-down youth, the website proclaims, "Whether in the fields or on the construction site, the youth could be seen pouring their sweat into the work, collectively laboring with the villagers." And evoking the Maoist policy of self-reliance, it characterizes the project as "using the power of the community's own products and resources to help peasants establish specialty cooperatives and develop a new-style rural collective economy."

At the same time, the Hope Town clearly reflects state priorities embraced much more recently, including market-friendly environmental and ethnic policies. The goal is to make Baise Hope Town "ecological, organic, green, and in harmony with the local natural environment," and most importantly of all to turn it into a "new socialist village possessed of agricultural development capability and with distinct local and ethnic characteristics"—Guangxi's ethnic minorities are part of its marketability in the new economy.[20] Blending Maoist self-reliance with Hu Jintao's call for a "scientific development worldview," the project is expected to "use a spirit of innovation to implement the scientific development worldview and help the nation by charting a new style and a new path in which enterprises use their own resources to actively participate in building a new socialist countryside."[21]

As in the Mao era, peasants in Baise Hope Town are celebrated for mastering the new skills associated with scientific approaches to agricul-

FIGURE 29. "Thanks to Wufeng, I Learned Scientific Breeding." Photograph taken by the author in Huarun Baise Hope Town.

ture; and as in the Mao era, these celebrations include propaganda posters mounted around village public areas. Of course, the messages—and the work they represent—also reflect the political and economic relationships of reform-era China. For example, one billboard depicts peasants proudly proclaiming that they have learned "scientific breeding" from the Huarun subsidiary Wufeng (figure 29). It also presents information about the two breeds of pigs raised in Baise Hope Town. On the left is the Bama Delectable pig (巴马香猪), a breed associated with the nearby Bama Yao Autonomous County (Yao is a minority ethnic group) and thus possessed of a certain cachet of tradition and "distinctive local and ethnic characteristics," not unlike the cachet that "heirloom breeds" now possess in the West. On the right is a new hybrid produced by crossing a Guangxi breed known as Luchuan pig (陆川猪) and the Danish Landrace pig (长白猪). The Danish Landrace has been one of the most important imported "improved varieties" since the 1960s; it has been promoted for its efficient conversion of feed to meat, but for flavor and other more subjective qualities, rural people prefer "local" (土 or 本地) breeds. The same is true of the chickens: during our visit, we saw "local" chickens on the loose that villagers raised for their own consumption (figure 30).

FIGURE 30. "Local" chickens for villagers' consumption wander through idyllic courtyards in Huarun Baise Hope Town. Photograph taken by the author.

The approach to agricultural research and extension found in Baise Hope Town builds on the top-down elements of the Mao-era system while at the same time seamlessly integrating the reform-era vision of market socialism. According to "A Village to Dream Of," "The Hope Town industries support the utilization of innovative, reformed, scientific methods in order to improve local crop varieties and animal breeds, raise agricultural productivity, and . . . develop agricultural produce rich in local flavor and compatible with the development of Huarun's industries." Still more tellingly, instead of a government agency providing extension services, the Huarun Corporation itself takes responsibility for choosing, vetting, and supplying the parent stock and sends its own "experts" to train peasants in "modern" breeding techniques, replacing the older system in which most peasants purchased young animals and then raised them for the market. The animals are then marketed specifically for Huarun's supermarket chains in Hong Kong and major mainland cities, where organic meat with "local" characteristics is in high demand. This type of "commercialization" (企业化) of extension services, such that they more effectively contribute to the market economy and involve more of "society" (rather than "the state"), is consistent with the state's own explicit policy orientation.[22]

My visit to another model village in nearby Tianyang County a few days later was even more evocative of an earlier era: I felt eerily as though I were repeating the experiences of American scientists who visited China in the 1970s, but translated into another time and another political economy. Like so many of the stops on their itineraries, my visit began with a rehearsed "introduction" by a government representative.[23] She was originally from another part of the county, had studied at the teachers' college in the county seat, and had been sent to this site after passing the government civil service exam. She called our attention to a billboard showing China's current leader Xi Jinping visiting the village and then brought us to a room displaying the organic products the village produced and sold, a room for cultural activities (with bookshelves and space for dancing), and the conference room where democratic decisions are made. Finally, she introduced us to an old man who was a former production team leader. This is where things broke down a bit: the interview was kindly arranged to accommodate my interest in learning about the past; it was apparently not part of the standard script, which was entirely about the village's present and future. We met the former production team leader in a large hall with an enormous television tuned to NBA basketball. He did his best to answer my questions but was clearly uncomfortable, not least because the government representative could not help chiming in here and there—including to admonish him not to put his bare feet up on his chair.

In the car on the way to our next stop, one of my traveling companions remarked that it would be wonderful if the whole Chinese countryside could be this nice, but another was quick to emphasize that this was a model; not only was it not representative, but given the level of investment, it was perhaps not even reproducible everywhere. Specifically, he noted that this was Xi Jinping's "spot" (or "point," 点), and "there aren't enough Xi Jinpings" in China to transform the entire country in this way. Here too the parallels with the Mao era are strong, since top leaders then as well had their own "spots" that received lavish state support and served as models for the programs they favored.[24]

We left that village for a tour of an extraordinary set of greenhouses displaying innovative agricultural technologies, part of a collaborative project between Baise and the Association for Southeast Asian Nations. Like some of the 1970s delegates, I felt inspired to be in the presence of such forward-looking, environmentally sensible research—though the greenhouses do not at all represent the Mao-era ethic of making due with "crude," locally available materials, but rather dazzle the eye with their impressive economic investment.

During lunch at another model village in the county, the feeling of repeating 1970s history grew even stronger. A special room had been set up for hosting guests. About a dozen well-dressed city people treated my companions and me to a generous meal of "local" food, with many kinds of meat and fish, and many dishes left unfinished, in a "farmer's house" where the farmers themselves were not in evidence. Despite the obviously theatrical character of the experience, there was genuine warmth and friendship all around and a sincere desire to create positive relationships on the basis of a shared hope for the future of the Chinese countryside, Sino-US relations, and the planet as a whole. Toward the end of the meal, our host toasted my collaborator Cao Xingsui and me together, saying that Cao was *tu* (native) and I was *yang* (foreign), and that we should be congratulated for "uniting *tu* and *yang*" (土洋结合) in this way.

The evocation of one of the most important slogans for the Maoist vision of science was the crowning touch on an experience richly resonant with an earlier era. But for all that, some very important elements of the Maoist vision of science were entirely absent. Perhaps ironically, the elements of Mao-era agricultural research and extension that best fit the demands of the market economy are the most top-down. The legacy of the more radical political values associated with the Mao-era scientific experiment movement lies elsewhere.

Peasant Participation in a New Political Language

Not everyone subscribes to the template for refashioning the Chinese countryside through "innovative" new partnerships between peasant cooperatives and market-oriented corporations. Whatever Wen Tiejun's role in the Huarun Hope Towns, his own work—and that of the New Rural Reconstruction movement more generally—emphasizes the need for more insulation from market forces and proposes a model for rural Chinese development alternative to that of global capitalism.[25] This perspective is also at play in the efforts of a transnational group of agronomists and social scientists who have organized the Participatory Plant Breeding project in maize-growing villages of Guangxi. The researchers hail from a number of Guangxi-based institutions in addition to the Center for Chinese Agricultural Policy (农业政策研究中心) of the Chinese Academy of Sciences and the Canadian government's International Development Research Centre; the project also has ties to CIMMYT, which was of central importance in developing green revolution technologies but now has its hand in breeding projects to salvage genetic diversity as a relatively small number of high-yield varieties have come to dominate the fields.

Like many of the agricultural experts I interviewed in 2012, the social scientists associated with the Center for Chinese Agricultural Policy bemoan the reform-era disintegration of what had in the Mao era been a robust system of agricultural extension. They call for a renewal of "public" (公共) extension services and criticize the blurring of the line between public extension and commercial enterprise.[26] However, unlike the more mainstream agricultural experts I interviewed, CCAP scholars and others associated with the Participatory Plant Breeding project do not—at least in their published writings—speak nostalgically of Mao-era agricultural extension; on the rare occasion that they mention it, they portray it as characterized by the same problematic top-down structure found in current extension activities.[27]

The Participatory Plant Breeding project has adopted the principles of "participatory action research" as the foundation for a new kind of agricultural research and extension system with the joint goals of protecting the environment, creating better varieties, and improving rural people's livelihoods. Nonetheless, its resonance with Mao-era agricultural science is perhaps even deeper than that found in the model villages discussed above. Moreover, where the model villages carry on the more top-down aspects of the Mao-era system, this project recalls specifically the bottom-up, *tu* science elements, such that its literature often reads like translations of Maoist discourse into the language of postsocialist, transnational, environmentalist social science. A 2006 article written by Chen Tianyuan and Huang Kaijian, collaborators in the Participatory Plant Breeding project who hail from the Guangxi Maize Research Institute, illustrates this clearly.

Unlike the directors of the Baise Hope Town, who call on the towns to "make the leap from a traditional to a modern style of agriculture,"[28] Chen and Huang have mixed feelings about modern agriculture. They credit the replacement of traditional agriculture for increasing output but assert that it has come with the heavy costs of lowered sustainability and loss of biological diversity.[29] Meanwhile, they criticize "top-down" (自上而下) systems of agricultural research and extension, in which "peasants are seen as merely receivers of research and not participants" and in which research is "concentrated in the laboratory / experiment station" and so unable to "effectively satisfy peasants' real needs." As they go on to explain, "The aloof and remote [高高在上] research system has developed much research, but most of the results (including the results of experiment stations) are difficult to put into practical use or are unable to serve agricultural production."[30] Similar to Mao-era critics of ivory-tower science, Chen and Huang point to the way research becomes "divorced" from the needs of production: "Conventional research places too much emphasis on the research questions themselves and not enough on resolving problems. The research questions get deeper and deeper the more they are researched . . . while solving problems becomes a matter for other people, often peasants themselves." They conclude that this "top-down" system is unsuitable to the new economic and social conditions of rural China: "Scientific research has been divorced from peasants' real situations, breeding goals have been divorced from market needs . . . and the resulting varieties cannot satisfy the market's need for high-quality specialty corn."[31]

Chen and Huang claim that in contrast to the "conventional" system, "participatory" (参与式) methods "not only create the conditions for peasants to actively participate, but at the same time strengthen the integration of scientific research departments with production." The Participatory Plant Breeding approach is said to be "intimately related to peasants' needs" and to foster "cooperation between researchers and peasants," while siting "the majority of the experiments in peasants' fields." Peasants do not just play "supporting roles" but act as genuine "collaborators." Peasants contribute to the breeding program the crop varieties their own families have developed: this results in varieties "better suited to the local environment" and more likely to spread widely among farmers; thus, the system both protects the environment and facilitates the dissemination of better varieties.[32]

The authors emphasize that "the key to using entirely new agricultural research methods—participatory methods—is how to mobilize the peasants, how to get peasants to actively participate, and for this we need to transform the outsiders' perspectives so that they adopt an attitude of equality and modestly learning from peasants, emphasizing 'local knowledge' [乡土知识], 'local talent' [乡村天才], and 'peasant experts' [农民专家]." And they have focused especially on women peasants, whose involvement in agriculture has increased as men have migrated to factory jobs. The project included five groups of peasant women, and the authors quoted the project's leader, Song Yiching, as saying, "From the start, the women were wholly enthusiastic and actively participated through the whole process. Some men at the beginning were a bit shocked by women's participation, but they accepted it very quickly."[33]

The Mao-era echoes ring strongly throughout the article, though in places they are translated into a new jargon—for example, "participatory" replaces "mass," "top-down" replaces "technocratic," and the term "environment" sometimes appears in place of "conditions" in the familiar call for research to "suit local conditions." More importantly, some of the most fundamental priorities of Maoist science are almost unchanged, especially the cultivation of *tu* knowledge and "peasant experts," the call for researchers to "learn modestly from peasants," the requirement that science "be put into practical use" and "serve production," and the need for women's enthusiastic participation and men's rapid accommodation of the new gender dynamics.

And yet the researchers involved in this project do not position their work in relation to the Mao-era scientific experiment movement. Perhaps this is because they are uncomfortable connecting their approach with the radical politics of the Mao era, and perhaps because they genuinely do not see the connections—it is certainly clear that they perceive their project to be in opposition to what has come before. In the English-language book *Seeds and Synergies*, other members of the project similarly characterize Mao-era extension as unequivocally "top-down."[34] Given the extensive damage incurred during the Mao era through the imposition of inappropriate models on locales without concern for local conditions (or "environment") and without consulting local people for their perspectives, the "top-down" label is perhaps justified. Nevertheless, the Participatory Plant Breeding project's critique is in fact consistent with critiques articulated during the Mao era: of ivory-tower science divorced from the needs of the peasants and production, of inappropriate technologies imposed from above, and of failures to recognize the knowledge that rural people possess. In a striking display of Maoist-inflected discourse, the authors of *Seeds and Synergies* highlight the "need to challenge most if not all traditional plant science assumptions, such as the belief that farmers are less knowledgeable than breeders."[35] Of course the authorities referenced are different: Mao himself in the Mao era, and Western social science today.

And there are other important differences between the Participatory Plant Breeding project and the Mao-era scientific experiment movement. Most obviously, the PPB is guided by a full-blown environmentalism that was only a whisper in the 1960s when Pu Zhelong and other Chinese scientists worked to find alternatives to chemical insecticides. More subtle, but even more important, is the difference between encouraging peasants to become involved in modern agricultural methods and recognizing the legitimacy and value of social networks and knowledge communities outside those organized by the state. The latter is what the Participatory Plant Breeding project promotes: they seek to acknowledge and protect the practices through which peasants select their own seeds and then circulate them in informal markets, and further to bring these networks into productive exchange with the formal research and extension system of scientists. This is clearly a significant departure from the Mao-era scientific experiment movement, which had at best a co-opting relationship, and at worst an openly hostile relationship, with cultural forms and social activities outside the official political economy.

Moreover, the rhetorical differences between socialist "propaganda" and postsocialist "jargon" have important social and political consequences. The Mao-era discourse of "mass science" was undeniably fiercer with its frontal attacks on capitalism, imperialism, and patriarchy, and for many that fierceness made it more inspiring than the discourse of "participatory action research" is ever likely to be. It is hard to imagine many posters depicting social scientists and communicating participatory action research slogans tacked up on the walls of Western activist organizations and college dorm rooms the way Maoist propaganda posters were in the 1970s—when, not coincidentally, participatory action research was growing in popularity among Western social scientists in leftist circles. But if it is less galvanizing, participatory action research is also far safer. Mao-era rhetoric was filled with "enemies," whereas post-Mao rhetoric has only "systems" or perhaps "forces" in need of transformation. In this way, it is consistent with the more pragmatic and instrumental approach that the postsocialist technocratic leadership has adopted with its "managed campaigns." And it is also consistent with the more violence-averse politics of Western activist movements today. Like so much else about the Mao era, what is needed is a way to recognize the continued relevance and lasting influence of radical approaches to agricultural science, while simultaneously viewing them in critical perspective.

The Food Sovereignty Movement in China

The Participatory Plant Breeding project is one strand of an expanding network of Chinese academics and activists committed to "food sovereignty"—an agricultural movement rooted in anticorporate, anti-imperialist activism in Latin America and South Asia, and already well established in places as close to China as South Korea and Taiwan. Food sovereignty activists are best known for their opposition to the use of genetically modified organisms in agriculture, but GMOs are just the most potent example of numerous technologies perceived to threaten ecological health, the economic interests of countries in the Global South, indigenous knowledge, and farmers' rights.

An open letter written in 2011 to Yuan Longping by Li Changping did much to spread awareness of food sovereignty, as newspapers and websites picked up the story that this New Rural Reconstruction advocate had "called upon Yuan Longping to give peasants back the right to freely select seeds."[36] Li began his letter by claiming to be a "fan" of Yuan Longping, stating his belief that Yuan "deserved" the appellation "father of hybrid rice" (which some other leftist critics have challenged), and crediting him with enabling 1.3 billion Chinese people to fill their bellies. And yet Li's purpose in writing the letter was to request that Yuan cease his efforts to "scale new peaks" in hybrid rice research and spend the remainder of his life breeding conventional rice varieties instead. Li denounced "geneticists and seed-industry capitalists" for pursuing "monopolistic profits by doing everything possible to exterminate peasants' conventional seeds." He related his experience attempting to purchase seed from a supply company and finding it impossible to buy "a single grain of conventional seed"; they were all "death-without-progeny" (断子绝孙) seeds—what activists in Anglophone countries often call "suicide seeds," because they contain a gene that renders them infertile, compelling farmers to return to the seed companies each year for new supplies, and for that reason they are implicated in the literal suicides of debt-stricken farmers in India.[37]

Li's concerns should be familiar to people around the world who pay attention to agriculture and food issues. First is food security: the Chinese state's and Chinese people's national security depends on conventional varieties, for what if a terrorist attack or natural disaster wiped out the seed corporations' storehouses? Second is environmental protection: conventional varieties require less chemical fertilizer and have greater natural resistance to pests. And third is peasant livelihood: in the event of a poor spring harvest, peasants need to be able to "turn over to autumn" (翻秋), replanting seeds from the spring harvest with the hope of a better late crop; moreover, the price of hybrid seeds has escalated out of proportion, now costing peasants twenty times the value of the equivalent amount of grain, whereas in the old days conventional rice seed cost just twice as much as the same weight in grain.

But for our purposes, Li's most important point may be his assertion that conventional varieties selected by peasants are perfectly capable of generating high yields. He specifically recalled a variety of rice popular during his time as village party secretary in the 1980s. A peasant named Hu had selected a variety that had been well received by local peasants (a criterion commonly encountered in Mao-era materials on local variety improvement) and had come to be known as "Hu Select" (胡选). However, according to Li, the popularity of the variety was a "thorn in the side" of the seed research and development agencies, who sought a monopoly on profits, and now Hu Select has disappeared. Li closed his letter by appealing to Yuan as a "serious scientist" who, unlike government officials and many others in the world of science, should be capable of "climbing down from the speeding chariot of commerce."

Li Changping's call for "peasant seed sovereignty" (农民种子主权) is part of China's recently emergent participation in the global food sovereignty movement (in Chinese, 食物主权 or 粮食主权). A growing network of Chinese academics and activists who identify with this movement are collaborating on a website called People's Food Sovereignty (人民食物主权, http://www.shiwuzq.com/) launched in 2013. They come from Mainland China, Taiwan, and Hong Kong, and their website (hosted in Hong Kong) provides a venue for a wide range of subjects—from articles redeeming the history of collectivism in China, to interviews with figures like Li Changping, to critiques of the "build the new socialist countryside" campaign's emphasis on "sending capital down to the villages." The site features two sections with titles reminiscent of Maoist epistemology: "Knowing and Practice" (认识与实践) and "People's Science" (人民科学). And it includes an article by Song Yiching calling for "scientific knowledge and rural traditional knowledge to come together, respond to one another, and ceaselessly carry on."[38] Song does not call this *yang* and *tu*, but the roots of her epistemology are unmistakable.

In their approach to the global decolonial movement to mobilize indigenous knowledge, some Chinese participants in the food sovereignty movement evince still deeper connections to Maoist discourse on *tu* science. In an article on microbial fertilizers, the website offers a critique of the ways in which indigenous knowledge is exploited for the profit of organic farming enterprises. The anonymous author writes, "Modern science does not serve peasants and their indigenous knowledge, but rather strips them of the traditional wisdom and skills they embody and makes off with the profits. They [scientists] collect the forest topsoil and take it to the lab to cultivate microbial strains, then after separating out each strain, they name them and patent them, then sell them back to the peasants." The author testifies, "What we want to say is that modern scientific knowledge and technology is not that far removed from peasant traditional and local wisdom, and moreover the only appropriate science and technology is that which applies science locally using methods that enhance and preserve peasant agency."[39] Here in particular the resonance with the Mao era is palpable. We may recall Jiangsu party secretary Xu Jiatun's 1965 declaration that there was no "Great Wall" between peasant experience and science, and that peasant experience was necessary for scientific advancement.[40] And we may further remember the Cultural Revolution–era allegations that some "technocratic" educated youth had become expropriators of peasant wisdom: they had "taken the technologies that the poor and lower-middle peasants slaved to teach them and secreted them away in their own sacks."[41]

The concept of "food sovereignty" itself bears no small resemblance to the idea of "self-reliance." Food sovereignty and self-reliance both emerged from resistance to colonialism, and thus both insist on the importance of maintaining internal strength and independence from outside forces. At the same time, food sovereignty shares with self-reliance an impressive flexibility, and so attracts people with diverse political interests. Especially telling is the way the emphasis on "security" appeals to nationalist impulses: Li Changping's reference to potential terrorist attack is particularly striking, but other influential food sovereignty advocates have expressed similar concerns about food imports undermining Chinese national interests.[42] It is not surprising, then, that the food sovereignty movement has become a global phenomenon in current decolonial struggles in much the same way that self-reliance inspired diverse groups around the world fighting imperialism during the Mao era.

China, Global Food Movements, and the Legacy of Mao-Era Scientific Farming

The work of researchers involved in Guangxi's Participatory Plant Breeding project and of the diverse participants in the People's Food Sovereignty website is exciting. Through these efforts, Chinese academics and activists are joining global movements to recognize indigenous agricultural knowledge for its ecological wisdom and sustainability, and to mobilize it against the onslaught of capitalist agricultural technologies. Akhil Gupta has rightly criticized the indigenous knowledge movement for its problematic assumptions about the purity of local people's knowledge and the tendency to want to preserve people in that supposed state of purity when they themselves may not desire it. However, following Gayatri Spivak, such "strategic essentialism" may be appropriate or even necessary to counter the extraordinary power of developmentalist discourse.[43] That is, even if indigenous identity is understood to be a social construct that carries potentially damaging stereotypes, it may still serve as a needed weapon against state and corporate efforts to "develop" traditional communities and their ecologies out of existence.

In China, voices critical of developmentalism have had a harder row to hoe than in some other places, especially South Asia and Latin America. Outside certain small circles, Chinese academics and peasants alike sing the praises of modernization and speak of peasant knowledge derisively—and in the case of peasant informants, with shame. The disparagement of peasant culture as backward was prevalent in the Mao era, and it served as a major justification for state intervention. However, this attitude coexisted with a simultaneous political need to celebrate peasant experience as a key foundation for agricultural scientific knowledge. The Mao-era policy of "raising *tu* and *yang* together" attempted to make ideological sense out of a profound contradiction in state attitudes toward peasant culture and what decolonial activists would call "indigenous knowledge." That contradiction prevented nativist, peasant-based *tu* science from mounting an effective challenge to developmentalism of the kind decolonial activists encourage through the indigenous knowledge movement. But Mao-era celebrations of *tu* science at least served to check what otherwise might have been wholesale disregard for long-standing forms of knowledge. Today, celebrations of peasant knowledge are much harder to find in official state policies than they were in the past; critical social scientists like those involved with the Participatory Plant Breeding project and People's Food Sovereignty movement are filling that gap.

So what is China's place in the larger, global histories of struggles over agricultural science and technology? It is not hard to understand why so many people in China and elsewhere view the history of agriculture as a linear process from a state of "backwardness" to one of "modernity": this teleology is apparently confirmed wherever one looks.[44] At one lunch I attended with agricultural experts, I heard someone speak at length of his trip to North Korea. He was astonished by just how much North Korea of today resembled China of the Cultural Revolution era—from the low level of technology, to the scarcity of consumer products, to the loudspeakers blaring political propaganda in every community. In the realm of agriculture, he noted especially the lack of any improved varieties or breeds—all they had were "native," "traditional" plants and livestock. He told his hosts repeatedly that if they would hire him, he could increase their production. They rejected his offer, saying that he was peddling "revisionist technology" they neither wanted nor needed. The others at our lunch table agreed that North Korea was "thirty years behind China."

The developmentalist narrative proves remarkably robust even in today's environmentalist climate. In Baise and Tianyang's model villages, organic farming—and even the marketing of "local" breeds—is touted as evidence of a more "modern" approach, or, in Hu Jintao's terms, a "scientific development worldview." This is the kind of rural reconstruction that easily accommodates organic methods if they serve market needs, but does not accommodate the tendency for an old peasant to put his bare feet up on a chair while chatting with a guest, which is a clear mark of "backwardness."

Ironically, even within progressive movements, China is at risk of being perceived as "backward" because it is "still" mired in developmentalism, especially if we view projects like Participatory Plant Breeding as directed by Western academic trends. Helpful here is Michael Hathaway's effort to chart the global manifestations of what he calls "winds"—he borrows the term from the Mao-era policy "winds" that shaped political work at the local level, but he uses it more generally to mean patterns of cultural influence. Hathaway recognizes the significance in contemporary China of Western cultural influences, and his case in point is specifically the shift to valuing the environmental knowledge of local people, rather than condemning their "slash and burn" agriculture and other allegedly destructive practices. He not only highlights the agency of Chinese actors in contributing to these paradigm shifts but also notes that prevailing cultural winds sometimes blow the other way, seen, for example, in the impact of Maoism on 1960s radical activism in the West.[45]

The limitation of the otherwise very compelling metaphor of "winds" is that it captures influences across geographical space far better than those across time. What I would add to this analysis is an appreciation for how Mao-era radicalism continues to serve as an intellectual resource for Chinese academics—and by extension, those who work with them—as they develop a critique of technocracy and intellectual elitism, even though they appear far more consciously tuned to the currently strong "West wind" of participatory action research. And it serves as a resource, again largely unspoken, in the Chinese movement for food sovereignty, which owes no small debt to Mao's philosophy of self-reliance.

To adopt another metaphor, we might see this history as the layers of soil in which current movements are planted. The past does not go away just because new layers are added: it gets plowed up, consciously or inadvertently, again and again to mix in new ways with the work of today.[46] Self-reliance, learning from peasants, making theory serve practice: Participatory Plant Breeding in China would not be the same without these legacies from the socialist Chinese past. Perhaps the richest element of the Mao-era layer of history is the insistence on pursuing agricultural science and technology in actively political ways. The man who coined the term "green revolution" conceived it as an alternative to red revolutions, but critics on the left—with Maoists in the lead—have never let that vision stand unchallenged. China's experience by no means suggests that the flaws of green revolution ideology can be magically cured by red revolution: the environmental consequences of chemical-intensive agriculture and the human costs of emphasizing increasing production over equitable distribution of food and labor have been felt as strongly in China as anywhere on the planet. But recognizing the Maoist resources that nourish today's agricultural movements will provide those movements with a clearer understanding of their own histories and thus a stronger, healthier root system to support future growth.

Acknowledgments

One of my goals in launching this project was to acknowledge and document the participation of diverse social actors in the production of knowledge. It is only fitting that I should recognize the many people who helped in the research and editing of this book, though of course any mistakes or omissions are solely my responsibility.

I have greatly benefited from the support of several intellectual communities. I could not ask for a more encouraging environment or better colleagues than I enjoy in the UMass Amherst History Department; I especially thank my chair, Joye Bowman, for her continuous support. My advisers at UCSD, Joe Esherick and Paul Pickowicz, and the cohort of graduate students they trained continue to act as a vibrant community that our geographical scattering has not disrupted. The UMass Social Thought & Political Economy Program has challenged me to make bolder connections between scholarship and activism. I gained tremendous inspiration—and learned many valuable lessons—from the visionaries of community agriculture involved in Montview Neighborhood Farm. And I am honored to be part of an ongoing community of scientists and activists interested in reviving the radical 1970s–1980s organization Science for the People, many of whom participated in the original organization and came together at UMass in 2014 for an inspiring retrospective and look ahead.

This book would not have been possible without the willingness of many people who experienced scientific farming in socialist-era China to share their stories. I cannot name them all here (those who wished to be named appear in the footnotes), but I would like to highlight a few to whom I am especially indebted. Cao Xingsui not only granted invaluable interviews on his experiences but organized a ten-day trip to multiple sites in Guangxi where

I was able to interview dozens of others. In Guangdong, Gu Dexiang and Mai Baoxiang spent many days answering my questions about Pu Zhelong and the history of insect control research and taking me to Sihui to obtain further interviews and historical materials. Gu Dexiang also provided much follow-up help for which I am deeply grateful. I don't know how I came to be so blessed as to have Ye Wa's help on yet another project, but once again speaking with her has been among the most meaningful and enjoyable parts of my research.

Zhang Li at the Institute for the History of Natural Sciences has been extraordinarily generous with her time and energy; not only has she given me lively conversation on the history of science in the PRC, but she has processed much red tape to facilitate my visits to China. I keep hoping there will be a way for me to repay her many kindnesses. Peng Guanghua at the National Agricultural University had the brilliant, course-changing idea of introducing me to Cao Xingsui, for which I will be forever grateful. Zhang Guren of Sun Yat-sen University and Zhang Su of the National Agricultural Museum spent much time facilitating my research in Guangdong and Guangxi respectively. Yang Yang, Boqiao, and Feng have made it possible, and fun, for me to get my work done in China.

I am also deeply grateful for the assistance of many librarians and archivists in the United States and in China, including at the University of Minnesota, the University of Illinois at Urbana-Champaign, Texas A&M University, and Sun Yat-sen University. I especially thank Sharon Domier whose support has once again made it possible to get a great deal accomplished without leaving Amherst. Also at UMass, I thank Duan Lei for his assistance during his semester as my research assistant. In Beijing, Zhang Xiaoyan expertly transcribed many interview recordings.

The feedback I have received at numerous conferences, workshops, and invited lectures has challenged me to consider alternative perspectives and hone my arguments. I thank the organizers of the Hixon-Riggs Forum on Science, Technology and Society (Zuoyue Wang); the "ReFocus" series of workshops on science and technology in modern China and India (Tong Lam, Jahnavi Phalkey, and Grace Shen); the East China Normal University "Symposium on Society and Culture in 1950s China" (Yang Kuisong); the Simon Fraser workshop "Between Revolution and Reform: China at the Grassroots, 1960–1980" (Jeremy Brown and Matthew Johnson); the Francis Bacon Conference "How the Cold War Transformed Science" (Erik Conway, John Krige, and Naomi Oreskes); and the "Knowing and Doing in Asian Agriculture" workshop (Jacob Eyferth and Don Harper). I also thank Michitake Aso, Xuan Geng, Chuck Hayford, Danian Hu, Eugenia Lean, and Zuoyue

Wang for organizing thought-provoking conference panels. And I thank Jacob Eyferth, Judith Farquhar, Jim Hevia, and their colleagues at the University of Chicago; Victor Ho, Elisabeth Köll, Liz Perry, and others at Harvard University; Zhang Li at the Institute for the History of Natural Sciences in Beijing; Cao Xingsui at the National Agricultural Museum in Beijing; and Dean Kinzley at the University of South Carolina for the chance to present parts of this research to constructively critical audiences.

Jeremy Brown, Shane Hamilton, Gail Hershatter, and Fabio Lanza read the entire manuscript and offered very helpful comments and corrections. I hope the end result justifies their extraordinarily generous efforts and lives up to their expectations.

Stefan Landsberger and Wang Zuoyue shared valuable sources I would not otherwise have found. Peter Lavelle and Gabriel Rosenberg shared unpublished writings. Madhumita Saha has been a wonderful partner on a collaborative paper destined to be published as part of the "ReFocus" project.

In addition, I am grateful to the following people who offered valuable suggestions along the way: Michael Ash, Francesca Bray, Madeleine Charney, Tim Cheek, Richard Chu, Jerry Dennerline, Michael Dietrich, Fa-ti Fan, Jacob Eyferth, Arunabh Ghosh, Michael Hathaway, Lijing Jiang, Tong Lam, Dick Levins, Laura Lovett, Dick Minear, Hiromi Mizuno, Brian Ogilvie, Lisa Onaga, Larry Owens, Zhaochang Peng, Steve Platt, Andy Rotter, Michael Schoenhals, Jared Schy, Steve Smith, Elena Songster, Bill Summers, Saul Thomas, Vinton Thompson, Rudolf Wagner, Woody Watson, Zhun Xu, Yuezhi Zhao, Wang Zheng, and Wang Zuoyue.

I received financial support for travel to China through a research grant from the D. Kim Foundation, a fellowship from the Marion and Jasper Whiting Foundation, a Franklin Research Grant from the American Philosophical Association, and a Faculty Research Grant from UMass Amherst. A subvention from the UMass Office of Research Development supported the color gallery.

I count myself highly fortunate to be publishing again with Karen Darling, Kelly Finefrock-Creed, Evan White, Tadd Adcox, and the rest of the team at the University of Chicago Press. I know I couldn't be in better hands, and I just love working with them. Paige Bridgens read the entire manuscript with me at the proofing stage and raised many stimulating questions for the next project.

My parents and in-laws frequently made space where I could write while the children enjoyed playing with their Oma and Opa, Grandma and Grandpa. My father, Victor Schmalzer, to whom this book is dedicated, was supportive until the very end: he died just six days after I submitted the edited manuscript. He was a book publisher of the first water and an avid gardener and composter. I like to think his influence can be seen in everything I do, and I hope this book is a worthy tribute to his memory.

I have always sworn I wouldn't be the kind of scholar who had to apologize to her family for neglecting them in order to write her books. I fear that in the last stages of completing this book, I have failed a bit on that end. Whether or not an apology is in order, a thank you certainly is: I am profoundly grateful to Winston, Ferdinand, and Anarres for all their love and support.

One apology is required. In order to make this book a reasonable length, something had to give, and I am sorry that the two chapters I cut (on American visitors to socialist China) had to be the ones most relevant to Fabio Lanza and my friends from Science for the People. All I can offer is a sincere effort to do that material justice elsewhere.

Notes

Introduction

1. In 1973, Jim Hightower offered a scathing indictment of the role of agribusiness in agricultural science research and extension, including charges of classism and institutionalized racism: "There has been more than a 'green revolution' out there—in the last thirty years there literally has been a social and economic upheaval in the American countryside. It is a protracted, violent revolution, and it continues today." *Hard Tomatoes, Hard Times*, 2. On environmental degradation and social consequences in the Global South, see Shiva, *The Violence of Green Revolution*. For a Marxist analysis of the science and business of plant breeding, see Kloppenburg, *First the Seed*; and Berlan and Lewontin, "The Political Economy." On the green revolution and the Cold War, see Perkins, *Geopolitics and the Green Revolution*; and Cullather, *The Hungry World*.

2. Examples are numerous. Especially influential and interesting works include Haldane, *The Marxist Philosophy and the Sciences*; Bernal, *The Social Function of Science*; and Haraway, "Primatology Is Politics by Other Means."

3. For a thoughtful discussion of recent scholarship that highlights this appetite, see Rogaski, "Addicted to Science."

4. Gaud, "AID Supports the Green Revolution."

5. Perkins, *Geopolitics and the Green Revolution*; Cullather, *The Hungry World*.

6. Kennedy, "Special Message to Congress."

7. "Zhengzhi jingji weiji riyi jiashen."

8. Examples of Soviet influence include the machine tractor stations (Miller, *One Hundred*) and scientific publications on various aspects of agricultural science (see, for example, a Soviet book in Chinese translation on biological control of insect pests, Jielianjia, *Nonglin haichong shengwu fangzhi*). On the influence of US agricultural technology and US agricultural experts in the Soviet Union, see Fitzgerald, "Blinded by Technology"; and J. L. Smith, *Works in Progress*.

9. For example, when John Lossing Buck began his agricultural reform project in North China in 1916, he organized his efforts around "a test farm for experiments, a demonstration farm for a model, and a school program for boys." Stross, *The Stubborn Earth*, 111. In a recent publication, Peter Lavelle demonstrates that the staff in China's first wave of agricultural experiment stations were more likely to train in Japan than in the United States. This changed dramatically during the 1920s. Lavelle also argues that the 1920s saw a relative decline in the

significance of research accomplished at experiment stations compared with that at universities. Lavelle, "Agricultural Improvement."

10. *Quotations from Chairman Mao Tse-tung*, 40.

11. Also often called "agricultural [农业] scientific experiment movement" or just "scientific experiment movement." For "experiment," 实验 and 试验 are used interchangeably.

12. "Zhuangzu guniang xue Dazhai: Kexue zhongtian duo gaochan." The use of the term "science" here may strike readers as unusual. Socialist-era Chinese sources demonstrated little of the concern in distinguishing between "science" and "technology" that has occupied scholars in the West. In Chinese, then and now, the words are very often used together in a term that might be translated as "sci-tech" (科技). At the same time, considering the use of pig manure fertilizer to represent "science" or "technology" at all is more consistent with the perspectives of scholars in science and technology studies than that of William Gaud or, indeed, of many others who may not only see science and technology as independent of social and political forces but also equate the terms with only the most "advanced" or "modern" methods. Scholars who study science and technology typically accept a far broader range of knowledge and practices within these categories and insist on their embeddedness in the social world. As Francesca Bray has conceptualized the term, technologies "are specific to a society, embodiments of its visions of the world and of its struggles over social order." Bray, *Technology and Gender*, 16.

13. Cullather, *The Hungry World*, 10.

14. Bräutigam, *Chinese Aid and African Development*, 1–2, 176–79. As in China, political leaders in West Africa recognized the usefulness of a philosophy that not only stoked anti-imperialist sentiment but also encouraged locales not to depend on aid from the central government. See chapter 5 of this book.

15. Mahoney, "Estado Novo, Homem Novo (New State, New Man)," 191. See also Cook, "Third World Maoism."

16. Saha and Schmalzer, "Science and Agrarian Modernization." On the "crisis of sovereignty," see Gupta, *Postcolonial Developments*, 35.

17. Gupta, *Postcolonial Developments*, 51.

18. Saha and Schmalzer, "Science and Agrarian Modernization."

19. Gupta, *Postcolonial Developments*, 172–76. In a different way, Madhumita Saha has complicated our understanding of the "indigenous" by focusing attention on the Indian state's own distinct use of the term in a manner that "had almost no epistemological connotation." Saha, "State Policy, Agricultural Research and Transformation of Indian Agriculture."

20. Han, "Rural Agriculture"; Xu, "The Political Economy of Agrarian Change"; Peng, "Decollectivization and Rural Poverty in Post-Mao China."

21. During his 1974 and 1977 visits to China, Norman Borlaug noted with pleasure the building of chemical fertilizer plants. Borlaug, Field Notebooks, China, no. 1 (1974), 49, 50, 60; Borlaug, Field Notebooks, China, no. 2 (1974), 51; Borlaug, Field Notebooks, China, no. 2 (1977), 42. Many observers remarked positively on night soil and manure collection. See, for example, Metcalf, "China Unleashes."

22. Stavis, *The Politics of Agricultural Mechanization in China*.

23. Both approaches could be, and were, trumpeted through the use of quotations from Chairman Mao. In his 1957 "Be Activists in Promoting the Revolution," Mao suggested that China depend on "intensive cultivation" to make it possible to feed one person with just one *mu* of land (and then quickly added the need for birth control). Mao, *Selected Works*, 5:486. In his 1959 "Six Questions on Agriculture," he declared mechanization to be the "way out" for Chinese agriculture. Stavis, *The Politics of Mechanization*, 129.

24. Shiva, *The Violence*, 72.

25. Harlan, "Plant Breeding and Genetics," 307.

26. "Until the second half of the twentieth century," millet was "the principal source of subsistence for the peasants and the urban poor of the north." Sorghum was an important supplement, since it tolerated fields prone to flooding where millet and wheat could not be sown. Li, *Fighting Famine in North China*, 90–99, especially 93–94.

27. Brown, "Spatial Profiling," 212.

28. Borlaug, Field Notebooks, China, no. 3 (1977), 83.

29. This is why Pan Yiwei sought improved irrigation for his village (chapter 7, page 181).

30. Coffey, "Fertilizers."

31. Mao, "Intra-Party Correspondence."

32. Stavis, *Making Green Revolution*, 44.

33. For an influential critique of the pesticide treadmill, see van den Bosch, *The Pesticide Conspiracy*, 17–35.

34. I heard from many people about the promotion of cover crops for green manure during the 1970s, most vividly from Cao Xingsui. The top level was alfalfa (green), the middle was a purple type of milk vetch known as 紫云英, and the bottom level was a red type of milk vetch called 红花草.

35. Melillo, "The First Green Revolution," 1028–60.

36. Fukuyama, *The End of History*.

37. As Aminda Smith says in her study of thought reform in China, "The official version is crucial, not *despite the fact* that it is idealized, but precisely because it is idealized." Smith, *Thought Reform*, 7, italics in the original.

38. Timothy Cheek offers a very thoughtful discussion of the significance of ideology and propaganda as understood by Chinese establishment intellectuals throughout his *Propaganda and Culture* (see especially pp. 13–20). See also Schurmann, *Ideology*.

39. Lynch, "Ideology," 199. I depart from Lynch, who seeks an approach that facilitates the identification and elimination of ideological elements in scientific knowledge, and who for that reason defines ideology narrowly and negatively, as "any aspect of knowledge, understood as a product or process, and ranging from the scientific to the mundane, that plays a causal role in maintaining or creating power disparities in society" (206–7). I change just the last few words of his definition, to read instead "maintaining or creating political power," in recognition that ideology can work to empower oppressed or subaltern groups and so decrease power disparities in society. The difference stems not just from a desire to recognize the possibility of progressive ideologies, but also (and perhaps more importantly) from a deep skepticism about the possibility of dissecting knowledge into purely "scientific" and "ideological" components. I define ideology more broadly and neutrally because I do not imagine the possibility of identifying and eliminating ideological aspects of scientific knowledge.

40. Helen Longino's discussion of science and ideology is useful here: "While eschewing the concept of a single truth or the hope of a singular epistemological blessing, we can nevertheless rank theories as to their acceptability, in particular their worthiness as bases for collective action to solve common problems." *Science as Social Knowledge*, 214.

41. Marx and Engels, *The German Ideology*; de Certeau, *The Practice of Everyday Life*.

42. Farquhar and Zhang, "Biopolitical Beijing," 310.

43. Hershatter, *The Gender of Memory*, 216. And on page 235, she returns to this theme: "New women were brought into being, not by state fiat, but by the labor of cadres, the women themselves, their village communities, and regional or national reading and listening publics."

44. On interpreting Mao-era propaganda posters, see Evans and MacDonald, *Picturing Power*.

45. Shen Dianzhong, *Sixiang chenfu lu*, 286.

46. This continuity in the role played by ideology across periods is noticeable to some people who lived through socialist and postsocialist times. As one of my interview subjects put it, "The format [during the Mao era] was set. The government told you how to write, and you gave your speeches to the government to edit. It's just like today when we have to talk about Hu Jintao's 'development perspective' [发展观]. You have to say each sentence [just as they want it]. That time was just the same. You had to talk about class struggle, scientific revolution, waging revolution in the countryside." Interview in Guangxi, June 2012.

47. Deng Xiangzi and Ye Qinghua, *Bu zai ji'e*, 66−67. A full discussion of this material can be found in chapter 3.

48. Xiaoping Fang has written about a parallel phenomenon in which people are revisiting the famous "barefoot doctors" program of the Cultural Revolution. Fang's analysis demonstrates the success of the barefoot doctors program but challenges several elements of the current portrayal, including the notion that the gains for rural health care were lost in the reform period. Fang, *Barefoot Doctors*.

49. Cao specifically told me that agricultural extension was one of Mao's "three great successes" (the other two being water control and national defense).

50. Interview appearing in Hinton, Barmé, and Gordon, *Morning Sun.*

51. Examples of nationally important places include the rural areas around Beijing, Huarong County, and Dazhai Brigade. In 2012, I visited a number of sites in Guangxi Province with the agricultural historian Cao Xingsui, whose story appears in chapter 4, and several others with a personal friend who remains anonymous.

52. Michel Bonnin recognizes the existence of rural educated youth but nonetheless reinforces the conflation of "educated youth" and "urban youth" in the title and content of his important book *The Lost Generation: The Rustication of China's Educated Youth (1968–1980)*. He also gives short shrift to youth participation in the scientific experiment movement. An exception to this pattern is Miriam Gross's 2010 dissertation on anti-schistosomiasis campaigns, which examines the role of rural educated youth and argues that their incorporation into scientific and public health work helped consolidate state power. Gross, "Chasing Snails," 626−64.

53. On Chinese science as transnational, see especially Wang, "The Cold War and the Reshaping"; and Wang, "Transnational Science during the Cold War." *Tu* science is discussed elsewhere as "mass science" or "self-reliant science." See Suttmeier, *Research and Revolution*; Schmalzer, *The People's Peking Man*; and Schmalzer, "Self-Reliant Science." Exploring what he calls "citizen science," Fan relates mass mobilization in China to forms of public participation found in Western capitalist countries. "'Collective Monitoring,'" 148−49.

54. In his study of the earthquake prediction program, Fa-ti Fan concludes that "Maoist mass science" was "mostly top-down despite its claim of the mass line." Fan, "'Collective Monitoring,'" 149. Without denying the tendency toward dogmatism, I place more emphasis than Fan does on the ways *tu* science served diverse social actors in resisting top-down policies and hegemonic forces more generally. Thomas Mullaney's study of Chinese typewriters offers another approach to the tensions implicit in mass science: he suggests that mass science could indeed translate the labor of thousands of workers into dramatic technological innovations and simultaneously produce, "ironically, an ever more tight-fitting, personal connection and commitment to the rhetorical apparatus of Maoism." Mullaney, "The Moveable Typewriter," 807.

55. On the complex uses to which China's distinct approach to green revolution was put in Africa, see Bräutigam, *Chinese Aid and African Development.*

56. On persecution, see, for example, Neushul and Wang, "Between the Devil and the Deep Sea"; Schneider, *Biology and Revolution*; and Hu, *China and Albert Einstein.* On agency, Thomas Mullaney has shown that the 1950s project of ethnic classification—just the type of project we usually assume to be heavily state directed—depended less on the imposition of communist perspectives and more on the intellectual paradigms and priorities embraced by Chinese social scientists, who relied on knowledge of Chinese ethnicity produced by Western scholars. See his *Coming to Terms with the Nation,* especially chapter 2. In *The People's Peking Man,* I highlighted a different aspect of Chinese scientists' agency by arguing that they were active participants in hegemonic efforts to eradicate forms of knowledge deemed "superstitious" and replace them with forms deemed properly socialist and scientific.

57. On deskilling of rural people in socialist China, see Eyferth, *Eating Rice from Bamboo Roots.* The literature on deskilling in agriculture is extensive. See chapter 4, notes 92–96.

58. In *Barefoot Doctors,* Xiaoping Fang has demonstrated that the famed barefoot doctor program of the Mao era, widely celebrated for its supposed promotion of traditional Chinese medicine, in fact produced unprecedented acceptance of Western medicine throughout the countryside. Similarly, the scientific experiment movement, for all its emphasis on "self-reliance" and "native methods," was extraordinarily effective in convincing rural Chinese people of the superiority of the new agricultural technologies that constituted the US geopolitical strategy known as the "green revolution." However, medicine and agriculture represent two very different cases, since "Traditional Chinese Medicine" gained recognition in the twentieth century as a coherent body of knowledge with specialized practitioners representing a unified tradition (Lei, *Neither Donkey*), whereas the diverse array of agricultural knowledge and practices did not enjoy systematic codification and did not have an organized group of professional practitioners advocating for its survival.

59. Ma Bo, quoted in Pan, *Tempered in the Revolutionary Furnace,* 128.

60. Greenhalgh, *Just One Child.*

61. Ferguson, *The Anti-Politics Machine,* xv.

62. Kloppenburg, *First the Seed,* 290.

63. Agricultural activists today often point to Cuba as a model, in ways eerily reminiscent of earlier decades' celebration of China. I have not yet seen the kind of thorough, source-based analysis necessary to make sense of the complicated history of Cuba's nationwide program of organic farming. That said, Richard Levins has provided an important and intriguing start from the perspective of a Marxist biologist in "How Cuba Is Going Ecological."

Chapter One

1. See, among many examples, Yao, "Chinese Intellectuals"; Dong Guangbi, *Zhongguo jin-xiandai kexue*; Williams, "Fang Lizhi's Big Bang"; Neushul and Wang, "Between the Devil"; and Schneider, *Biology and Revolution.* To some extent, the pendulum narrative follows the "two-line" analysis advanced by Mao-era radicals themselves; the post-Mao antiradical historiography reverses the signs and depoliticizes the rhetoric. My understanding of Liu Shaoqi, Deng Xiaoping, and others as "technocrats" follows Andreas, *Rise of the Red Engineers.* See also Fan, "'Collective Monitoring.'" This chapter borrows from Schmalzer, "Self-Reliant Science."

2. Zweig, *Agrarian Radicalism,* 192. Zweig sees peasants in Popkin's terms as economically rational actors, which he seems to equate with an interest in economic development/modern-

ization, whereas in his view the radicals were committed to ideology over material concerns and sought to impose these values on the peasants.

3. The first reference to "scientific farming" (科学种田) in *People's Daily* is from 22 July 1961; in 1965 there were eleven references. This periodization is similar to that suggested by earlier observers more sympathetic to Chinese socialism. For example, Benedict Stavis marked 1960−1962 as the key watershed when China embarked on a technological transformation of agriculture still going strong when he visited China in the early 1970s. Stavis, *Making Green Revolution*. See also Kuo, *The Technical Transformation of Agriculture*.

4. In his study of science education films in the PRC, Matthew Johnson has similarly noted the important commonalities that underlay the various Cold War political ideologies of modernization and development; he dates the emergence of this "shared global culture—technocratic, homogenizing and nationally directed—which transcended nationally specific forms" to the early twentieth century. Johnson, "The Science," 31.

5. Gilman, "Modernization Theory," 48−49.

6. J. L. Smith, *Works in Progress*, especially 6−12, 120−21.

7. Bray, "Chinese Literati," 301.

8. Rowe, "Political, Social and Economic Factors," 29.

9. Perdue, *Exhausting the Earth*, 131−35.

10. For an expanded discussion of how this slogan came about, see Tan Shouzhang, *Mao Zedong*, 103−4. Note that the push for "close planting" has developed a negative reputation (and certainly if pushed to extremes makes little sense), but the basic premise is sound: many of the new varieties require less space than older varieties do, and farmers had to be convinced to plant them more closely together to improve harvests.

11. Steve Smith, "Local Cadres," 1021. See also Schmalzer, *The People's Peking Man*, 281−82; Perry, *Anyuan*, e.g., 9, 44, 244−46.

12. The best account of US influence on the modernization of Chinese agriculture remains Stross, *The Stubborn Earth*.

13. J. K. King, "Rice Politics," 458.

14. J. L. Buck, "Missionaries Begin," 78.

15. Examples abound. See, e.g., Xu Jiatun, "Shixian nongye kexue," 6.

16. Republican-era Chinese approaches to agricultural extension were also directly influential on the history of the green revolution as it unfolded in Taiwan. In a recent journal article, James Lin argues that in Taiwan agricultural modernization blended the technocratic methods of the green revolution with social organizing through farmers' associations and cooperatives, and he shows how this approach built on the Republican-era work of James Yen and others. Given their shared roots in 1930s and 1940s rural reconstruction, a deeper comparison of the Chinese and Taiwanese experiences with the green revolution is clearly warranted. Lin, "Sowing Seeds."

17. Yang, "Promoting Cooperative Agricultural Extension," 55.

18. Ibid., 60, 57.

19. On the US visitors' perceptions of Chinese agricultural science, see Schmalzer, "Speaking about China."

20. Kuhn, "Political and Cultural Factors," 66.

21. Heilmann, "From Local Experiments," 13−14.

22. Secord, "Knowledge in Transit."

23. Mitchell, *Rule of Experts*, 52.

24. Meisner, *Li Ta-chao*.

25. Stuart Schram cautions that the links between Yan'an and later Maoist perspectives on economic development represent "existential continuity" but "no intellectual continuity in terms of detailed policy formulations, and certainly no unbroken chain of development in Mao's own thinking." I do not seek to argue this point beyond the level of "existential continuity"; it is sufficient, I think, to recognize the lasting impact of both the experience of Yan'an and the heroic narrative woven around the "Yan'an Way." Schram, *The Thought of Mao Tse-tung*, 93; Selden, *The Yenan Way*.

26. Reardon-Anderson, *The Study of Change*, 323.

27. Schneider, *Biology and Revolution*, 105. Note that Le Tianyu's family name has previously been incorrectly Romanized as "Luo" (in Pinyin) and "Lo" (in Wade-Giles) by a number of scholars, including me.

28. Two excellent discussions of this episode can be found in Reardon-Anderson, *The Study of Change*, 352–59; and Schneider, *Biology and Revolution*, 104–8.

29. Interestingly, as Mao began pushing harder for Chinese self-reliance, Soviet leaders were beginning to admit defeat on their long-standing commitment to maintaining agricultural self-sufficiency; in 1963, Khrushchev signed a deal with Kennedy to import US wheat. J. L. Smith, *Works in Progress*, 19.

30. For examples of these various uses of the term 洋专家, see "Tu zhuanjia he yang zhuanjia" (which speaks of "洋专家, who have received formal higher education") and "Zhonghua ernü duo qizhi."

31. Anderson, "Introduction: Postcolonial Technoscience," 644.

32. As Lisa Rofel says, "Socialist power operates by positing within workers and peasants a subaltern consciousness that the state then authorizes and represents. For Chinese Marxists, subaltern agency was the major means of constituting modernity." Rofel, *Other Modernities*, 27.

33. This alternative vision for science that Mao helped create was not the only way Chinese intellectuals could have responded to the challenge posed by Western science. As Tong Lam has argued, Republican-era Chinese intellectuals largely differed from their Indian counterparts in their attitude toward modern science and technology. While Indian intellectuals, according to Gyan Prakash, rejected the universality of modern science and instead sought to construct "another reason" indigenous to India, Chinese intellectuals widely accepted Western science as universal. Lam, *A Passion for Facts*, 115.

34. As Walter Mignolo puts it, the goal of decolonialism is a "de-linking from the rhetoric of modernity and the logic of coloniality." He seeks a "de-colonial epistemic shift" that "brings to the foreground other epistemologies, other principles of knowledge and understanding and, consequently, other economy, other politics, other ethics." Mignolo, "Delinking," 453.

35. Zhou Yun, "Cong genben shang," 3.

36. Xinhua she, "Nongcun tiandi guangluo," 4.

37. Recent years have seen a boom in scholarly publications on the Great Leap Forward. Perhaps the most useful is the volume edited by Kimberley Ens Manning and Felix Wemheuer, *Eating Bitterness*. Works for more general audiences, which offer fiercer condemnations of the collectivist economic system as a whole, include Yang Jisheng, *Tombstone*; and Dikötter, *Mao's Great Famine*.

38. Schmalzer, "Breeding a Better China."

39. MacFarquhar, *The Origins of the Cultural Revolution*, vol. 2; and MacFarquhar, *The Origins of the Cultural Revolution*, vol. 3, especially p. 286.

40. Stavis, *Making Green Revolution*, 174–76; Dangdai Zhongguo, *Dangdai Zhongguo de nongye*, 570–71; Kuo, *The Technical Transformation*, 22.

41. Stavis, *Making Green Revolution*, 161.

42. This the first such explicit reference in the *People's Daily* to what would, beginning in 1978, become Deng Xiaoping's famous Four Modernizations platform. "Chanming nongye kexue," 1. In January, Zhou Enlai had referenced the "Four Modernizations" at the Conference on Scientific and Technological Work held in Shanghai but had not explicitly spelled out what they comprised. See "Zai Shanghai juxing," 1.

43. Lü Xinchu and Gu Mainan, "Shi kexuejia daxian shenshou," 2.

44. "Yige shehui zhuyi jiaoyu yundong hou," 42–49.

45. "Ba puji xiandai nongye kexue," 1.

46. " 'Yangbantian' shi nongye kexue," 1. On the larger politics of mass science, see Schmalzer, *The People's Peking Man*.

47. Stavis, *Making Green Revolution*, 164–65; Kuo, *The Technical Transformation*, 23; Dangdai Zhongguo, *Dangdai Zhongguo de nongye*, 571.

48. Here demonstration is 示范.

49. "Banhao sanjiehe de yangbantian, cujin nongke kexue shiyan yundong," 2. "In September 1962, the Tenth Plenum of the Eighth Central Committee emphatically declared that we must strengthen scientific and technological research with special attention to research in agricultural science and technology. In May 1963 Comrade Mao Zedong further called for . . . the three great revolutionary movements" (ibid.). On the significance of "backbones," see Walder, "Organized Dependency."

50. Here demonstration is 样板 again.

51. "Gao ju Mao Zedong sixiang hongqi gengjia guangfan," 9.

52. Jin Shanbao, "Yangbantian," 15.

53. Guangdong sheng, "Jieshao yige nongcun," 7.

54. Interview in Guangxi, June 2012.

55. "Gao ju Mao Zedong sixiang hongqi gengjia," 2.

56. Miriam Gross notes that an important difference between the anti-schistosomiasis campaigns of the Great Leap era and those of the mid-1960s lay in the dramatic increase in the availability of rural youth who had received schooling. Gross, "Chasing Snails," 624–25.

57. See, for example, Fyfe, "Reading Children's Books"; Kohlstedt, *Teaching Children Science*; Melanie Keene, " 'Every Boy & Girl a Scientist' "; and Owens, "Science 'Fiction.' "

58. "Beijing shi nongcun kexue shiyan xiaozu," 20–21.

59. Zhonggong Fuqing xian Yinxi dadui, "Tuchu zhengzhi, kexue zhongtian" [Give prominence to politics, farm in a scientific way], in Henan sheng, *Quanguo nongcun*, 7.

60. MacFarquhar notes that in December 1964 Mao invited Chen Yonggui of Dazhai to his birthday party but also that this was before the official call to study Dazhai. MacFarquhar, *The Origins of the Cultural Revolution*, 3:423–24. The first mention of the movement in the *People's Daily* was in September 1965, with only two articles until August 1966. The movement took off in 1967.

61. On some of the fascinating twists and turns in the history of Dazhai as political symbol, see Friedman, "The Politics of Local Models."

62. For a sense of the frustrations one locale experienced over the course of multiple campaigns taking Dazhai as a model, see Friedman, Pickowicz, and Selden, *Revolution, Resistance, and Reform*. On creative uses of the Dazhai model to further local interests, see ibid., 71–72, 79.

63. Henan sheng, *Banhao siji nongye*.

64. Lin Biao was the chief choreographer of Mao's cult of personality and widely recognized as Mao's second-in-command during the early Cultural Revolution. In 1971, he lost Mao's con-

fidence and on 13 September died in a plane crash while attempting to escape the country with his family.

65. Henan sheng, *Banhao siji nongye*, 2–3.

66. Cenxi xian, "Zajiao shuidao."

67. Teiwes and Sun, "China's New Economic Policy"; Meisner, *Mao's China and After*.

68. "Hua Zhuxi zhangduo," 1.

Chapter Two

1. Borlaug, Field Notebooks, China, no. 4 (1977), 25. This chapter borrows from Schmalzer, "Insect Control."

2. Esherick, *Ancestral Leaves*, 263–75.

3. Liu Chongle jinianguan [Liu Chongle commemorative museum], accessed 4 June 2015, http://www.yiqin.com/memorial/intro.html?m_id=18318.

4. On this important theme in recent Chinese history, see, for example, Williams, "Fang Lizhi's Big Bang"; Hu, *China and Albert Einstein*; and Miller, *Science and Dissent*. During the May Fourth era (c. 1915–1925), Chinese intellectuals called for modernization, national sovereignty, enlightenment, social transformation, and above all "science and democracy."

5. Interview with anonymous member of Science for the People, 2008. Pu Zhelong is discussed in Science for the People, *China*, 155–64.

6. Mai Shuping, "Shengwu huanbao," 10; Su Shixin, "Gaoshan yangzhi," 1–3.

7. US influence in early twentieth-century Chinese science is relatively well represented in the literature on science in modern China. See Schneider, "The Rockefeller Foundation," 1217; Bullock, *An American Transplant*; and P. Buck, *American Science*.

8. Liu, Letter to William A. Riley.

9. Wu, Letter to William A. Riley.

10. G. Y. Shen, *Unearthing the Nation*; Reardon-Anderson, *The Study of Change*.

11. H. C. Chiang, "I Am Happy"; Mai Shuping, "Shengwu huanbao," 10.

12. H. C. Chiang, "I Am Happy," 278.

13. Gu Dexiang and Feng Shuang, *Nan Zhongguo*, 80–81.

14. Pu distinguished himself among his peers in entomology at Minnesota. Clarence Mickel, at that time the chief of the Entomology Division, ranked Pu second of the ten Chinese graduate students then in residence when Mickel applied for continued funding for Pu from the China Institute of America in 1948. Mickel, Letter from Mickel to China Institute of America.

15. Chiang's subsequent experiences, including his return to China in 1975, are recounted in Schmalzer, "Insect Control."

16. Wang, "Transnational Science." On "Western-trained scholars and patriotic returnees," see also Cheek, *The Intellectual*.

17. Interviews with Gu Dexiang form an essential part of the evidence for this chapter. Born in 1936 in Malaysia, Gu returned in childhood to his ancestral home in Meizhou City, Guangdong. He graduated from secondary school in 1954 and received his BA from the Biology Department of Sun Yat-sen University in 1958, where he has remained throughout his career. He became a member of the Chinese Communist Party in 1959. His career is closely tied to that of Pu Zhelong, especially the research at Big Sand, where he and Pu spent twenty years beginning in 1973. Since Pu's death, Gu has played the leading role in organizing memorial volumes and other projects to celebrate Pu's legacy. "Suiyue de henji: Gu Dexiang jiaoshou jishi" [Traces of time: Professor Gu Dexiang's chronicle], accessed 3 January 2015, http://ch.sysu.edu.cn/gdx/experience/index.html.

18. Pu was eventually compensated two hundred yuan for these losses. "Guangzhou diqu chu."

19. Gu Dexiang and Feng Shuang, *Nan Zhongguo*, 80–81. Incidentally, Qianyang was where Yuan Longping did his groundbreaking research on hybrid rice—see chapter 3.

20. "Song jiaoyu ban."

21. Interviews in Guangdong, 2010.

22. Wang, "The Cold War."

23. Xinhua she, "Rang kunchongxue," 3. In a recent interview, Gu Dexiang gave much the same explanation, with less strident rhetoric. (Indeed, he may well have been the source for the *People's Daily*'s 1977 account.) Foreign scientists could recommend insecticides, but because it was impossible for China to produce or import them, there was no choice but to rely on "native methods" (土办法, e.g., using ducks to catch insects) and biological controls such as ducks, parasitic wasps, and bacteria. Interview with Gu Dexiang, June 2010.

24. The Soviet Union had pledged only sixteen thousand tons of chemical fertilizer for 1964, and though Cuba had sent requests to some "capitalist countries," they were unwilling to supply (presumably because of pressure from the United States). Deng Xiaoping supported a plan to provide an immediate twenty to thirty thousand tons of relief supplies and to attempt to buy some extra for Cuba from a Western European organization during their next round of negotiations. "Guanyu Guba yaoqiu." China also provided several varieties of insecticide, including DDT, BHC, and Dipterex. See "Gei Guba tigong."

25. Zhongguo kexueyuan, *Zhongguo zhuyao*, 15.

26. Xia Yunfeng, "Nongmin weishenme," 2.

27. Wu Tingjie, " 'Liuliuliu' bu shi," 6.

28. Dangdai Zhongguo, *Dangdai Zhongguo de nongye*, 478–79.

29. This need was recognized by Chinese scientists already in the 1930s and 1940s. Yun-pei Sun and Ming-tao Jen conducted research in this area at the University of Minnesota, and Ting Wu (Mrs. T. Shen) applied for a fellowship from the China Foundation (funded by the Boxer Indemnity) to conduct similar research (I am unsure whether she was successful). Shepard, Letter from H. H. Shepard.

30. Kogan, "Integrated Pest Management," 245. The article in question was Michelbacher and Bacon, "Walnut Insect."

31. Stern, "The Integrated."

32. Qi Zhaosheng, Zhang Zepu, and Wang Dexiu, "Mian hongzhizhu."

33. Pan Chengxiang, "The Development," 4n9.

34. Pu Zhelong, Zhu Jinliang, and Wu Weiji, "Ganshu xiaoxiang."

35. See, for example, Pu Zhelong, Mai Xiuhui, and Huang Mingdu, "Liyong pingfu."

36. This figure does not include military advisers. Goikhman, "Soviet-Chinese Academic," 282; Stiffler, " 'Three Blows."

37. Goikhman, "Soviet-Chinese Academic."

38. Mai Shuping, "Shengwu huanbao."

39. American Insect Control Delegation, *Insect Control*, 91.

40. Pan Chengxiang argues that Soviet influence on PRC insect control was slight, whereas US influence was strong owing to Republican-era connections. Pan Chengxiang, "The Development," 4 and 7.

41. Zhu Ruzuo and Hu Yongxi, "Chiyanfeng shenghuo." Note that the large majority of citations are to US publications, but a few Soviet sources also appear.

42. Interview with Gu Dexiang, June 2010.

43. Gu Dexiang and Feng Shuang, *Nan Zhongguo*, 91–93.

44. Grishin, Letter from Grishin to Pu Zhelong, 2 November 1959; Grishin, Letter from Grishin to Pu Zhelong, 20 April 1961; Pu Zhelong, Draft letter from Pu Zhelong to Grishin, 14 June 1961; Pu Zhelong, Draft letter from Pu Zhelong to Grishin, 1 November 1961.

45. "Zhongguo kunchong," 5.

46. Zhu Ruzuo and Hu Cui, "Nongzuowu haichong."

47. "Zhiwu baohu xuehui." The report did, however, highlight the importance of prevention, and in that respect presaged the policy of "integrated control with prevention foremost" that would be adopted in 1974. It is also notable for its very critical discussion of Great Leap–era developments and its emphasis on rebuilding infrastructure and personnel as a necessary foundation for improving plant protection.

48. Ma Shijun, "Xulun" [Introduction], in Zhongguo kexueyuan dongwu yanjiusuo, *Zhongguo zhuyao haichong*, 1–2.

49. This is clearly evident in the 1974 edition (1976 printing) of *Zhibaoyuan shouce*.

50. Interview with Gu Dexiang, June 2010. On the politics of insect control science and chemical corporations in the United States, see van den Bosch, *The Pesticide Conspiracy*; and Schmalzer, "Insect Control."

51. Xinhua she, "Zai renmin"; interview with Mai Baoxiang, June 2010.

52. For a comparison, see Metcalf, "Changing Role," 219–20.

53. Mai Baoxiang, "Pu Zhelong jiaoshou zai Dasha." See also "Dasha gongshe chonghai zonghe fangzhi." Mai Baoxiang has shared with Sun Yat-sen University portions of his diary from the 1970s dealing with Pu Zhelong's activities.

54. 白天燕了田间飞，夜晚啾啾闻蛙鸟，沿途蟑蟑装楼房. Interview with Gu Dexiang, June 2010.

55. Stross, *The Stubborn Earth*, 206.

56. Mai Baoxiang, "Pu Zhelong jiaoshou zai Dasha"; Pu Zhelong, *Haichong shengwu*, 236. See also Guangdong sheng Sihui xian Dasha gongshe, "Dasha gongshe shuidao."

57. Perry, *Anyuan*.

58. Pu had no party affiliation. "Guangzhou diqu chuli."

59. Note that Li Cuiying also served as a representative from the world of education on the Chinese People's Political Consultative Conference.

60. Cheng, "Insect Control," 269.

61. Interview with Gu Dexiang, June 2010.

62. Pu Zhelong and Liu Zhifeng, "Woguo zhequ"; Pu Zhelong and Liu Zhifeng, "Chiyanfeng daliang."

63. Xinhua she, "Zhongshan daxue pin."

64. Li Shimei has fared less well at the hands of historians. For example, one article concludes the episode was ultimately disappointing, since Li's research did not contribute in any substantial way to general knowledge. Xue Pangao, "Dui tu zhuanjia." However Gu Dexiang remembers him as being extremely experienced and competent: "他是很有办法的，因为他是农民出身，很有经验，所以聘请农民专家来搞研究." Interview with Gu, June 2010. And in later years Li Shimei remained on the faculty of the Chinese Academy of Science's Entomological Research Institute and coauthored an occasional paper; in 1989, he was on the other side of the professional / lay divide, appearing in *People's Daily* as a scientist supporting another amateur (or "fan," 迷) recognized for his accomplishments in termite control. Cheng Guansen, "Ta mizui."

65. Interview with Gu Dexiang, June 2010.

66. Zhongshan daxue, "Mianxiang qunzhong." In tune with the revolutionary tenor of the times, the article did not mention the names of specific individuals, attributing all the work to the Revolutionary Committee or to "teachers and students" (师生). However, this was clearly Pu's project.

67. See Pan, *Tempered.*

68. Xinhua she, "Zai renmin jiaoshi."

69. "Song jiaoyu ban."

70. Interview with Gu Dexiang, June 2010; Gu Dexiang and Feng Shuang, *Nan Zhongguo*, 28.

71. Interviews with Gu Dexiang, June 2010.

72. In a 2004 interview discovered in 2014, Xi spoke at length about his experiences as a sent-down youth: "The first test was the flea test. When I first got there, I couldn't stand the fleas. I don't know if they still have them. My skin was very allergic to them, and they bit my skin into swathes of red sores that became blisters that burst. It was so painful you didn't want to live, but after three years my skin was as tough as ox meat and horse hide and wasn't afraid of being bitten." Buckley, "An Interview."

73. Interview with Gu Dexiang, June 2010.

74. Mai Baoxiang, "Pu Zhelong jiaoshou zai Dasha."

75. Interview with former plant protection specialist at Big Sand, June 2010.

76. Mai Baoxiang, "Pu Zhelong jiaoshou zai Dasha."

77. Guangdong sheng Sihui xian Dasha gongshe, "Dasha gongshe shuidao."

78. Pu Chih Lung and Tsui-Ying Lee, Letter to Dr. and Mrs. Richards.

79. Xinhua she, "Mei diguo"; "Meiguo qinlüezhe jinxing."

80. On Chinese allegations of US use of germ warfare, see Rogaski, "Nature, Annihilation, and Modernity"; Chen, "History of Three Mobilizations"; and Esherick, *Ancestral Leaves*, 243–44. The historical evidence points strongly to the conclusion that the Chinese state fabricated evidence to bolster the charges in addition to pressuring scientists to make allegations against their former colleagues.

81. On Zhou Yiliang, who was also educated abroad and later participated in criticisms of his former Chinese colleagues and Western academics, see Cheek, *The Intellectual.*

82. Zhonggong Guangdong shengwei Tongzhanbu, Qiaowu zu, Letter to the Zhongshan daxue geweihui Zhenggongzu.

83. Guangdong sheng geweihui banshizu waishi bangongshi.

84. Visitors included Arthur Galston and Ethan Signer (who arrived even before Kissinger's landmark journey), Science for the People, the 1975 US insect control delegation, the 1975 Swedish insect control delegation, the 1976 International Rice Research Institute delegation, and numerous others. See Schmalzer, "Speaking about China"; and Schmalzer, "Insect Control."

85. Borlaug, Field Notebooks, China, no. 2 (1974), 38.

86. Chiang, "I Am Happy," 289.

87. Schmalzer, "Insect Control." The tremendous enthusiasm among foreign observers for Chinese efforts in integrated pest control and also in biogas (using methane produced through composting for energy) inspired Rudolf Wagner to conduct a very useful survey of the available Chinese literature on these subjects. See Wagner, "Agriculture."

88. Mai Baoxiang, "Pu Zhelong jiaoshou zai Dasha." This was apparently Professor M. J. Way.

89. Brinck, *Insect Pest.* Although the Swedish delegation did not go to Big Sand, they attended a lecture by Pu Zhelong at Sun Yat-sen University in which he spoke at length about the

work there (39–43). On the generally very positive impressions foreign visitors had of Chinese science during the late Cultural Revolution, see Berner, *China's Science.*

90. Johansson, "Mao and the Swedish." On Chinese efforts to enroll Mexico, see in the same volume, Rothwell, "Transpacific Solidarities."

91. "Report of IRRI Team Visit," 10.

92. Pu Chih-lung, "The Biological Control."

93. Pu Zhelong, "Kexue yanjiu."

94. In at least one case we can see strong evidence of direct copying from one biographical article to another, but the phenomenon is much broader and more subtle than that. Compare Su Shixin's "Gaoshan yangzhi" and Luo Lixin's "Laizi ling yige."

95. Su Shixin, "Gaoshan yangzhi."

96. Luo Lixin, "Laizi ling yige," 70. "Ancestral country" (祖国) is a common way of referring to China.

97. Su Shixin, "Gaoshan yangzhi," 2.

98. Gu Dexiang and Feng Shuang, *Nan Zhongguo,* 30–31; Su Shixin, "Gaoshan yangzhi," 3; Luo Lixin, "Laizi ling yige," 72.

99. Su Shixin, "Gaoshan yangzhi"; Luo Lixin, "Laizi ling yige."

100. Su Shixin, "Gaoshan yangzhi," 2.

101. Luo Lixin, "Laizi ling yige," 72.

Chapter Three

1. Like so many institutions, Anjiang Agricultural School underwent name changes, and for much of the Mao era, it was known as Qianyang Agricultural School.

2. Yuan acknowledges the inspiration that hybrid corn and sorghum provided his work, along with the previous achievements in hybrid rice made by scientists in Japan and the United States. Yuan Longping and Xin Yeyun, *Yuan Longping koushu,* 54.

3. Yuan Longping, "Shuidao de yongxing."

4. In a 1986 article, "The Political Economy of Hybrid Corn," Berlan and Lewontin argued that hybrid corn was not more productive than open-pollinated varieties and attributed its development rather to the commercial interests of agribusiness. It would be fascinating, but beyond my abilities at this point, to compare and contrast this account with a political economy of the development of hybrid rice in socialist China.

5. Hunan sheng, "Xuanyu shuidao," 8.

6. "Yong Mao zhuxi"; Anhui sheng, "Liyong shuidao." For the use of the theory of contradiction to guide breeding work in soybeans, see Wang Jinling et al., "Hunhe geti."

7. "Peiyu shuidao"; Liaoning sheng, "Shuidao zazhong"; "Shuidao xiongxing"; Shuidao xiongxing, "1972 nian shuidao"; Guizhou nongxueyuan, "Shuidao xiongxing"; "Wosheng zhaokai."

8. Guangdong sheng keji ju, "Zazhong youshi."

9. Xinjiang Weiwuer, "Yebai gengxing."

10. Zhao'an xian, "Shuidao nantaigeng."

11. Xiong Weimin and Wang Kedi, *Hecheng yi ge.*

12. Using the China Academic Journals database, I found more than a dozen such references from 1971 to 1976.

13. For Li Bihu, see Jiangxi sheng, *Zenyang zhonghao,* 11.

14. Interviews in Guangxi, June 2012.

15. "Shuidao xiongxing buyuxi de xuanyu," 5.

16. Zhejiang sheng, *Jiji shizhong*, 6.

17. Guangxi shiyuan, "Kaizhan shuidao."

18. Jiangxi sheng, *Zenyang zhonghao*.

19. Xiangzhou xian shuidao, "Jiaqiang lingdao."

20. Guangdong sheng Hainan, "Yi jieji douzheng."

21. Teiwes and Sun, "China's New Economic Policy"; Meisner, *Mao's China and After*, 428–32.

22. I have found only one mention of Hua in any of the Mao-era articles on hybrid rice available through the China Academic Journals database, and that is an oblique reference to a speech he gave at a national agricultural conference in 1973. "Wosheng zhaokai shuidao."

23. Xinhua she, "Zajiiao shuidao."

24. "Hua zhuxi zhangduo." I highlight the dating error to show that the story of Hua's involvement was at this point not well fixed.

25. "Lüse wangguo."

26. *Hua zhuxi zai Hunan*, 182–89.

27. "Di wu ci quanguo zajiao," 2–3.

28. Yuan Longping, Li Bihu, and Yin Huaqi, "Tantan zajiao shuidao."

29. Yuan Longping, "Zajiao shuidao peiyu."

30. Xie Changjiang, *Yuan Longping yu zajiao*, 158 (quote); Xie Changjiang, *Zajiao shuidao zhi fu*; Xie Changjiang, *Yuan Longping*.

31. Luo Runliang and Wu Jinghua, *Lüse shenhua*, 98, 125ff.

32. Quan Yongming, "Qiantan Yuan Longping keyan zhong maodun guandian de yunyong" [Thoughts on the application of the perspective of contradiction in Yuan Longping's scientific research], in Qi Shuying and Wei Xiaowen, *Yuan Longping zhuan*, 411–14.

33. Yuan Longping and Xin Yeyun, *Yuan Longping koushu*, 234–35.

34. Deng Xiangzi and Ye Qinghua, *Bu zai ji'e*, 195. Translation from the English version of this book: Deng Xiangzi and Deng Yingru, *The Man Who Puts*, 200. The original article was Yuan Longping, "Cun cao."

35. Deng Xiangzi and Ye Qinghua, *Bu zai ji'e*, 79–85.

36. Qi Shuying and Wei Xiaowen, *Yuan Longping zhuan*, 141–47.

37. Yuan Longping and Xin Yeyun, *Yuan Longping koushu*, 241.

38. Xie Changjiang, *Zajiao shuidao zhi fu*, 75, 89.

39. Yao Kunlun, *Zoujin*, 156, 217; Xie Changjiang, *Yuan Longping*, 121–22.

40. See, for example, Yao Kunlun, *Zoujin*, 241; Nie Leng and Zhuang Zhixia, *Lüse wangguo*, 138–39; Qi Shuying and Wei Xiaowen, *Yuan Longping zhuan*, 183–84.

41. One key exception is Luo Runliang and Wu Jinghua, *Lüse shenhua*, 92, which talks about the three-in-one teams that integrated cadres, technicians, and peasants, and the further integration of experiment, modeling, and extension. However, elsewhere this book depicts the Cultural Revolution in the typical manner as ten years of turmoil. Another exception is Xie Changjiang, *Zajiao shuidao zhi fu*, 41, which also highlights the integration of research and production.

42. Qi Shuying and Wei Xiaowen, *Yuan Longping zhuan*, 71, 142.

43. Nie Leng and Zhuang Zhixia, *Lüse wangguo*, 34; Luo Runliang and Wu Jinghua, *Lüse shenhua*, 60.

44. Yuan Longping and Xin Yeyun, *Yuan Longping koushu*, 102.

45. Deng Xiangzi and Ye Qinghua, *Bu zai ji'e*, 86–87. Translation from Deng Xiangzi and Deng Yingru, *The Man Who Puts*, 94.

46. Nie Leng and Zhuang Zhixia, *Lüse wangguo*, 109.

47. Yuan Longping and Xin Yeyun, *Yuan Longping koushu*, 117–18.

48. Ibid., 70.

49. Qu Chunlin, Xiang Biao, and Wang Jue, "Zajiao shuidao wenhua," 160.

50. Yao Kunlun, *Zoujin*, 40.

51. The passage in question is clearly translated from the 1950 edition of the English-language original. The relevant passage can be found in Sinnott, Dunn, and Dobzhansky, *Principles of Genetics*, 327–28.

52. Yuan Longping and Xin Yeyun, *Yuan Longping koushu*, 51.

53. Gowen, *Heterosis*, especially 55, 173.

54. Dong Bingya, "Yumi yichuan"; Schneider, *Biology and Revolution*, 71. On Republican-era use and translation in 1943 of the first edition of Sinnott's textbook, see Pi Yan et al., "Guonei gaoxiao."

55. Yang Shouren et al., "Xianjingdao zajiao."

56. Xu Guanren and Xiang Wenmei, "Liyong xiongxing"; Zhongguo kexueyuan, *Zenyang zhong*.

57. "Zajiao yuzhong de qunzhong."

58. See, for example, Siddiqi, *The Red Rockets' Glare*. For a recent reassessment of Lysenko's impact on Soviet agriculture, see J. L. Smith, *Works in Progress*.

59. The key work on Lysenkoism in China is Schneider's *Biology and Revolution*.

60. Qi Shuying and Wei Xiaowen, *Yuan Longping zhuan*, 76–79; Deng Xiangzi and Ye Qinghua, *Bu zai ji'e*, 48–51, 51–54.

61. Qi Shuying and Wei Xiaowen, *Yuan Longping zhuan*, 125ff.

62. Xie Changjiang, *Yuan Longping yu zajiao*, 23.

63. Qi Shuying and Wei Xiaowen, *Yuan Longping zhuan*, 51.

64. Deng Xiangzi and Ye Qinghua, *Bu zai ji'e*, 66–67. In another version of the first episode, Yuan was compelled to apologize to his wife for spending money meant for the family on his research; finding the discarded pots solved the problem. Nie Leng and Zhuang Zhixia, *Lüse wangguo*, 69–71.

65. Yao Kunlun, *Zoujin*, 79.

66. Deng Xiangzi and Ye Qinghua, *Bu zai ji'e*, 121. Translation from Deng Xiangzi and Deng Yingru, *The Man Who Puts*, 127–28. The use of the rope is corroborated in many places, including Yuan Longping and Xin Yeyun, *Yuan Longping koushu*, 151. An interviewee who grew up in rural Zhejiang remembers people using this method when beginning the local production of hybrid rice.

67. On this phenomenon more generally, see Schmalzer, *The People's Peking Man*, 124–25.

68. Qi Shuying and Wei Xiaowen, *Yuan Longping zhuan*, 104.

69. Ibid., 97.

70. Yao Kunlun, *Zoujin*, 153–54.

71. CCTV, "Women dou you."

72. Qi Shuying and Wei Xiaowen, *Yuan Longping zhuan*, 193.

73. Taiji Mao, "Yuan Longping."

74. Han, *The Unknown*; Andreas, "Leveling the Little Pagoda."

75. Qi Shuying and Wei Xiaowen, *Yuan Longping zhuan*, 177.

76. Li Guihong and Xiao Jian, "Lengshui yuzhong."

77. Borlaug, Letter to Richard Critchfield, 94–95; Critchfield, "China's Miracle Rice."

78. Epstein, *Dossier.*

79. Hammer, "On a Vast."

80. Deng Xiangzi and Deng Yingru, *The Man Who Puts,* 205–6, 223.

81. Livingstone, *Putting Science,* 134.

Chapter Four

1. In translating the Chinese word *nongmin* (农民) as "peasant," I am consciously choosing not to follow a noble movement among some China scholars to adopt the more respectful term "farmer" (see, e.g., Hayford, "The Storm over the Peasant"). I employ "peasant" precisely because the word is more loaded in English and so is a better translation of the very loaded Chinese term.

2. We may compare this with the way gay and lesbian people in some countries have claimed the epithet "queer," or more recently women have taken up "slut walks" to protest rape, though there is an obvious difference in that it was the Chinese state rather than the peasants themselves choosing the terminology.

3. "Report of IRRI Team Visit," 59.

4. Rowe, "Political, Social and Economic," 29.

5. Shanxi Yanbeiqu, "Yanbei zhuanshu"; Xie Jieyin, "Woyang xian Zhaowo xiang"; Fenghuang xian, "Laonong fangzhi."

6. Sun Xiuchun, "Yushi xiang."

7. Xu Jiatun, "Shixian nongye kexue," 3–4.

8. I am grateful to Peter Lavelle for sharing his as yet unpublished paper, "Imperial Texts in Socialist China." Among many relevant articles, see Li Baochu, "Yi, Lüfei zuowu."

9. Zhu Xianli, Chen Jisheng, and Ren Huiru, *Nongyan zhujie,* 2–3.

10. He Minshi and Wang Jianxun, *Guangzhou minjian chengyu,* 5.

11. Zhu Xianli, Chen Jisheng, and Ren Huiru, *Nongyan zhujie,* 7–8.

12. The campaign to eliminate the "four pests" is among the Mao-era campaigns best known outside China. When it began in 1958, it targeted rats, sparrows, flies, and mosquitos. However, when the loss of sparrows resulted in surging locust populations, sparrows were replaced with bed bugs. As with so much else in Mao-era science, many Western observers—including scientists—were initially enthusiastic about the possibilities that such mass mobilization offered, but they turned to mock the efforts when the political tide shifted and such quintessential examples of Maoism were repudiated. Schmalzer, "Insect Control"; Shapiro, *Mao's War,* 86–89.

13. Zhu Xianli, Chen Jisheng, and Ren Huiru, *Nongyan zhujie,* 7–10.

14. Zhu Dehui and Zhu Xianli, *Nongyan li de kexue.*

15. Li Qun, "*Qimin yaoshu.*"

16. Zhejiang sheng Huangyan xian, "Pinxia zhongnong." See also Sun Changzhou, "Guantian shaobing"; "Guantian ji"; and Xinhua she, "Qunzhong huanying."

17. Zhejiangsheng Huangyan xian, "Pinxia zhongnong."

18. Jiangsu sheng Jianhu xian, *Tianjia wuxing.* The *People's Daily* article in question wrongly identified the publication date as 1977 and even more egregiously represented the book as arising from the work of a returned educated youth. Sun Changzhou, "Guantian shaobing."

19. Wang Yuying et al., "Lunzuo, jianzuo"; Beijing nongye daxue, "Jianzuo tanzhong."

20. Zhu Dehui and Zhu Xianli, *Nongyan li de kexue daoli*, 108–11.

21. References to intercropping abound in Cultural Revolution–era sources. Borlaug, Field Notebooks, China, no. 3 (1977), 5.

22. Interviews in Qinzhou and Baise, June 2012. Qinzhou is an area of southern Guangxi Province that in the 1960s was part of Guangdong.

23. Mao, *Selected Works*, 5:486.

24. Xinhua she, "Fazhan nongye."

25. Interview with Yi Ruoxin, June 2012.

26. The Chinese state-produced web encyclopedia *Baike* provides an account of this, which appears to be cribbed from a newspaper interview with a former junior colleague of Huang's named Jiang Yijun (江奕君). It is, however, possible that the newspaper article cribbed the information from *Baike* and attributed it, word for word, to Jiang Yijun. For the *Baike* article, see http://baike.baidu.com/view/321423.htm (accessed 18 July 2014). For the newspaper article, see "Huang Yaoxiang 'banweigan shuidao zhi fu.'" Note that the specific area credited is Shantou (汕头), part of the Chaoshan region.

27. Many Chinese journal articles discuss "Chaoshan culture." For an English-language example, see Y. Z. Wang and Y. T. Chen, "The Eco-unit Settlement."

28. Xinhua she, "Fazhan nongye."

29. Kuhn, "Political and Cultural Factors," 66. Note that the section quoted here did not appear in the delegation's formal report. For more on this, see page 252, note 14.

30. Interview in Guangxi, June 2012.

31. Interview with Cao Xingsui, June 2012.

32. See also Schmalzer, *The People's Peking Man*.

33. Judith Farquhar reports similar testimony from peasants in Shandong who credit the extension work of agricultural technicians for long-sought improvements in agricultural production. Farquhar, *Appetites*, 83.

34. Zhonggong Huarong xian weiyuanhui [Chinese Communist Party Committee of Huarong County], "Jiaqiang lingdao, yikao qunzhong, banhao siji nongye kexue shiyan wang" [Strengthen leadership, rely on the masses, build the four-level agricultural scientific experiment network], in Henan sheng, *Banhao siji nongye*, 29 and 39.

35. This is similar to Xiaoping Fang's findings on the barefoot doctor movement in *Barefoot Doctors*.

36. *Gongnong famingjia xiaozhuan*.

37. Zhonggong Nantong, "Nongmin zijue."

38. "Gao ju Mao Zedong sixiang hongqi gengjia," 4.

39. Zhonggong Huarong xian, "Jiaqiang lingdao, yikao qunzhong," in Henan sheng, *Banhao siji nongye*, 33–37; Nanzhao xian geming weiyuanhui, ed., "Quandang dongshou cengceng zhua, qun ban kexue kai honghua" [The whole party initiates and each level grasps, mass science produces red flowers], in Henan sheng, *Banhao siji nongye*, 49–50.

40. Interview with Tan Chengping, in Qinzhou, June 2012.

41. Interview in Qinzhou, June 2012.

42. The campaign to study Dazhai's experience in agriculture did not begin until 1965, but it is possible that Mao's promotion of Dazhai as an example of self-reliance would have led the party secretary to have been discussed in those terms. MacFarquhar, *The Origins of the Cultural Revolution*, 3:423–24.

43. "Yi suo pinxia zhongnong guanli de xinxing xuexiao" [A new-style school run by poor and lower-middle peasants], in Henan sheng, *Banhao siji nongye*, 68–78.

44. On state efforts to promote women's participation in cotton production, see Hershatter, *The Gender of Memory*, 214–15; and Gao Xiaoxian, " 'The Silver Flower Contest.' "

45. Zhang Gui? [last character illegible], "Gao ju Mao Zedong sixiang hongqi, wei geming banhao mianhua" [Raise high the red flag of Mao Zedong thought, grow cotton for the revolution], in Henan sheng, *Banhao siji nongye*, 24–29.

46. "Sunan nongmin."

47. "Zuotan xin qingkuang." On Mao-era science education films, see M. Johnson, "The Science."

48. Tang Ruifu, "Chen Yongkang"; "Report of IRRI Team Visit," 126–30.

49. Lü Xinchu and Gu Mainan., "Shi kexuejia daxian shenshou."

50. Li Zhensheng, "Dalaocu."

51. Li Zhensheng, "Mao Zhuxi zhexue."

52. Ningwu xian liangzhong fanzhi chang, "Yunyong bianzheng weiwulun, Chuangkai shanyao yuzhong men" [Using dialectical materialism, open the door to breeding sweet potatoes], in Shanxi sheng Xinxian diqu, *Xinxian diqu nongye*, 49–56.

53. Yangchun xian kexue jishu xiehui, "Yige tuchu zhengzhi jianku fendou de kexue shiyan xiaozu" [A scientific experiment group gives priority to politics and struggles arduously], in Henan sheng, *Quanguo nongcun*, 31.

54. Interview with Cao Xingsui, June 2012.

55. Kuhn, "Political and Cultural Factors," 66. Note that the section quoted here did not appear in the delegation's formal report.

56. Interview in Youjiang, June 2012.

57. Interview with Cao Xingsui, June 2012.

58. Stavis, *The Politics of Agricultural*, especially p. 170.

59. See, e.g., "Fenxi xingshi"; and "Xiaomai ruhe."

60. Xiyang xian weiyuanhui, "Yi dang de jiben luxian wei gang renzhen tuiguang Dazhai kexue zhongtian jingyan" [Taking the fundamental line of the party as the key link, extend the scientific farming experience of Dazhai], in Mianyang xian, *Qunzhongxing nongye kexue*, 7–8.

61. Bai Zhaoqing and He Wentong, "Minbei shanqu."

62. Wu Jun, "Yu shuidao tu zhong."

63. Beijing nongye daxue and Shandong nongxueyuan, *Nongye huaxue*, 128.

64. Pingyang xian nonglin ju, "Pingyang xian 1964."

65. "He'erbin shi nongye."

66. Cao Longgong, "Tan Chen Fu de."

67. Dazhai dadui, "Cong Dazhai gaitu."

68. Bray, *Science and Civilization*, 293.

69. F. H. King, *Farmers of Forty*, 10. King's 1895 book *The Soil: Its Nature, Relations, and Fundamental Principles of Management* was one of many Western works on technical subjects translated into Chinese and published by the Jiangnan Arsenal during the last fifty years of the Qing dynasty. Lavelle, "Agricultural Improvement," 336.

70. Hua xian Jiang Qizhang, "Pinxia zhongnong."

71. "Shai tian" [Sun drying fields], *Baidu baike*, accessed 7 May 2015, http://baike.baidu.com/view/127244.htm.

72. Interview in Guangxi, June 2012. He further pointed to the lack of scientific approaches to controlling insects and fertilizer.

73. Interview in Qinzhou, June 2012.

74. Interview in Guangxi, June 2012. At first the production team leader suggested the min-

ing technology was introduced around 1966 or 1967, but when pressed, he said it was during the Four Cleanups Movement, which would have been a few years earlier.

75. Edward D. Melillo has documented the human suffering involved in the laborious mining of nitrogen-rich guano fertilizer, including by Chinese laborers in Peru, during what he calls the "first green revolution." Melillo, "The First Green Revolution."

76. Interview in Qinzhou, June 2012.

77. Interviews in Guangxi, June 2012.

78. On Mao-era efforts to transform rural gender relations, see, among others, K. A. Johnson, *Women, the Family*; and Hershatter, *The Gender of Memory*.

79. Hershatter, *The Gender of Memory*.

80. See, for example, Naquin, *Millenarian Rebellion*.

81. "Jianchi kexue shiyan." For another example of women working in a group of women, see Xinxian Qicun gongshe Qicun dadui geming weiyuanhui, "Yanzhe Mao Zhuxi de geming luxian da gao qunzhongxing de nongye kexue shiyan yundong" [Following Chairman Mao's revolutionary line, pursue the mass rural scientific experiment movement in a big way], in Shanxi sheng Xinxian diqu, *Xinxian diqu nongye*, 39.

82. Fogang xian, "Jiji zuzhi funü."

83. Gao Xiaoxian, "'The Silver Flower Contest'"; Hershatter, *The Gender of Memory*, 214–34; on "badly remunerated production drives," see Eyferth, "Women's Work," 386. Soviet history offers an interesting parallel in the cultivation of women swineherds called *svinarki*; they were encouraged through competitions to improve piglet survival and weight gain, while especially inept or lazy *svinarki* would be "shamed" in the newspapers. Jenny Leigh Smith credits the improvements *svinarki* brought to swine management for some of the success Lysenkoist programs enjoyed in the post-WWII period. Smith, *Works in Progress*, 10.

84. Zhang Kui?, "Gao ju Mao Zedong sixiang," 24–29. Hershatter states that for women, "Songs provided a personal connection to state initiatives such as land reform, the literacy drive, and the Marriage Law, linking the formal messages imparted in political meetings to the work of daily life." *The Gender of Memory*, 102.

85. Gao Xiaoxian has emphasized that stories about these labor models served as vessels for extension of agricultural technologies far more than media for advancing women's rights and interests. Gao, "'The Silver Flower Contest,'" 600.

86. Murphy, "Changes in Family," 227.

87. Interview in Nanjing, June 2012.

88. Interview with Cao Xingsui, June 2012.

89. "Gao ju Mao Zedong sixiang hongqi gengjia guangfan."

90. Jiangxi sheng Nancheng xian Guanzhen si jiemei shouyi zhan fuzhanzhang Wu Lanxian, "Bu pa gan 'chou' shi, ganyu huan xintian" [Not fearing "smelly" things, courageously changing the world], in Henan sheng, *Quanguo nongcun*, 17–23.

91. Braverman, *Labor*.

92. Fitzgerald, "Farmers Deskilled."

93. Vandeman, "Management in a Bottle."

94. See Grossman, *The Political Ecology*; and Cooke, "Expertise, Book Farming." On reskilling, see Apple, "Curriculum."

95. Stone, "Agricultural Deskilling," 85.

96. Eyferth, *Eating Rice*.

97. "Gao ju Mao Zedong sixiang hongqi gengjia," 9.

98. Eyferth, *Eating Rice*.

99. Richard Levins argues that this is what we see in Cuban agriculture, where "socialist social arrangements and ideological priorities made ecological development an almost 'natural' correlate of the economic and social development and of the commitment to improving the quality of life as the primary goal of development." Levins, "How Cuba," 23.

100. Haraway, *Simians*, 68.

101. Schmalzer, "Speaking about China."

Chapter Five

1. J. K. King, "Rice Politics," 458.

2. In an examination of forestry agents in Senegal, Giorgio Blundo adopts the same title to question Scott's depiction of the state as "a unified source of intentions, policies, and coherent plans" and to "enquire whether civil servants 'see' in the same way as the state." Blundo, "Seeing Like a State Agent."

3. Jean Oi has analyzed cadre-peasant relations in terms of patron-clientelism; Li Huaiyin has challenged this assessment, pointing to the limits of cadre power especially in the face of active efforts on the part of peasants to check abuses. Oi, *State and Peasant*; Li Huaiyin, *Village China*.

4. Steve Smith, "Local Cadres," 1010.

5. Except where noted, the information on Mai's experiences comes from interviews with him in June 2010.

6. He Minshi and Wang Jianxun, *Guangzhou minjian chengyu*, 21.

7. Mai Baoxiang, "Pu Zhelong jiaoshou zai Dasha."

8. Dasha gongshe geweihui, "Dasha gongshe jiji tuiguang."

9. "Sihui xian Dasha gongshe 1975 nian shuidao bingchonghai."

10. "Dasha gongshe chonghai zonghe fangzhi."

11. "Sihui xian Dasha gongshe 1975 nian shuidao bingchonghai."

12. "Dasha gongshe chonghai zonghe fangzhi."

13. "Sihui xian Dasha gongshe 1975 nian shuidao bingchonghai."

14. Kuhn, "Political and Cultural Factors," 66. The published version of this essay is shorter and differs somewhat in wording; it also is not attributed directly to Kuhn. I use the draft since it is more likely to represent Kuhn's own interpretation. See *Plant Studies in the People's Republic of China*, 162–67. Miriam Gross has also highlighted the difference between experiment and what she calls "performance" in public health campaigns. She considers scientific performances to have been "unnecessary" and "odd" in that they served political rather than scientific purposes, but she emphasizes their usefulness to the state in that "scientific performances permit villagers to conduct experiments according to totally standardized methods that achieve identical results and inevitably reaffirm Party guidance." While Gross focuses on how performances, cloaked as experiments, served to strengthen party control, I highlight the ways state and nonstate actors used the practices and rhetoric of experiment to resolve disagreements and to struggle over crucial questions of science and power. Gross, "Chasing Snails," 656, 660.

15. Gail Hershatter discusses the *dundian* system at some length in *The Gender of Memory*.

16. Here as elsewhere, I have used the full-text search feature of the *People's Daily* database to identify when terms emerge and when they enter widespread use.

17. Xu Jiatun, "Shixian nongye kexue," 8.

18. Jin Shanbao, "Yangbantian fazhan le nongye," 14.

19. Xu Jiatun, "Shixian nongye kexue," 2.

20. Jin Shanbao, "Yangbantian fazhan le nongye," 15.

21. Interview in Guangxi, June 2012.

22. Interview in Qinzhou, June 2012. The peasant referred to was probably Luo Jiagui (罗家贵).

23. Interview in Youjiang, June 2012.

24. "Ronggeqing dadui dangzhibu." This state-produced document is an especially clear account of a party secretary's role in nurturing youth and encouraging them in scientific farming. Ye Wa, who was a sent-down youth in Shaanxi, spoke to me of the tremendous importance of the support given her by her production team leaders—both an older man and the younger man who replaced him. Interview, March 2012.

25. Cao Xingsui's comments during a group interview in Qinzhou, June 2012.

26. Shanxi sheng Xinxian diqu, *Xinxian diqu nongye*, 15. For another example, see *Nongcun zhishi qingnian*, 35.

27. Interview with Yi Ruoxin, June 2012.

28. Interview with Pan Yiwei, June 2012.

29. Interview with Cao Xingsui, June 2012.

30. Interviews in Qinzhou and Baise, June 2012.

31. Interview in Qinzhou, June 2012.

32. Ibid.

33. Interview with Cao Xingsui, June 2012.

34. Interview in Youjiang, June 2012.

35. Henan sheng, *Quanguo nongcun*, 10.

36. Schmalzer, "Speaking about China."

37. Interview with Pan Yiwei, June 2012. For one of many examples of state rhetoric on the subject, see Zhang Dianqi, "Shehui zhuyi shi."

38. Friedman, Pickowicz, and Selden, *Revolution, Resistance, and Reform*, 178.

39. Zhongguo kexueyuan caizheng bu, "Zengbo qunzhongxing kexue shiyan." In this document, the amount allotted by the Chinese Academy of Sciences to the entire province of Guangdong for mass scientific experiment activities was sixty thousand yuan.

40. Sheng nongkezhan weiyuanhui, "Guanyu shenqing qunzhongxing."

41. Nanzhao xian geming weiyuanhui, ed., "Quandang dongshou cengceng zhua, qun ban kexue kai honghua" [The whole party initiates and each level grasps, mass science produces red flowers], in Henan sheng, *Banhao siji nongye*, 41–53.

42. "Gao ju Mao Zedong sixiang hongqi gengjia," 11.

43. Peng Baoshan, "Kaizhan qunzhong yundong," 7.

44. Wu Xianzhi, " 'Si zi yi fu.' "

45. "Gao ju Mao Zedong sixiang hongqi gengjia," 9.

46. Zhonggong Huarong xian weiyuanhui [Chinese Communist Party Committee of Huarong County], "Jiaqiang lingdao, yikao qunzhong, banhao siji nongye kexue shiyan wang" [Strengthen leadership, rely on the masses, build the four-level agricultural scientific experiment network], in Henan sheng, *Banhao siji nongye*, 38.

47. The critical literature on this issue is vast. For just a few examples, see Berlan and Lewontin, "The Political Economy"; Lewontin, "Agricultural Research"; and Shiva, " Seeds of Suicide."

48. Xiangpu xian. "Jiaqiang lingdao"; Shaoyang di, *Wei geming zhong hao*.

49. Peng Baoshan, "Kaizhan qunzhong yundong," 8. See also Guangdong sheng Sihui xian Dasha gongshe, "Dasha gongshe shuidao."

50. Interview in Guangxi, June 2012.

51. Ibid.

52. Ibid.

53. Ibid.

54. Ibid.

55. Chan, Madsen, and Unger, *Chen Village*, 239–40.

56. Interview with Ye Wa, March 2012.

57. Gongqingtuan zhongyang qingnongbu, *Wei geming gao nongye*, 4.

58. "Gao ju Mao Zedong sixiang hongqi gengjia," 11.

59. Xinxian Qicun gongshe Qicun dadui geming weiyuanhui, "Yanzhe Mao Zhuxi de geming luxian da gao qunzhongxing de nongye kexue shiyan yundong" [Following Chairman Mao's revolutionary line, pursue the mass rural scientific experiment movement in a big way], in Shanxi sheng Xinxian diqu, *Xinxian diqu nongye*, 34–35.

60. Liaoning sheng, "Shuidao zazhong youshi."

61. Xiyang xian weiyuanhui, "Yi dang de jiben luxian wei gang renzhen tuiguang Dazhai kexue zhongtian jingyan" [Taking the fundamental line of the party as the key link, extend the scientific farming experience of Dazhai], in Mianyang xian, *Qunzhongxing nongye kexue*, 8–9.

62. Interview with Mai Baoxiang, June 2010.

63. "Yige shehui zhuyi jiaoyu yundong hou," 44.

64. Xiangzhou xian, "Kexue zhongtian duo gaochan," 36–38.

65. "Yige shehui zhuyi jiaoyu yundong hou," 44.

66. Interview in Qinzhou, June 2012.

67. Zhonggong Fuqing xian Yinxi dadui, "Tuchu zhengzhi, kexue zhongtian" [Give prominence to politics, farm in a scientific way], in Henan sheng, *Quanguo nongcun*, 5.

68. Yangchun xian kexue jishu xiehui, "Yige tuchu zhengzhi jianku fendou de kexue shiyan xiaozu" [A scientific experiment group gives priority to politics and struggles arduously], in Henan sheng, *Quanguo nongcun*, 34.

69. "Gao ju Mao Zedong sixiang weida hongqi, da gao kexue shiyan, si nian liangshi zengchan yi bei ban" [Holding high the red flag of Mao Zedong Thought and pursuing scientific experiment in a big way, increasing grain production 1.5 times in four years], in Henan sheng, *Quanguo nongcun*, 11.

70. Hua xian Jiang Qizhang, "Pinxia zhongnong"; Zhonggong Fuqing xian Yinxi dadui, "Tuchu zhengzhi, kexue zhongtian" in Henan sheng, *Quanguo nongcun*, 4–5. Archival sources are replete with discussions of peasants' "superstitions." See S. A. Smith, "Talking Toads"; Steve Smith, "Local Cadres"; and Gross, "Chasing Snails." On the Republican-era state's antisuperstition campaigns, see Nedostup, *Superstitious Regimes*.

71. Dasha gongshe geweihui, "Dasha gongshe jiji tuiguang."

72. "Gao ju Mao Zedong sixiang weida hongqi, da gao kexue shiyan," 15–16.

73. Henan sheng, *Quanguo nongcun*, 10.

74. Pu Zhelong and Liu Zhifeng, "Woguo zhequ qunzhong."

75. Yangchun xian kexue jishu xiehui, "Yige tuchu zhengzhi jianku fendou de kexue shiyan xiaozu," in Henan sheng, *Quanguo nongcun*, 34.

76. Zhonggong Fuqing xian Yinxi dadui, "Tuchu zhengzhi, kexue zhongtian," in Henan sheng, *Quanguo nongcun*, 4–5.

77. Interview in Youjiang, June 2012.

78. Interview in Qinzhou, June 2012.

79. Luo Zhongbi (罗仲弼), survey response (see sources).

80. Shapin, *A Social History of Truth.*

81. "Yige shehui zhuyi jiaoyu yundong hou," 43.

82. Ibid., 44.

83. Interview with Cao Xingsui, June 2012.

84. Schmalzer, "Breeding a Better China," 14–15.

85. Interview with Zhao Yuezhi, March 2012.

86. Interview with Cao Xingsui, June 2012. Cao credits this practice with preserving black rice and other traditional varieties from extinction.

87. See chapter 1.

88. Deborah Fitzgerald has made a similar point about agricultural science in the United States: "The peculiar needs of the rural population . . . have obliged agricultural scientists to consider science itself as a negotiation between natural and human forces." *The Business of Breeding,* 3.

Chapter Six

1. Shen Dianzhong, *Sixiang chenfu,* 10.

2. Ibid., 249.

3. Hinton, Barmé, and Gordon, *Morning Sun;* Chan, *Children of Mao,* 61–62.

4. Han, *The Unknown,* 29; Bernstein, *Up to the Mountains,* 61–62.

5. Mao, *Selected Works,* 5:264.

6. See the sixth quotation in chapter 30, "Youth."

7. This phrase appeared unprompted in a survey response submitted by Chen Haidong and in an interview conducted in Qinzhou, June 2012.

8. Goldman, *China's Intellectuals,* 135–38; Williams, "Fang Lizhi's Big Bang," bk. 2, 679; Schmalzer, *The People's Peking Man,* 124–25.

9. "Lingdao zhishi qingnian."

10. Lu Youshang, "Guangkuo tiandi."

11. Interview with Pan Yiwei, June 2012.

12. *Kexue zhongtian,* 32.

13. *Nongcun zhishi qingnian,* 3.

14. Interview with Cao Xingsui, June 2012.

15. Interview with Chen Yongning, June 2012.

16. On the significance of books to urban educated youth in the countryside, and their broader hankering for "culture" in all forms, see Bonnin, *The Lost Generation,* 261–65.

17. Interview with Pan Yiwei, June 2012; Shen Dianzhong, *Sixiang chenfu lu,* 265.

18. Interview in Baise, June 2012.

19. *Kexue zhongtian,* 10.

20. Dai, *Balzac.*

21. Interview in Guangxi, June 2012.

22. Bonnin, *The Lost Generation,* 344–49.

23. Link, "The Limits"; Kong, "Between Undercurrent."

24. Zhang Yang, *"Di'erci woshou" wenziyu,* 149.

25. Interview with Ye Wa, March 2012.

26. Interview with Chen Yongning, June 2012.

27. Interview with Pan Yiwei, June 2012.

28. Interview in Guangxi, June 2012.

29. Essay shared by interview subject in Guangxi, June 2012.

30. Interview with Cao Xingsui, June 2012.

31. Interview with Huang Shaoxiong, June 2012.

32. Gongqingtuan zhongyang, *Wei geming*, 4.

33. Xinhua she, "Quanguo nongcun."

34. 说理论天花乱坠，论实际稀泥软蛋.

35. Xinxian Qicun gongshe Qicun dadui geming weiyuanhui, "Yanzhe Mao Zhuxi de geming luxian da gao qunzhongxing de nongye kexue shiyan yundong" [Following Chairman Mao's revolutionary line, pursue the mass rural scientific experiment movement in a big way], in Shanxi sheng Xinxian diqu, *Xinxian diqu nongye*, 36.

36. "Lingdao zhishi qingnian," 4.

37. Seybolt, *The Rustication*, 60–63. This is a translation of the Chinese collection entitled *Reqing guanhuai xiaxiang*. See also Heilongjiang sheng, "Bai ying dadou," 4; and Zhang Renpeng, "Houlu duizhang," 26.

38. "Young Girl Fulfills," 8–9.

39. *Nongcun zhishi qingnian*, 25.

40. E.g., ibid., 29, 50.

41. *Nongcun zhishi qingnian*, 5, 21. Arunabh Ghosh provides a helpful discussion of the scholarly literature to date on "Maoist social investigation as method" and demonstrates its tensions with established Marxist approaches to social science research based on statistical analysis. Ghosh, "Making It Count," 52–58.

42. *Kexue zhongtian*, 3, 12, 19–20.

43. Yunnan shengchan, "Jinjina shumiao," 6–7.

44. *Kexue zhongtian*, 29, 37–40.

45. Interview in Guangxi, June 2012; interview with Chen Yongning, June 2012. Chen clarified, however, that he thought the real problem was that people did not want to take on the experiment duties because it was extra work, and it looked so "idiotic" to bring lights out into the field at night for insect monitoring.

46. *Kexue zhongtian*, 29, 37–40.

47. Ibid., 14–16.

48. *Nongcun zhishi qingnian*, 36.

49. Gongqingtuan Shaanxi, *Dazhai Xiyang*, 175; Zhonggong Hunan sheng, "Women shi zenyang," 2. Peasants from revolutionary classes also reportedly used rhyme to say supportive things about scientific experiment. Hukou xian, "Hukou xian yingyong."

50. Gongqingtuan Shaanxi, *Dazhai Xiyang*, 171.

51. Gongqingtuan zhongyang qingnongbu, *Wei geming gao nongye*, 8.

52. Wang Chunling, "Jianjue zou."

53. Shen Dianzhong, *Sixiang chenfu*, 3–7; Lafargue, "Reminiscences of Marx," 23.

54. Shen Dianzhong, *Sixiang Chenfu*, 10, 360.

55. Zhang Yang, *"Di'erci woshou" wenziyu*, 91–99, 104, 105, 129, 405–7.

56. Link, "The Limits," 158.

57. On the transformation of one village in Dongguan, see Saich and Hu, *Chinese Village, Global Market*.

58. Interview in Baise, June 2012.

59. Interview in Guangxi, June 2012.

60. Ibid.

61. On protests by sent-down youth in Yunnan, see Yang, "'We Want.'"

62. Guojia nongken, *Yong yu pandeng*, e.g., 46.

63. Ibid., 37–38.

64. Ibid., 5.

65. Paola Iovene has recently offered an intriguing comparison of late 1950s and late 1970s science fiction in China. Perhaps surprisingly, Great Leap–era science fiction "envisioned a future laborless world" not unlike that described in early post-Mao works. Literature of the intervening period, including the Cultural Revolution, did not share these visions. Moreover, Iovene notes that late 1970s and early 1980s science fiction was distinctive for its "help in promoting a separation between mental and manual labor." Iovene, *Tales of Futures Past*, 27–29, 34.

66. Shan Ren, "Xiangcun."

67. Mao Zedong, "Mao Zhuxi gei Mao Anying."

68. The slogan "Springtime for science" emerged from the March 1978 National Conference on Science. *People's Daily* printed a series of drawings on this theme to celebrate the conference, including one highlighting children in connection with an "earnest hope" for the future of science. Yang Yuepu, "Kexue zhi chun."

69. Gao Shiqi, "Chuntian."

70. Zhou Peiyuan, "Kexue de weilai."

Chapter Seven

1. Interview with Pan Yiwei, Nanning, June 2012.

2. Bernstein suggests that rural youth represented the majority of young people who participated in the scientific experiment movement (*Up to the Mountains*, 224–25), and my sense of the sources agrees. On the complexity of interpreting statistics on sent-down and returned educated youth, see ibid. 22–32.

3. Qin, "The Sublime"; Zheng, "Images, Memories."

4. "'Yinhua' kai."

5. Bernstein, *Up to the Mountains*, 22.

6. Guo Xiulian, "Sanjie qingnian."

7. Tuan shengwei, "Guangdong Yangchun Sanjie."

8. Yangchun xian kexue jishu xiehui, "Yige tuchu zhengzhi jianku fendou de kexue shiyan xiaozu" [A scientific experiment group gives priority to politics and struggles arduously], in Henan sheng, *Quanguo nongcun*, 32.

9. Tuan shengwei, "Guangdong Yangchun Sanjie."

10. *Nongcun zhishi qingnian*, 21–22, 49, 64, 67. See also Gongqingtuan zhongyang, *Wei geming*, 40.

11. Lianjiang xian fulian, "Shengchan douzheng."

12. *Nongcun zhishi qingnian*, 19–20. Emphasis added.

13. Interview with Ye Wa, June 2012.

14. Shi Weimin, *Zhiqing riji*, 160–61.

15. Shen Dianzhong, *Sixiang chenfu*, 249, 255.

16. Interview with Chen Yongning, Nanning, June 2012.

17. Here it may be useful to remember that in the conversation recounted in chapter 6, Chen argued for the importance of top-down state administration rather than youth agency in the successes of scientific farming. Thus, I would not expect him to overemphasize the significance of his own initiative, and I am even more inclined to trust his statements about the role of choice in his trajectory.

18. Interview with Mai Baoxiang, June 2010.

19. Interview with Luo Zhongbi, June 2010.

20. Luo Zhongbi, survey response.

21. Interview with Cao Xingsui, June 2012.

22. Interview with Pan Yiwei, Nanning, June 2012.

23. Yue, *The Mouth that Begs*, 156, 182.

24. Chan, Madsen, and Unger, *Chen Village*, 238.

25. Shi Weimin, *Zhiqing riji*, 274. It was the boys who went fishing; the girls cleaned the fish.

26. Fan, " 'Collective Monitoring,' " 136.

27. Interview with Zhao Yuezhi, March 2012.

28. Yang Qiuying, survey response.

29. Bai Di, "Wandering Years in the Cultural Revolution," in Zhong, Zheng, and Di, *Some of Us*, 92.

30. Lihua Wang, "Gender Consciousness in My Teen Years," in Zhong, Zheng, and Di, *Some of Us*, 121.

31. Wang Zheng, "Call Me Qingnian but Not Funü," in Zhong, Zheng, and Di, *Some of Us*, 37.

32. Interview with Cao Xingsui, June 2012.

33. Hanson, "Trip diary," 165–66. Haldore Hanson was then the director-general of Norman Borlaug's home institution, CIMMYT, but had a distinguished past life as a journalist in war-torn China of 1935–1938 with sympathies for the communist revolution. See also Borlaug, Field Notebooks, China, no. 3 (1977), 72, 78.

34. Sun Zhongchen, "Gao hao kexue."

35. Zhonggong Nanweizi, "Yi jieji douzheng," 9.

36. Interview with Chen Yongning, Nanning, June 2012.

37. Pan Yiwei, "Yijiu qiqi." Pan provided me with the text of the article, which he indicated had been revised somewhat since the original publication.

38. Interview in Guangxi, June 2012. Wang Xiaodong is a pseudonym.

39. Interview with Ye Wa, March 2012.

40. Interview with Chen Yongning, Nanning, June 2012.

41. Interview with Pan Yiwei, Nanning, June 2012.

42. I am repeatedly struck in these interviews by how clearly people remember exactly how much things cost, an indication that every penny mattered dearly.

43. Interview with Cao Xingsui, June 2012.

44. Tapping personal connections was not a tactic limited to educated youth. When hybrid rice first came on the scene, the head of one county-level agricultural science institute in Guangxi immediately ran to the Agricultural Academy and found someone from his home village who would give him seed quickly. This gave his county a head start, such that they extended hybrid rice faster than anyone else in the region, and by 1978, 80 percent of their rice was already converted to hybrid. Interview in Qinzhou, June 2012.

45. See also Chan, Madsen, and Unger, *Chen Village*, 95.

46. Gongqingtuan zhongyang, *Wei geming*, 5.

47. Zhonggong Hunan sheng, "Women shi zenyang."

48. *Kexue zhongtian*, 3, 12, 19–20, 27–28, 31, 64; "Lingdao zhishi qingnian."

49. *Nongcun zhishi qingnian*, 37.

50. Interview with Ye Wa, March 2012; e-mail correspondence with Ye Wa, August 2014.

51. Interview with Ye Wa, March 2012.

52. Shen Dianzhong, *Sixiang chenfu*, 249–50.

53. Ibid., 257–60.

54. Ibid., 297.

55. She Shiguang, *Dangdai Zhongguo de qingnian*, 293–94.

56. Dangdai Zhongguo, *Dangdai Zhongguo de nongye*, 571–72.

57. This is a central principle of the strong programme in the sociology of scientific knowledge, which emerged in the 1970s and was especially promoted by scholars at the University of Edinburgh.

58. Ye Weili makes a similar point in *Growing Up*, 1.

59. Hightower, *Hard Tomatoes*, 124.

60. Rosenberg, *The 4-H Harvest*, 6.

61. Stross, *The Stubborn Earth*; Thomson, *While China*; Schmalzer, "Breeding a Better." More recently, Seung-joon Lee has argued for a reconsideration of Republican-era efforts to bring science and technology to bear on the problem of feeding the nation. Rather than focusing on government incompetence, as many previous studies have done, he takes seriously the Guomindang's "forward-looking attitude." However, he concludes that Guomindang programs failed because they overly emphasized "quantification and simplification," a verdict consistent with the notion that technocratic impulses stymied reform efforts. Lee, *Gourmets*, 135.

62. Interview with Zhao Yuezhi, March 2012.

63. Pan Gang, "Siying qiye."

Epilogue

1. A good critical overview of this literature can be found in Bramall, "Origins"; Peng, "Decollectivization"; and Zhun Xu, "The Chinese Agriculture."

2. Bramall, "Origins"; Zhun Xu, "The Chinese Agriculture."

3. Qualitative evidence from earlier chapters supports this conclusion. While statistical data on the effectiveness of agricultural extension in the Mao era is not available, studies conducted in the late 1980s and early 2000s testify to the positive impact of participation in agricultural extension activities on farmers' willingness to adopt new technologies. Delman, "'We Have to Adopt"; Huang Jianmin, Hu Ruifa, and Huang Jikun, "Jishu tuiguang." Justin Yifu Lin conducted research on factors influencing household adoption of hybrid rice and concluded that education was the most important variable; however, contact with extension services was not observable in his study. Lin, "Education and Innovation," 720.

4. Although agriculture represents a small fraction of the total GDP, and industry was the leading sector in post-Mao economic growth, according to Chris Bramall, "agriculture played the key role in the growth process between 1978 and 1984 in most parts of China, and in poor regions, agriculture was of critical importance throughout the post-1978 era." Bramall, *Sources*, 56–75, quotation on 74. Note also that the widespread assumption among economists that given a strong enough industrial sector agricultural produce can always be imported does not account for the compelling political reasons that states have for maintaining some degree of self-sufficiency, especially in grain.

5. Although hunger rates have fallen dramatically in China in recent decades, the Food and Agriculture Organization terms the current situation in China, among other countries, as a "double burden of malnutrition," with 8 percent of schoolchildren in China undernourished and 23 percent suffering from obesity. It further highlights disparities among regions, ethnicities, and socioeconomic groups, with rates of stunting among children in poor, rural areas

reaching as high as 29 percent. Food and Agriculture Organization, *The Double Burden*, 40, 12. Judith Farquhar has observed that the writings of former sent-down youth are most responsible for creating the theme of Cultural Revolution hunger in post-Mao literature: "Their narratives of hunger both place shortages firmly in the past (although there are plenty of people in China today without enough to eat) and privilege the phases that have been especially important to intellectuals in a public discourse that is easily dominated by the educated and the powerful." Farquhar, *Appetites*, 82.

6. Ho, Zhao, and Xue, "Access and Control," 358–60.

7. Economy, *The River*.

8. Interview with Chen Yongning, June 2012.

9. Jacobsen, "Stung by Bees."

10. Interviews in Guangxi, 2012.

11. Bramall, "Origins"; Zhun Xu, "The Chinese Agriculture"; Peng, "Decollectivization."

12. Day, *The Peasant*; Hale, "Reconstructing"; Lammer, "Imagined."

13. Day, *The Peasant*, 168.

14. Perry, "From Mass Campaigns."

15. Day, *The Peasant*, 166–67.

16. Zhongguo fupin jijinhui, "Meng kaishi."

17. Ibid. Huarun's Hope Towns resonate with the broader "dreaming green" vision of contemporary Chinese eco-cities that Julie Sze analyzes in *Fantasy Islands*.

18. Day, *The Peasant*, 168.

19. Elizabeth Perry has also documented efforts on the part of state agents to distinguish the new socialist countryside program from Mao-era campaigns and has provided a similar analysis of rhetorical and policy resonances between the current commitment to "building a new socialist countryside" and Mao-era efforts. Perry, "From Mass Campaigns," 36, 38–42.

20. The population of Baise is 80 percent Zhuang ethnicity, and Guangxi is officially a "Zhuang autonomous region" rather than a province.

21. Zhongguo fupin jijinhui, "Meng kaishi."

22. Chen Xiwen, *Zhongguo zhengfu*, 164.

23. On the history of these visits, see chapter 2; Schmalzer, "Speaking about China"; and Schmalzer, "Insect Control."

24. Brown, *City Versus Countryside*, 200.

25. Day, *The Peasant*, 108. See also Hale, "Reconstructing."

26. Zhang Li et al., "Opening Our Eyes: Renewing the Chinese Public Extension System," in Song and Vernooy, *Seeds and Synergies*, 85–111; Li Liqiu et al., "Jianli guojia."

27. This assessment of both the post-Mao and Mao-era extension systems is found also in a Western analysis, though the evidence comes only from interviews conducted in the 1980s: Delman, "'We Have to Adopt.'" Delman has elsewhere expressed his support for greater privatization of the Chinese extension system, specifically the transfer of extension responsibilities to nongovernmental organizations: "Continued state ownership of the extension services and most of the dairy plants and the imposition of control over potentially free and democratic farmers' organizations testify to a party-state reluctant to give up control and promote market-based institutional innovation in dialogue with the operators within the industry." Delman, "Cool Thinking?," 4.

28. Zhongguo fupin jijinhui, "Meng kaishi."

29. Chen Tianyuan and Huang Kaijian, "Canyushi zhiwu," 490.

30. Ibid., 491–92.

31. Ibid., 493–94.

32. Ibid., 491.

33. Ibid., 493–94. Note that Song Yiching publishes in English as well as Chinese and has chosen this Romanization for her name rather than the Pinyin, which would render it Song Yiqing.

34. Zhang Li et al., "Opening Our Eyes," in Song and Vernooy, *Seeds and Synergies,* 87.

35. Song and Vernooy, *Seeds and Synergies,* 9.

36. Jinghua shibao, "Sannong zhuanjia." Li Changping's letter has been reproduced in many places on the web, including here: Li Changping, "Zhi Yuan Longping."

37. See, e.g., Shiva, "Seeds of Suicide."

38. Song Yiqing, "Nongmin liuzhong."

39. "Wei shenme women."

40. See chapter 4, page 103.

41. See chapter 6, page 165.

42. Yan Hairong, "Cong dadou." John Perkins has illuminated the profound role of national security interests throughout the history of the green revolution and highlights the way "this legacy still hangs over all efforts to reform agriculture in order to make it more sustainable." *Geopolitics,* 264.

43. Gupta, *Postcolonial Developments,* 172–76, 229–30.

44. For a critique of developmentalism in agriculture from the perspective of a Marxist biologist, see Levins, "Science and Progress."

45. Hathaway, *Environmental Winds,* 8–33.

46. Helpful also is the metaphor Elizabeth Perry employs in her analysis of the changing ways the Chinese Communist Party has "mined" Chinese cultural resources, including its own "revolutionary tradition." Her moving conclusion calls upon future generations "not to forget or falsify the past but to mine the revolutionary inheritance in ways that encourage its inspiring vision to triumph over its appalling violence." Perry, *Anyuan,* 296.

Sources

Archives

Archives in China vary widely in accessibility, and timing makes a big difference. I visited several archives in Guangdong in the summer of 2010, when a recent political incident involving the use of archival material had made archivists there very uncertain about how much access to provide, even to well-connected Chinese people.

The Guangdong Provincial Archives was the most "modern" and ostensibly "open" of the archives I visited, and it offered an elegant computer search interface on the full index. However, I was not allowed to make photocopies or scans of any materials, and I was allowed to see and take notes on only about 20 percent of the materials I requested. I could not determine the rationale behind the selection: some obviously "political" materials were allowed, while others that would seem from their titles to have been very mundane were not.

The Sihui Municipal Archives were housed in a much more "local" institution that had clearly not been renovated in many decades. I was introduced to the archivist by a person with excellent political connections, and so the archivist was willing to allow that person and a few others who had accompanied me to look through the bound indexes (with me looking over their shoulders) to identify materials relevant to my topic and have them copied. There was no censorship of the materials in either the selection or the copying. However, concerns about the propriety of delivering those materials to me meant that I left with only digital photographs of the photocopies, and I was able to take the photographs only after all archival document numbers had been removed (thus protecting the archivists from identification should the documents be confiscated on my way through customs).

The archivists at Sun Yat-sen University were very helpful and provided copies of a number of important documents on Pu Zhelong and his work at Big Sand Commune. However, I was not able to obtain document numbers for these materials.

The other archives listed below presented no unusual obstacles.

Bibliographic information for individual archival documents is included in the main section of the bibliography.

ARCHIVAL COLLECTIONS CONSULTED

Beijing Municipal Archives
Committee on Scholarly Communication with the People's Republic of China Papers. Global Resource Center, George Washington University.
Entomology Papers. Division of Entomology and Economic Zoology, University of Minnesota Archives.
Guangdong Provincial Archives
Norman E. Borlaug Papers. UMedia Archive, University of Minnesota.
Robert Metcalf Papers. University of Illinois Archives.
Sihui Municipal Archives (Guangdong)
Sun Yat-sen University Archives
Waijiaobu (Foreign Affairs Ministry) Archives (Beijing)

ABBREVIATIONS

BSN. *Beijing shi nongcun kexue shiyan xiaozu jiji fenzi huiyi wenjian* [Documents from the Conference of Activists in Beijing Municipal Rural Scientific Experiment Groups].
NEB. Norman E. Borlaug Papers. UMedia Archive, University of Minnesota, http://umedia.lib .umn.edu /.

Interviews

Although this book is by no means an ethnography, interviews with a variety of people who encountered scientific farming in socialist China (including scientists, agricultural technicians, educated youth, and peasants) form an important subset of the source materials. All interviews with Chinese-speaking subjects were conducted in Chinese; most of these occurred in China, but I conducted interviews in Toronto with one former sent-down youth and one former rural youth, both of whom now reside in North America. Some of the interview subjects preferred to be named, while others requested that their names be withheld. To the best of my ability I have honored the former preferences by citing their names in the notes; of course, I have always been careful to conceal the identity of interview subjects who wished to be anonymous. Some of the interviews I conducted in China occurred in group settings. In most of these the interviewees appeared uninhibited in their responses, but I could not be absolutely sure that they would have felt comfortable publicly expressing reservations about being named. For that reason, I have taken the safer course of withholding names for subjects interviewed in most group settings. I have also withheld names for comments heard outside of formal interviews (for example, in cars, at meals, etc.). Even when the interview was private and permission clearly given, I have opted to withhold names in cases where I suspect that a specific story or opinion shared might bring trouble or embarrassment to the interviewee.

Survey

Big Sand (大沙) Commune in Sihui County, Guangdong, was for more than twenty years the site of Pu Zhelong's renowned research on biological control of insect pests. Like much of the surrounding area, however, it has experienced extensive industrialization and urbanization in recent decades; the villages simply do not exist in their previous form anymore, and the people have scattered more even than is the case elsewhere in the rapidly changing country. For this reason, in addition to interviewing a few former residents of Big Sand during my visit to Guangdong in 2010, I asked Pu Zhelong's colleague Gu Dexiang to circulate a survey to former educated youth about their experiences participating in the work. I received detailed responses from four people; their stories are featured in chapters 6 and 7.

Works Cited

WORKS WITH NO AUTHOR

"Ba puji xiandai nongye kexue jishu jianli zai qunzhong de jichu shang" [Build dissemination of modern agricultural science and technology on the foundation of the masses]. *Renmin ribao*, 21 May 1964, 1.

"Banhao sanjiehe de yangbantian, cujin nongke kexue shiyan yundong" [Organizing three-in-one demonstration fields and promoting the agricultural scientific experiment movement]. *Renmin ribao*, 28 March 1965, 2.

"Beijing shi nongcun kexue shiyan xiaozu jiji fenzi huiyi jianbao" [Brief report on the Conference of Activists in Beijing Municipal Rural Scientific Experiment Groups]. 15 November 1965. BSN, Beijing Municipal Archives, 2.22.31, 20–21.

"Chanming nongye kexue gongzuo renwu" [Explaining agricultural science and technology work assignments]. *Renmin ribao*, 22 February 1963, 1.

"Dasha gongshe chonghai zonghe fangzhi, cong 1972 nian wanzao xiaomianji kaishi shiyan, dao 1973 nian zai Anren dadui kaizhan" [Big Sand Commume integrated control of insect damage, from a small area experiment in the late crop of 1972 to its development in 1973 in Anren Brigade]. 1982? Sihui Municipal Archives.

"Di wu ci quanguo zajiao shuidao keyan xiezuo hui zonghe jianbao (chubao)" [Summary report on the Fifth National Coordinating Conference on Scientific Research on Hybrid Rice]. *Hunan nongye keji* 1977.3: 1–5.

"Fenxi xingshi zhengjia ganjin" [Analyze circumstances and increase vigor]. *Renmin ribao*, 17 January 1959, 3.

"Gao ju Mao Zedong sixiang hongqi gengjia guangfan shenru de kaizhan nongcun qunzhongxing kexue shiyan yundong (cao)" [Hold high the red flag of Mao Zedong thought in order to increase, broaden, and deepen the development of the mass scientific experiment movement (draft)]. 15 November 1965. BSN, Beijing Municipal Archives, 2.22.31.

"Gei Guba tigong nongyao ziliao shi" [On the provision of pesticide materials to Cuba]. 12 October 1961–21 February 1962. Waijiaobu Archives, 111-00444-10.

Gongnong famingjia xiaozhuan: "Tu zhuanjia" sai guo "yang zhuanjia" [Short biographies of worker and peasant innovators: "Native experts" rival "foreign experts"]. Jiangsu renmin chubanshe, 1958.

"Guangdong sheng Sihui xian Dasha gongshe 1975 nian shuidao bingchonghai zonghe fangzhi gongzuo zongjie" [Summary report of integrated control of rice insect pests and diseases in Guangdong Province, Sihui County, Dasha Commune, 1975]. 1975. Sihui Municipal Archives.

"Guangzhou diqu chuli bei chachao hu dingxing fucha, buchang biao: Pu Zhelong" [Guangzhou district form for investigating and compensating households that suffered confiscation]. 17 March 1987. Sun Yat-sen University Archives.

"Guantian ji" [On weather management]. *Renmin ribao*, 18 January 1973, 4.

"Guanyu Guba yaoqiu women gou huafei wenti de qingshi" [Request for instructions on Cuba's request for us to sell them chemical fertilizer]. 12–16 December 1963. Waijiaobu Archives, 111-00471-20.

"He'erbin shi nongye zhiye xuexiao yang xinxing nongmin" [Cultivating new-style peasants at the Harbin Agricultural Vocational School]. *Renmin ribao*, 23 November 1965, 2.

"Hua Zhuxi zhangduo bu mihang" [Chairman Hua will steer the boat and not drift off course]. *Renmin ribao*, 2 April 1977, 1.

"Huang Yaoxiang 'banweigan shuidao zhi fu' yinling diyici lüse geming" [Huang Yaoxiang, 'father of semi-dwarf rice' pioneered the first green revolution]. *Jiangmen ribao*, 16 November 2012. http://www.jmnews.com.cn/c/2012/11/16/09/c_1287424.shtml.

"Jianchi kexue shiyan wunian de laotaitai Dan Liangyu" [Dan Liangyu, an old lady who for five years has persisted in scientific experiment]. 15 November 1965. BSN, Beijing Municipal Archives, 2.22.31, 50–52.

Kexue zhongtian de nianqing ren [Youth in scientific experiment]. Beijing: Zhongguo qingnian chuban she, 1966.

"Lingdao zhishi qingnian jiji kaizhan kexue shiyan" [Leading educated youth to actively develop scientific experiment]. *Renmin ribao*, 16 October 1972, 4.

"Lüse wangguo de yi ke xin xing: Woguo zajiao shuidao yanjiu de zhongda tupo" [A New Star in the Green Kingdom: China's Great Breakthrough in Hybrid Rice Research]. *Renmin ribao* 1978.7.21: 3.

"Meiguo qinlüezhe jinxing xijunzhan de zuixing laibudiao" [The biological warfare crimes of the American aggressors are undeniable]. *Renmin ribao*, 4 May 1952.

Nongcun zhishi qingnian kexue shiyan jingyan xuanbian [Selected experiences of rural educated youth in the scientific experiment movement]. Beijing: Renmin chubanshe, 1974.

"Peiyu shuidao xiongxing buyu baochixi de xin fangan: xuanze zijiao fa" [New plan for breeding male-sterile maintainer lines in rice: Methods of selection and self-fertilization]. *Keji jianbao* 1972.5: 18.

Plant Studies in the People's Republic of China: A Trip Report of the American Plant Studies Delegation. Washington, DC: National Academy of Sciences, 1975.

"Report of IRRI Team Visit, October 7–27, 1976." Commissions, Conferences, Councils and Symposia, 1952–2001, China. NEB. http://purl.umn.edu/106463.

Reqing guanhuai xiaxiang zhishi qingnian de chengzhang [Have a warm concern for the maturation of sent-down educated youth]. Beijing: Renmin chubanshe, 1973.

"Ronggeqing dadui dangzhibu shi zenyang peiyang keji xiaozu zhichi shinian yuzhong de" [How the party branch secretary of Ronggeqing brigade cultivated the science and technology group to support ten years of breeding (work)]. 15 November 1965. BSN, Beijing Municipal Archives, 2.22.31, 62–65.

"Shuidao xiongxing buyuxi de xuanyu ji yingyong" [Selection and use of male-sterile lines in rice]. *Nongye keji tongxun* 1972.Z1: 4–7.

"Song jiaoyu ban" [To the Education Office]. 10 January 1972. Sun Yat-Sen University Archives.

"Sunan nongmin Chen Yongkang shuidao fengshou" [Southern Jiangsu peasant Chen Yongkang's bumper rice harvest]. *Huadong xinwen huibian* 1951.12: 66.

"Tu zhuanjia he yang zhuanjia de jingyan huiliu" [Converging experiences of tu experts and yang experts]. *Renmin ribao*, 14 July 1958, 1.

"Wei shenme women bu zhijie mai youjifei huo junzhong? Zizhi junzhong de" [Why don't we directly buy organic fertilizer or microbial strains? Homemade microbial strains]. 26 May 2014. http://www.shiwuzq.com/food/rights/science/2014/0526/276.html.

"Wosheng zhaokai shuidao xin pinzhong xuanyu ji zazhong youshi liyong yanjiu xiezuo zuotanhui" [Sichuan Coordinating Conference on Research in Selection of New Rice Varieties and Heterosis]. *Sichuan nongye keji* 1973.1: 8–11.

"Xiaomai ruhe pingheng shifei" [How to apply fertilizer evenly in wheat]. *Renmin ribao*, 24 April 2005, 6.

"'Yangbantian' shi nongye kexue wei shengchan fuwu de zhuyao zhendi" ["Demonstration fields" are the battlefront for agricultural science in the service of the people]. *Renmin ribao*, 25 October 1964, 1.

"Yige shehui zhuyi jiaoyu yundong hou chengzhang qilai de keji xiaozu: Tongxian Xiji gongshe Zhaoqing dadui keji xiaozu" [A science and technology group formed since the Socialist Education Movement: The Tong County Xiji Commune Zhaoqing Brigade Science and Technology Group]. 15 November 1965. BSN, Beijing Municipal Archives, 2.22.31, 42–49.

"'Yinhua' kai zai dahuang" ['Silver flowers' bloom in the great (northern) waste]. *Renmin ribao*, 20 December 1972, 2.

"Yong Mao zhuxi de guanghui zhexue sixiang zhidao xiongxing buyu yanjiu" [Using Chairman Mao's brilliant philosophical thought to guide research in male infertility]. *Liaoning nongye kexue* 1971.4: 33.

"Young Girl Fulfills Geological Prospecting Task by Several Times." *Survey of China Mainland Press* 2136 (7 November 1959): 8–9.

"Zai Shanghai juxing de kexue jishu gongzuo huiyi shang Zhou Enlai chanshu kexue jishu xiandaihua de zhongda yiyi" [At the Shanghai Science and Technology Work Conference, Zhou Enlai elaborates on the great significance of scientific and technological modernization]. *Renmin ribao*, 31 January 1963, 1.

"Zajiao yuzhong de qunzhong yundong zhengzai woguo pengbo xingqi" [The mass movement for hybrid breeding is flourishing in China]. *Gansu nongye keji jianxun* 1971.5: 18–19.

"Zhengzhi jingji weiji riyi jiashen" [The political and economic crisis deepens daily]. *Renmin ribao*, 25 October 1969, 5.

Zhibaoyuan shouce [Plant protector handbook]. 1974. Reprint, Shanghai: Shanghai renmin chubanshe, 1976.

"Zhiwu baohu xuehui guanyu fangzhi bingchonghai wenti de baogao" [Report from the Plant Protection Conference on problems in the control of plant diseases and pests] (1 August 1962). In Nie Rongzhen. *Nie Rongzhen keji wenxuan*, 349–59. Beijing: Guofang gongye chubanshe, 1999.

"Zhongguo kunchong xuehui zhaokai xueshu taolunhui" [Chinese Entomology Association launches academic symposium]. *Renmin ribao*, 1 February 1962, 5.

"Zhonghua ernü duo qizhi" [China's sons and daughters have high aspirations]. *Renmin ribao*, 10 June 1969, 4.

"Zhuangzu guniang xue Dazhai: Kexue zhongtian duo gaochan" [Zhuangzu girls study Dazhai: Scientific farming reaps big harvests]. *Guangxi nongye kexue* 1975.7: 32–35.

"Zuotan xin qingkuang xia de dianying gongzuo" [A discussion of film work under new conditions]. *Renmin ribao*, 1 May 1957, 7.

WORKS WITH ONE OR MORE AUTHORS

American Insect Control Delegation. *Insect Control in the People's Republic of China: A Trip Report of the American Insect Control Delegation, Submitted to the Committee on Scholarly Communication with the People's Republic of China.* Washington, DC: National Academy of Sciences, 1977.

Anderson, Warwick. "Introduction: Postcolonial Technoscience." *Social Studies of Science* 32.5/6 (2002): 643–58.

Andreas, Joel. "Leveling the Little Pagoda: The Impact of College Examinations, and Their Elimination, on Rural Education in China." *Comparative Education Review* 48.1 (2004): 1–47.

———. *Rise of the Red Engineers: The Cultural Revolution and the Origins of China's New Class.* Stanford, CA: Stanford University Press, 2009.

Anhui sheng Wuhu diqu nongkesuo. "Liyong shuidao zazhong youshi de 'liangxifa.'" *Keji jianbao* 1972.20: 21–23.

Apple, Michael W. "Curriculum and the Labor Process: The Logic of Technical Control." *Social Text* 5 (1982): 108–25.

Bai Zhaoqing and He Wentong. "Minbei shanqu qunzhong de 'sitou' shifei fa" [The 'four-head' method of fertilizer application among the masses of norther Fujian's mountain region]. *Turang* 1960.4: 25.

Beijing nongye daxue and Shandong nongxueyuan, eds. *Nongye huaxue* [Agricultural chemistry]. Vol. 1. Bejing: Nongye chubanshe, 1961.

Beijing nongye daxue gengzuoxue jiaoyanzu. "Jianzuo tanzhong dayou kewei" [Intercropping and interplanting have great potential]. *Renmin ribao*, 3 January 1961, 7.

Berlan, Jean-Pierre, and Richard Lewontin. "The Political Economy of Hybrid Corn." *Monthly Review* 38 (1986): 35–47.

Bernal, J. D. *The Social Function of Science.* London: Routledge, 1939.

Berner, Boel. *China's Science through Visitors' Eyes.* Lund: Research Policy Program, 1975.

Bernstein, Thomas. *Up to the Mountains and Down to the Villages: The Transfer of Youth from Urban to Rural China.* New Haven, CT: Yale University Press, 1977.

Blundo, Giorgio. "Seeing Like a State Agent: The Ethnography of Reform in Senegal's Forestry Services." In *States at Work: Dynamics of African Bureaucracies*, edited by Thomas Bierschenk and Jean-Pierre Olivier de Sardan, 69–90. Leiden: Brill, 2014.

Bonnin, Michel. *The Lost Generation: The Rustication of China's Educated Youth (1968–1980).* Translated by Krystyna Horko. Hong Kong: The Chinese University Press, 2013. [Originally published in French in 2004.]

Borlaug, Norman. Field Notebooks, China, no. 1. 1974. Field Notebooks and Appointment Books, 1948–2000. NEB. http://purl.umn.edu/106358.

———. Field Notebooks, China, no. 2. 1974. Field Notebooks and Appointment Books, 1948–2000. NEB. http://purl.umn.edu/106357.

———. Field Notebooks, China, no. 2. 1977. Field Notebooks and Appointment Books, 1948–2000. NEB. http://purl.umn.edu/106370.

———. Field Notebooks, China, no. 3. 1977. Field Notebooks and Appointment Books, 1948–2000. NEB. http://purl.umn.edu/106369.

———. Field Notebooks, China, no. 4. 1977. Field Notebooks and Appointment Books, 1948–2000. NEB. http://purl.umn.edu/106372.

———. Letter to Richard Critchfield. 22 May 1992. Chronological Correspondence; Correspondence, 1954–2006, 94–96. NEB. http://purl.umn.edu/106268.

Bramall, Chris. "Origins of the Agricultural 'Miracle': Some Evidence from Sichuan." *China Quarterly* 143 (Sep. 1995): 731–55.

———. *Sources of Chinese Economic Growth, 1978–1996.* New York: Oxford University Press, 2000.

Bräutigam, Deborah. *Chinese Aid and African Development: Exporting Green Revolution.* New York: St. Martin's Press, 1998.

Braverman, Harry. *Labor and Monopoly Capital.* New York: Monthly Review Press, 1975.

Bray, Francesa. "Chinese Literati and the Transmission of Technological Knowledge: The Case of Agriculture." In *Cultures of Knowledge: Technology in Chinese History,* edited by Dagmar Schäfer, 299–325. Leiden: Brill, 2011.

———. *Science and Civilization in China.* Vol. 6, pt. 2, *Agriculture.* Cambridge: Cambridge University Press, 1984.

———. *Technology and Gender: Fabrics of Power in Late Imperial China.* Berkeley: University of California Press, 1997.

Brinck, Per, ed. *Insect Pest Management in China.* Stockholm: Ingenjörsvetenskapsakademien, 1979.

Brown, Jeremy. *City Versus Countryside in Mao's China: Negotiating the Divide.* Cambridge: Cambridge University Press, 2012.

———. "Spatial Profiling: Seeing Rural and Urban in Mao's China." In *Visualizing Modern China: Image, History, and Memory, 1750–Present,* edited by James Cook et al., 203-18. Lanham, MD: Lexington Books, 2014.

Buck, J. Lossing. "Missionaries Begin Agricultural Education in China." *Millard's Review* 14 (Sept. 1918): 78–79.

Buck, Peter. *American Science and Modern China, 1876–1936.* New York: Cambridge University Press, 1980.

Buckley, Chris. "An Interview With Xi, Long Before He Was China's Leader." *New York Times,* 12 June 2014. http://sinosphere.blogs.nytimes.com/2014/06/12/an-interview-with-xi-long-before-he-was-chinas-leader/.

Bullock, Mary B. *An American Transplant: The Rockefeller Foundation and Peking Union Medical College.* Berkeley: University of California Press, 1980.

Cao Longgong. "Tan Chen Fu de 'dili chang xin lun'" [On Chen Fu's "On rejuvenating fertility"]. *Renmin ribao,* 2 November 1965, 5.

CCTV. "Women dou you yishuang shou" [We each have a pair of hands]. Accessed 5 February 2014. http://www.cctv.com/program/ddgr/20030627/100701_4.shtml.

Cenxi xian nongyeju. "Zajiao shuidao shizhong tihui" [Experiences in experimental planting of hybrid rice]. *Guangxi nongye kexue* 1976.2: 24–25.

Chan, Anita. *Children of Mao: Personality Development and Political Activism in the Red Guard Generation.* Seattle: University of Washington Press, 1985.

Chan, Anita, Richard Madsen, and Jonathan Unger. *Chen Village: The Recent History of a Peasant Community in Mao's China.* Berkeley: University of California Press, 1984.

Cheek, Timothy. *The Intellectual in Modern Chinese History.* Cambridge: Cambridge University Press, 2015.

———. *Propaganda and Culture in Mao's China: Deng Tuo and the Intelligentsia.* Oxford: Clarendon Press, 1997.

Chen, Shiwei. "History of Three Mobilizations: A Reexamination of the Chinese Biological Warfare Allegations against the U.S. in the Korean War." *Journal of American-East Asian Relations* 15 (2009): 213–47.

Chen Tianyuan and Huang Kaijian. "Canyushi zhiwu yuzhong yu kechixu liyong shengwu duoyangxing—yi Guangxi yumi wei li" [Participatory plant breeding and sustainable use of biological diversity: The case of maize in Guangxi]. *Zhongguo nongxue tongbao* 22.7 (2006): 490–94.

Chen Xiwen. *Zhongguo zhengfu zhinong zijin shiyong yu guanli: tizhi gaige yanjiu* [The use and management of Chinese state funds to support agriculture: Research on system reform]. Taiyuan: Shanxi jingji chubanshe, 2004.

Cheng, Tien-Hsi. "Insect Control in Mainland China." *Science* 140.3564 (1963): 269–77.

Cheng Guansen. "Ta mizui zai baimi wanggong" [He is fascinated by the termite's palace]. *Renmin ribao*, 16 April 1989, 6.

Chiang, H. C. "I Am Happy to Be an Entomologist." *Zhonghua kunchong* 13.2 (1993): 275–92.

Coffey, Brian F. "Fertilizers to the Front: HAER and U.S. Nitrate Plant No. 2." *Journal of the Society for Industrial Archeology* 23.1 (1997): 25–42.

Cook, Alexander C. "Third World Maoism." In *A Critical Introduction to Mao*, edited by Timothy Cheek, 288–312. Cambridge: Cambridge University Press, 2010.

Cooke, Kathy J. "Expertise, Book Farming, and Government Agriculture: The Origins of Agricultural Seed Certification in the United States." *Agricultural History* 76.3 (Summer 2002): 524–45.

Critchfield, Richard. "China's Miracle Rice: A New Rivalry with the West." *International New York Times*, 17 June 1992. http://www.nytimes.com/1992/06/17/opinion/17iht-edcr.html.

Cullather, Nick. *The Hungry World: America's Cold War Battle against Poverty in Asia*. Cambridge, MA: Harvard University Press, 2010.

Dai Sijie. *Balzac and the Little Chinese Seamstress*. Translated by Ina Rilke. New York: Anchor Books, 2002.

Dangdai Zhongguo congshu bianji weiyuanhui, ed. *Dangdai Zhongguo de nongye* [Contemporary Chinese agriculture]. Beijing: Zhongguo shehui kexue chubanshe, 1992.

Dasha gongshe geweihui. "Dasha gongshe jiji tuiguang huaxue chucao jieshao" [Introduction to Big Sand Commune's active extension of chemical herbicide]. 1973. Sihui Municipal Archives.

Day, Alexander F. *The Peasant in Postsocialist China: History, Politics, and Capitalism*. Cambridge: Cambridge University Press, 2013.

Dazhai dadui "san jiehe" keyan zu. "Cong Dazhai gaitu de shijian kan turangxue lilun de fazhan" [The development of soil science theory from the perspective of Dazhai's soil improvement practices]. *Renmin ribao*, 4 November 1975, 2.

de Certeau, Michel. *The Practice of Everyday Life*. Translated by Steven Rendall. Berkeley: University of California Press, 1984.

Delman, Jørgen. "Cool Thinking? The Role of the State in Shaping China's Dairy Sector and Its Knowledge System." *China Information* 17.2 (2003): 1–35.

———. "'We Have to Adopt Innovations': Farmers' Perceptions of the Extension-Farmer Interface in Renshou County." In *From Peasant to Entrepreneur: Growth and Change in Rural China*, edited by E. B. Vermeer, 83–103. Wageningen, Neth.: Pudoc, 1992.

Deng Xiangzi and Deng Yingru. *The Man Who Puts an End to Hunger: Yuan Longping, "Father of Hybrid Rice."* Beijing: Foreign Languages Press, 2007.

Deng Xiangzi and Ye Qinghua. *Bu zai ji'e: Shijie de Yuan Longping* [Never again famine: The world's Yuan Longping]. Changsha: Hunan wenyi chubanshe, 2007.

Dikötter, Frank. *Mao's Great Famine: The History of China's Most Devastating Catastrophe, 1958–1962*. London: Bloomsbury, 2010.

Dong Bingya. "Yumi yichuan yuzhong zhuanjia: Li Jingxiong" [Corn genetics and breeding specialist Li Jingxiong]. *Zhongguo keji shiliao* 15.3 (1994): 54–61.

Dong Guangbi. *Zhongguo jinxiandai kexue jishu shi lungang* [Outline discussion of science and technology in modern and contemporary Chinese history]. Changsha: Hunan jiaoyu chubanshe, 1992.

Economy, Elizabeth. *The River Runs Black: The Environmental Challenge to China's Future*. 2nd ed. Ithaca, NY: Cornell University Press, 2010.

Epstein, Edward Jay. *Dossier: The Secret History of Armand Hammer.* New York: Random House, 1996.

Esherick, Joseph W. *Ancestral Leaves: A Family Journey through Chinese History.* Berkeley: University of California Press, 2011.

Evans, Harriet, and Stephanie MacDonald, eds. *Picturing Power in the People's Republic of China: Posters of the Cultural Revolution.* Lanham, MD: Rowman & Littlefield, 1999.

Eyferth, Jacob. *Eating Rice from Bamboo Roots: The Social History of a Community of Handicraft Papermakers in Rural Sichuan, 1920–2000.* Cambridge, MA: Harvard University Asia Center, 2009.

———. "Women's Work and the Politics of Homespun in Socialist China, 1949–1980." *International Review of Social History* 57.3 (2012): 365–91.

Fan, Fa-ti. "'Collective Monitoring, Collective Defense': Science, Earthquakes, and Politics in Communist China." *Science in Context* 25.1 (2012): 127–54.

Fang, Xiaoping. *Barefoot Doctors and Western Medicine in China.* Rochester, NY: University of Rochester Press, 2012.

Farquhar, Judith. *Appetites: Food and Sex in Postsocialist China.* Durham, NC: Duke University Press, 2002.

Farquhar, Judith, and Qicheng Zhang. "Biopolitical Beijing: Pleasure, Sovereignty, and Self-Cultivation in China's Capital." *Cultural Anthropology* 20.3 (2005): 303–27.

Fenghuang xian Longtan xiang youcai fengchan gongzuozu. "Laonong fangzhi youcai donghai de cuoshi" [Old peasants' methods for preventing freezing in rapeseed plant]. *Zhongguo nongye kexue* 1958.9: 483.

Ferguson, James. *The Anti-Politics Machine: "Development," Depoliticization and Bureaucratic Power in Lesotho.* Cambridge: Cambridge University Press, 1990.

Fitzgerald, Deborah. "Blinded by Technology: American Agriculture in the Soviet Union, 1928–1932." *Agricultural History* 70.3 (Summer 1996): 459–86.

———. *The Business of Breeding: Hybrid Corn in Illinois, 1890–1940.* Ithaca, NY: Cornell University Press, 1990.

———. "Farmers Deskilled: Hybrid Corn and Farmers' Work." *Technology and Culture* 34.2 (April 1993): 324–43.

Fogang xian Lingkuang fudaihui. "Jiji zuzhi funü canjia nongye kexue shiyan" [Actively organize women to participate in agricultural scientific experiment]. July 1973 [no day specified]. Guangdong Provincial Archives, 233-3-0017-105~107.

Food and Agriculture Organization of the United Nations. *The Double Burden of Malnutrition: Case Studies from Six Developing Countries.* Rome: FAO, 2006. ftp://ftp.fao.org/docrep/fao/009/a0442e/a0442e.zip.

———. *Guidelines for Integrated Control of Rice Insect Pests.* Rome: FAO, 1979.

Friedman, Edward. "The Politics of Local Models, Social Transformation and State Power Struggles in the People's Republic of China: Tachai and Teng Hsiao-p'ing." *China Quarterly* 76 (1978): 873–90.

Friedman, Edward, Paul Pickowicz, and Mark Selden. *Revolution, Resistance, and Reform in Village China.* New Haven, CT: Yale University Press, 2005.

Fukuyama, Francis. *The End of History and the Last Man.* New York: Free Press, 1992.

Fyfe, Aileen. "Reading Children's Books in Late Eighteenth-Century Dissenting Families." *The Historical Journal* 43.2 (June 2000): 453–73

Gao Shiqi. "Chuntian" [Spring]. *Shaonian kexue* 1979.1: 6–7.

Gao Xiaoxian. "'The Silver Flower Contest': Rural Women in 1950s China and the Gendered Division of Labour." Translated by Yuanxi Ma. *Gender and History* 18.3 (2006): 594–612.

Gaud, William. "AID Supports the Green Revolution." Address before the Society for International Development, 8 March 1968. Washington, DC?: Agency for International Development?, 1968.

Ghosh, Arunabh. "Making It Count: Statistics and State-Society Relations in the Early People's Republic of China, 1949–1959." PhD diss., Columbia University, 2014.

Gilman, Nils. "Modernization Theory, the Highest Stage of American Intellectual History." In *Staging Growth: Modernization, Development, and the Global Cold War*, edited by David C. Engerman, Nils Gilman, Mark H. Haefele, Michael E. Latham, 47–80. Amherst: University of Massachusetts Press, 2003.

Goikhman, Izabella. "Soviet-Chinese Academic Interactions in the 1950s: Questioning the 'Impact-Response' Approach." In *China Learns from the Soviet Union, 1949–Present*, edited by Thomas Bernstein and Hua-yu Li, 275–302. Lanham, MD: Lexington Books, 2010.

Goldman, Merle. *China's Intellectuals: Advise and Dissent.* Cambridge, MA: Harvard University Press, 1981.

Gongqingtuan Shaanxi sheng wei. *Dazhai Xiyang qingnian gongzuo jingyan.* Beijing: Zhongguo qingnian chubanshe, 1977.

Gongqingtuan zhongyang qingnong bu, ed. *Wei geming gao nongye kexue shiyan* [Agricultural scientific experiment for the revolution]. Beijing: Zhongguo qingnian chubanshe, 1966.

Gowen, John W. *Heterosis: A Record of Researches Directed toward Explaining and Utilizing the Vigor of Hybrids.* Ames: Iowa State College Press, 1952.

Greenhalgh, Susan. *Just One Child: Science and Policy in Deng's China.* Berkeley: University of California Press, 2008.

Grishin. Letter from Grishin to Pu Zhelong (translated into Chinese). 2 November 1959. Sun Yat-sen University Archives.

———. Letter from Grishin to Pu Zhelong (translated into Chinese). 20 April 1961. Sun Yat-sen University Archives.

Gross, Miriam. "Chasing Snails: Anti-schistosomiasis Campaigns in the People's Republic of China." PhD diss., University of California, San Diego, 2010.

Grossman, Lawrence. *The Political Ecology of Bananas: Contract Farming, Peasants, and Agrarian Change in the Eastern Caribbean.* Chapel Hill: University of North Carolina Press, 1998.

Gu Dexiang and Feng Shuang, eds. *Nan Zhongguo shengwu fangzhi zhi fu: Pu Zhelong* [The father of biological control in southern China: Pu Zhelong]. Guangzhou: Zhongshan daxue chubanshe, 2012.

Guangdong sheng. "Jieshao yige nongcun kexue shiyan xiaozu" [Introducing a rural scientific experiment group]. 23 November 1969. Guangdong Provincial Archives, 306-A0.02-7-28.

Guangdong sheng geweihui banshizu waishi bangongshi [External Affairs Office of the Guangdong Provincial Revolutionary Committee Administrative Group]. Memo to the Waijiaobu libinsi [Foreign Ministry Protocol Department]. 20 July 197? [illegible]. Sun Yat-sen University Archives.

Guangdong sheng Hainan Lizu Miaozu zizhu zhou kejiju qingbaosuo. "Yi jieji douzheng wei gang, dali tuiguang zajiao shuidao" [With class struggle as the key link, vigorously extend hybrid rice]. *Yichuan yu yuzhong* 1976.4: 23.

Guangdong sheng keji ju. "Zazhong youshi liyong he shengwu fangzhi liangxiang huizhan jinzhan qingkuang" [Progress on the Two Campaigns of Heterosis and Biological Control]. 28 May 1973. Guangdong Provincial Archives, 306-A0.02–41–85.

Guangdong sheng Sihui xian Dasha gongshe geweihui et al., ed. "Dasha gongshe shuidao haichong zonghe fangzhi" [Integrated control of rice insect pests in Big Sand Commune]. *Zhongshan daxue xuebao ziran kexue ban* 1976.2: 23–33.

Guangxi shiyuan shengwuxi shuidao "sanxi" keyan xiaozu. "Kaizhan shuidao zazhong youshi liyong jichu lilun yanjiu de yixie tihui" [Some experiences developing basic theoretical research on heterosis in rice]. *Guangxi shiyuan daxue xuebao* 1975.1: 10–12.

Guizhou nongxueyuan zuowu yuzhong jiaoyanzu. "Shuidao xiongxing buyu yanjiu jiankuang" [The status of research on male sterility in rice]. *Guizhou nongye kexue* [title on cover differs: *Nongye keji*] 1973.1: 3–5.

Guo Xiulian. "Sanjie qingnian kaizhan kexue shiyan duo gaochan huodong de huigu" [Reflections on the activities of the youth of Sanjie who developed scientific experiment and achieved high production]. In *Yangchun wenshi ziliao*, ed. Guangdong sheng Yangchun shi zhengxie wenshi ziliao weiyuanhui, vol. 18, 81–87. Yangchun xian: Yangchun xian zhengxie wenshi zu, 1996.

Guojia nongken zong ju kejiao ju. *Yong yu pandeng de nianqing ren: Guoying nongchang zhishi qingnian kexue shiyan de shiji* [Young people bravely scaling the heights: The scientific experiment achievements of educated youth on state farms]. Beijing: Nongye chubanshe, 1979.

Gupta, Akhil. *Postcolonial Developments: Agriculture in the Making of Modern India*. Durham, NC: Duke University Press, 1998.

Haldane, J. B. S. *The Marxist Philosophy and the Sciences*. London: Allen and Unwin, 1938.

Hale, Matthew A. "Reconstructing the Rural: Peasant Organizations in a Chinese Movement for Alternative Development." PhD diss., University of Washington, 2013.

Hammer, Armand. "On a Vast China Market." *Journal of International Affairs* 39.2 (1986): 19–25.

Han, Dongping. "Rural Agriculture: Scientific and Technological Development during the Cultural Revolution." In *Mr. Science and Chairman Mao's Cultural Revolution: Science and Technology in Modern China*, edited by Chunjuan Nancy Wei and Darryl E. Brock, 281–30. Lanham, MA: Lexington Books, 2013.

———. *The Unknown Cultural Revolution: Revolution: Education Reforms and Their Impact on China's Rural Development*. New York: Garland, 2000.

Hanson, Haldore. "Trip diary." U.S. Wheat Sudies Follow-up; China; Commissions, Conferences, Councils and Symposia, 1952–2001, 33–205. NEB. http://purl.umn.edu/106466.

Haraway, Donna. "Primatology Is Politics by Other Means." *Proceedings of the Biennial Meeting of the Philosophy of Science Association*, 1984, 489–524.

———. *Simians, Cyborgs, and Women: The Reinvention of Nature*. New York: Routledge, 1991.

Harlan, Jack R. "Plant Breeding and Genetics." In *Science in Contemporary China*, edited by Leo A. Orleans, 295–312. Stanford, CA: Stanford University Press, 1980.

Hathaway, Michael. *Environmental Winds: Making the Global in Southwest China*. Berkeley: University of California Press, 2013.

Hayford, Charles. "The Storm over the Peasant: Orientalism, Rhetoric and Representation in Modern China." In *Contesting the Master Narrative: Essays in Social History*, edited by Shelton Stromquist and Jeffrey Cox, 150–72. Iowa City: University of Iowa Press, 1998.

He Minshi and Wang Jianxun. *Guangzhou minjian chengyu nongyan tongyao* [Guangzhou folk sayings, agricultural maxims, and nursery rhymes]. Guangzhou: Guangzhou shi qunzhong yishuguan, [preface dated 1963].

Heilmann, Sebastian. "From Local Experiments to National Policy: The Origins of China's Distinctive Policy Process." *China Journal* 59 (2008): 1–30.

Heilongjiang sheng Binxian Xinlisi dui keyan xiaozu. "Bai ying dadou wang de xuanyu" [The selection of white-breast soybean king]. *Nongye keji tongxun* 1973.12: 4.

Henan sheng geming weiyuanhui kexue jishu weiyuanhui, ed. *Banhao siji nongcun kexue shiyan wang* [Build the four-level agricultural scientific experiment network]. N.p. 1975.

Henan sheng Nanyang zhuanqu kexue jishu xiehui, ed. *Quanguo nongcun kexue shiyan yundong jingyan huiji* [Collection of national experiences in the agricultural science experiment movement]. Vol. 1. N.p., 1966.

Hershatter, Gail. *The Gender of Memory: Rural Women and China's Collective Past.* Berkeley: University of California Press, 2011.

Hightower, Jim. *Hard Tomatoes, Hard Times: A Report of the Agribusiness Accountability Project on the Failure of America's Land Grant College Complex.* Cambridge, MA: Schenkman, 1973.

Hinton, Carma, Geremie R. Barmé, and Richard Gordon, dir. *Morning Sun.* Brookline, MA: Long Bow Group, 2005.

Ho, Peter, Jennifer H. Zhao, and Dayuan Xue. "Access and Control of Agrobiotechnology: Bt Cotton, Ecological Change and Risk in China." *Journal of Peasant Studies* 36.2 (2009): 345–64.

Hu, Danian. *China and Albert Einstein: The Reception of the Physicist and His Theory in China, 1917–1979.* Cambridge, MA: Harvard University Press, 2005.

Hua xian Jiang Qizhang. "Pinxia zhongnong yao dang kexue shiyan de chuangjiang" [Poor and lower-middle peasants must become pathbreakers for scientific experiment]. 1965. Guangdong Provincial Archives, 235-1-365-047~049.

Hua zhuxi zai Hunan [Chairman Hua in Hunan]. Beijing: Renmin chubanshe, 1977.

Huang Jianmin, Hu Ruifa, and Huang Jikun. "Jishu tuiguang yu nongmin dui xin jishu de xiuzheng caiyong" [Agricultural technology extension and farmers' modification of new technology]. *Zhongguo ruan kexue* 2005.6: 60–66.

Hukou xian nongye ju. "Hukou xian yingyong shengwu zhichong qude xin fazhan" [Hukou County has achieved new developments in the use of biology to control insect pests]. *Jiangxi nongye keji* 1976.8: 9–10.

Hunan sheng Qianyang diqu nongxiao keyanzu. "Xuanyu shuidao xiongxing buyu baochixi de yidian tihui" [Some experiences in selecting a male-infertile maintainer line in rice]. *Nongye keji tongxun* 1972.10: 8–9.

Iovene, Paola. *Tales of Futures Past: Anticipation and the Ends of Literature in Contemporary China.* Stanford, CA: Stanford University Press, 2014.

Jacobsen, Rowan. "Stung by Bees." *Newsweek.* 23 June 2008. Viewed 15 January 2009. http://www.newsweek.com/id/141461?tid=relatedcl.

Jiangsu sheng Jianhu xian *Tianjia wuxing* xuanze xiaozu. *Tianjia wuxing* [The farmer's five phases]. Beijing: Zhonghua shuju, 1976.

Jiangxi sheng Yichun diqu nongyeju et al., ed. and pub. *Zenyang zhonghao zajiao shuidao* [How to plant hybrid rice]. 1975.

Jielianjia, H. A. *Nonglin haichong shengwu fangzhi.* Shanghai: Shanghai kexue jizhu chubanshe, 1957. [The author is Russian; however, the original Russian name is unknown. The name provided here is a Chinese transliteration.]

Jin Shanbao. "Yangbantian fazhan le nongye shengchan cujin le nongye kexue geminghua: Zai quanguo nongye kexue shiyan gongzuo huiyi shang de fayan" [Model fields have developed agricultural production and furthered the revolutionization of agricultural science: Speech at the National Conference on Agricultural Scientific Experiment Work]. *Zhongguo nongye kexue* 1965.4: 12–15.

Jinghua shibao. "Sannong zhuanjia zhixin Yuan Longping: Huyu huan nongmin ziyou xuanze zhongzi quanli" [Three-rural expert writes letter to Yuan Longping: Give peasants back the right to freely select seeds]. 28 April 2011. http://www.chinanews.com /sh /2011/04-28/3003303.shtml.

Johansson, Perry. "Mao and the Swedish United Front against USA." In *The Cold War in Asia: The Battle for Hearts and Minds*, edited by Zheng Yangwen, Hong Liu, and Michael Szonyi, 217–40. Leiden: Brill, 2010.

Johnson, Kay Ann. *Women, the Family, and Peasant Revolution in China.* Chicago: University of Chicago Press, 1983.

Johnson, Matthew D. "The Science Education Film: Cinematizing Technocracy and Internationalizing Development." *Journal of Chinese Cinemas* 5.1 (2011): 31–53.

Keene, Melanie. "'Every Boy & Girl a Scientist': Instruments for Children in Interwar Britain." *Isis* 98.2 (June 2007): 266–89.

Kennedy, John F. "Special Message to Congress on Urgent National Needs, 25 May 1961." Speech Files, Papers of John F. Kennedy, Presidential Papers, President's Office Files. Accessed 19 January 2014. http://www.jfklibrary.org/Asset-Viewer/Archives/JFKPOF-034-030.aspx.

King, Franklin Hiram. *Farmers of Forty Centuries, Or, Permanent Agriculture in China, Korea and Japan.* Madison, WI: Mrs. F. H. King, 1911.

King, John Kerry. "Rice Politics," *Foreign Affairs* 31.3 (April 1953): 453–60.

Kloppenburg, Jack Ralph. *First the Seed: The Political Economy of Plant Biotechnology, 1492–2000.* 2nd ed. Madison: University of Wisconsin Press, 2004.

Kogan, Marcos. "Integrated Pest Management: Historical Perspectives and Contemporary Developments." *Annual Review of Entomology* 43 (1998): 243–70.

Kohlstedt, Sally Gregory. *Teaching Children Science: Hands-On Nature Study in North America, 1890–1930.* Chicago: University of Chicago Press, 2010.

Kong, Shu-yu. "Between Undercurrent and Mainstream: Social Energy and the Production of Hand-Copied Literature during and after the Cultural Revolution." Unpublished paper, delivered at the workshop "Between Revolution and Reform: China at the Grassroots, 1960–1980," Simon Frasier University, 2010.

Kraus, Richard. *Pianos and Politics in China: Middle-Class Ambitions and the Struggle over Western Music.* New York: Oxford University Press, 1989.

Kuhn, Philip. "Political and Cultural Factors Affecting Agricultural Development" (Kuhn's draft, marked "Confidential Norman Borlaug") [archive cuts off]. Background; China; Commissions, Conferences, Councils and Symposia, 1952–2001, 52–76. NEB. http://purl .umn.edu /106459.

Kuo, Leslie T. C. *The Technical Transformation of Agriculture in Communist China.* New York: Praeger, 1972.

Lafargue, Paul. "Reminiscences of Marx." In *Marx and Engels Through the Eyes of Their Contemporaries,* 22–39. Moscow: Progress Publishers, 1972.

Lam, Tong. *A Passion for Facts: Social Surveys and the Construction of the Chinese Nation State, 1900–1949.* Berkeley: University of California Press, 2011.

Lammer, Christof. "Imagined Cooperatives: An Ethnography of Cooperation and Conflict in New Rural Reconstruction Projects in a Chinese Village." PhD diss., University of Vienna, 2012.

Lavelle, Peter. "Agricultural Improvement at China's First Agricultural Experiment Stations." In *New Perspectives on the History of Life Sciences and Agriculture,* edited by D. Phillips, S. Kingsland, 323-44. Switzerland: Springer International Publishing, 2015.

———. "Imperial Texts in Socialist China: Republishing Agricultural Treatises in the Early Maoist Era." Paper presented at the History of Science Society Annual Meeting, Montreal, Canada, 5 November 2010.

Lee, Seung-joon. *Gourmets in the Land of Famine: The Culture and Politics of Rice in Modern Canton.* Stanford, CA: Stanford University Press, 2011.

Lei, Sean Hsiang-lin. *Neither Donkey nor Horse: Medicine in the Struggle over China's Modernity.* Chicago: University of Chicago Press, 2014.

Levins, Richard. "How Cuba Is Going Ecological." *Capitalism, Nature, Socialism* 16.3 (2005): 7–25.

———. "Science and Progress: Seven Developmentalist Myths in Agriculture." *Monthly Review* 38.3 (July–August 1986): 13–20.

Lewontin, Richard. "Agricultural Research and the Penetration of Capital." *Science for the People* 14.1 (1982): 12–17.

Li, Lillian M. *Fighting Famine in North China: State, Market, and Environmental Decline, 1690s–1990s.* Stanford, CA: Stanford University Press, 2007.

Li Baochu. "Yi, Lüfei zuowu" [One: Green manure crops]. *Zhejiang nongye kexue* 1961.9: 451–53.

Li Changping. "Zhi Yuan Longping Laoshi de yifeng xin gongkai xin" [An open letter to Li Yuan Longping]. 28 April 2011. http://www.snzg.net/article/2011/0428/article_23504.html.

Li Guihong and Xiao Jian. "Lengshui yuzhong, qianwu guren: Luo Xiaohe qiaojie zajiao shuidao nanti" [Breeding in cold water (i.e., without support from others), going where no one has gone before]. *Sanxiang dushi bao*, 15 June 2005. Viewed 26 October 2011. http://www.hn.xinhuanet.com/misc/2005-06/15/content_4442720.htm.

Li Huaiyin. *Village China under Socialism and Reform: A Micro-History, 1949–2008.* Stanford, CA: Stanford University Press, 2009.

Li Liqiu, Hu Ruifa, Liu Jian, and Feng Yan. "Jianli guojia gonggong nongye jishu tuiguang fuwu tixi" [Establishing a national public agricultural technology extension service system]. *Zhongguo keji luntan* 2003.6: 125–28.

Li Qun. "*Qimin yaoshu* he fajia sixiang" [*Essential technologies for the common people* and Legalist thought]. *Renmin ribao*, 16 September 1974, 3.

Li Zhensheng. "Dalaocu zui pei gao keyan" [Country bumpkins are a perfect match for scientific research]. *Nongcun kexue shiyan* 1976.4: 2–5.

———. "Mao Zhuxi zhexue sixiang shi wo peiyu yumidao de jin yaoshi" [Mao Zedong philosophical thought was my golden key in breeding corn-rice]. *Yichuan xuebao* 2.3 (1975): 187–93.

Lianjiang xian fulian. "Shengchan douzheng, kexue shiyan de jieguo, bian wo geng re'ai nongcun" [Results of the struggle for production and scientific experiment have deepened my love for the countryside]. 11 June 1966. Guangdong Provincial Archives, 233-2-0332-23~29.

Liaoning sheng nongkeyuan shuidao yanjiu shi. "Shuidao zazhong youshi liyong yanjiu chubu zongjie" [Initial summary of research on hybrid vigor in rice]. *Liaoning nongye kexue* 1972.11–12 (Z2): 3–4.

Lin, James. "Sowing Seeds and Knowledge: Agricultural Development in Taiwan and the World, 1925–1975." *East Asian Science, Technology and Society* 9 (2015): 1–23.

Lin, Justin Yifu. "Education and Innovation Adoption in Agriculture: Evidence from Hybrid Rice in China." *American Journal of Agricultural Economics* 73.3 (1991): 713–23.

Link, Perry. "The Limits of Cultural Reform in Deng Xiaoping's China." *Modern China* 13.2 (April 1987): 115–76.

Liu, C. L. Letter to William A. Riley. 28 July 1935. Box 10. Folder "China—Miscellaneous, 1932–1935." Collection 938, Entomology Papers, Division of Entomology and Economic Zoology, University of Minnesota Archives.

Livingstone, David N. *Putting Science in Its Place: Geographies of Scientific Knowledge.* Chicago: University of Chicago Press, 2003.

Longino, Helen E. *Science as Social Knowledge: Values and Objectivity in Scientific Inquiry.* Princeton, NJ: Princeton University Press, 1990.

Luo Lixin. "Laizi ling yige shehui de baogao: Ji zhuming kunchong xuejia Pu Zhelong jiaoshou" [Report from another society: Remembering the famous entomologist Professor Pu Zhelong]. *Gaojiao tansuo* 1989.1: 69–72.

Luo Runliang and Wu Jinghua. *Lüse shenhua jieshi: Lun Yuan Longping keji chuangxin* [Explaining the Green Legend: On Yuan Longping's Scientific-Technological Innovation]. Guangzhou: Guangdong keji chubanshe, 2003.

Lü Xinchu and Gu Mainan. "Shi kexuejia daxian shenshou de shihou le: Quanguo nongye kexue jishu gongzuo huiyi ceji" [This is the time for scientists to do their all: Notes from the National Agricultural Science and Technology Work Conference]. *Renmin ribao,* 6 April 1963, 2.

Lu Youshang. "Guangkuo tiandi dayou zuowei" [In the vast land, great achievements are possible]. *Kexue shiyan* 1976.7: 27.

Lynch, William T. "Ideology and the Sociology of Scientific Knowledge." *Social Studies of Science* 24.2 (1994): 197–227.

MacFarquhar, Roderick. *The Origins of the Cultural Revolution.* Vol. 2, *The Great Leap Forward, 1958–1960.* New York: Columbia University Press, 1983.

———. *The Origins of the Cultural Revolution.* Vol. 3, *The Coming of the Cataclysm, 1961–1966.* New York: Columbia University Press, 1997.

Mahoney, Michael. "Estado Novo, Homem Novo (New State, New Man): Colonial and Anticolonial Development Ideologies in Mozambique, 1930–1977." In *Staging Growth: Modernization, Development, and the Global Cold War,* edited by David C. Engerman, 165–98. Amherst: University of Massachusetts Press, 2003.

Mai Baoxiang. "Pu Zhelong jiaoshou zai Dasha de rizi" [Professor Pu Zhelong's days at Big Sand]. *Zhongshandaxue xinwenwang.* Accessed 8 May 2015. Presented in three parts: http://news2.sysu.edu.cn/theory01/124756.htm, http://news2.sysu.edu.cn/theory01/124755 .htm, http://news2.sysu.edu.cn/theory01/124754.htm.

Mai Shuping. "Shengwu huanbao di yi ren: Pu Zhelong" [Pioneer of biological conservation: Pu Zhelong]. *Jiankang da shiye* 2002.3: 10–11.

Manning, Kimberley Ens, and Felix Wemheuer, eds. *Eating Bitterness: New Perspectives on China's Great Leap Forward and Famine.* Vancouver: UBC Press, 2011.

Mao Tse-tung [Mao Zedong]. "Intra-Party Correspondence," 11 October 1959. In *Selected Works of Mao Tse-tung,* vol. 8. Accessed 7 May 2015. https://www.marxists.org/reference/archive/ mao/selected-works/index.htm.

———. *Quotations from Chairman Mao Tse-tung.* Beijing: Foreign Languages Press, 1966.

———. *Selected Works of Mao Tse-tung.* Vol. 5. Beijing: Foreign Languages Press, 1977.

Mao Zedong. "Mao Zhuxi gei Mao Anying, Mao Anqing tongzhi de xin" [A letter from Chairman Mao to Mao Anying and Mao Anqing]. *Shaonian kexue* 1979.1: 3.

Marx, Karl, and Friedrich Engels. *The German Ideology: Part One with Selections from Parts Two and Three, Together with Marx's "Introduction to a Critique of Political Economy."* New York: International Publishers, 1970.

Meisner, Maurice. *Li Ta-chao and the Origins of Chinese Marxism.* Cambridge, MA: Harvard University Press, 1967.

———. *Mao's China and After.* 3rd ed. New York: The Free Press, 1999.

Melillo, Edward D. "The First Green Revolution: Debt Peonage and the Making of the Nitrogen Fertilizer Trade, 1840–1930." *American Historical Review* 117.4 (October 2012): 1028–60.

Metcalf, Robert L. "Changing Role of Insecticides in Crop Protection." *Annual Review of Entomology* 25 (1980): 219–56.

————. "China Unleashes Its Ducks." *Environment* 18.9 (1976): 14–17.

Mianyang xian kexue jishu weiyuanhui. *Qunzhongxing nongye kexue shiyan ziliao huibian* [Collected materials on mass agricultural scientific experiment]. N.p., 1974.

Michelbacher, A. E., and O. G. Bacon. "Walnut Insect and Spider-Mite Control in Northern California." *Journal of Economic Entomology* 45.6 (1952): 1020–27.

Mickel, Clarence. Letter from Mickel to China Institute of America. 27 May 1948. Box 56. Folder "Pu, Chih Lung." Collection 938, Entomology Papers, Division of Entomology and Economic Zoology, University of Minnesota Archives.

Mignolo, Walter D. "Delinking: The Rhetoric of Modernity, the Logic of Coloniality and the Grammar of De-coloniality." *Cultural Studies* 21.2–3 (March/May 2007): 449–514.

Miller, H. Lyman. *Science and Dissent in Post-Mao China: The Politics of Knowledge.* Seattle: University of Washington Press, 1996.

Miller, Robert F. *One Hundred Thousand Tractors: The MTS and the Development of Controls in Soviet Agriculture.* Cambridge, MA: Harvard University Press, 1970.

Mitchell, Timothy. *Rule of Experts: Egypt, Techno-Politics, Modernity.* Berkeley: University of California Press, 2002.

Mullaney, Thomas S. *Coming to Terms with the Nation: Ethnic Classification in Modern China.* Berkeley: University of California Press, 2011.

————. "The Moveable Typewriter: How Chinese Typists Developed Predictive Text during the Height of Maoism." *Technology and Culture* 53.4 (2002): 777-814.

Murphy, Eugene T. "Changes in Family and Marriage in a Yangzi Delta Farming Community, 1930–1990." *Ethnology* 40.3 (2001): 213–35.

Naquin, Susan. *Millenarian Rebellion in China: The Eight Trigrams Uprising of 1813.* New Haven, CT: Yale University Press, 1976.

Nedostup, Rebecca. *Superstitious Regimes: Religion and the Politics of Chinese Modernity.* Cambridge, MA: Harvard University Asia Center, 2009.

Neushul, Peter, and Zuoyue Wang. "Between the Devil and the Deep Sea: C. K. Tseng, Mariculture, and the Politics of Science in Modern China." *Isis* 91.1 (2000): 59–88.

Nie Leng and Zhuang Zhixia. *Lüse wangguo de yiwan fuweng* [Billionaire of the green kingdom]. Beijing: Huayi chubanshe, 2000.

Oi, Jean C. *State and Peasant in Contemporary China: The Political Economy of Village Government.* Berkeley: University of California Press, 1989.

Owens, Larry. "Science 'Fiction' and the Mobilization of Youth in the Cold War." *Quest: The History of Spaceflight Quarterly* 14.3 (2007): 52–57.

Pan, Yihong. *Tempered in the Revolutionary Furnace: China's Youth in the Rustication Movement.* Lanham, MD: Lexingtoon Books, 2003.

Pan Chengxiang. "The Development of Integrated Pest Control in China." *Agricultural History* 62.1 (Winter 1988): 1–12.

Pan Gang. "Siying qiye tigao le nongmin shangping yishi" [Privately owned businesses raise peasants' commercial consciousness]. *Renmin ribao*, 12 April 1988, 2.

Pan Yiwei. "Yijiu qiqi shiluo de daxue meng" [1977: The lost dream of college]. *Nanguo zaobao*, 4 July 2007, 42.

Peng, Zhaochang. "Decollectivization and Rural Poverty in Post-Mao China: A Critique of the Conventional Wisdom." PhD diss., University of Massachusetts, Amherst, 2013.

Peng Baoshan. "Kaizhan qunzhong yundong da gao baijiangjun de shengchang he yingyong" [Develop a mass movement to produce and use *Beauvaria* on a large scale]. *Nongcun kexue shiyan* [Rural agricultural experiment] 1976.5: 6–8.

Perdue, Peter. *Exhausting the Earth: State and Peasant in Hunan, 1500–1850.* Cambridge, MA: Council on East Asian Studies, Harvard University Press, 1987.

Perkins, John H. *Geopolitics and the Green Revolution: Wheat, Genes, and the Cold War.* New York: Oxford University Press, 1997.

Perry, Elizabeth. *Anyuan: Mining China's Revolutionary Tradition.* Berkeley: University of California Press, 2012.

———. "From Mass Campaigns to Managed Campaigns: 'Constructing a New Socialist Countryside.'" In *Mao's Invisible Hand: The Political Foundations of Adaptive Governance in China*, edited by Sebastian Heilmann and Elizabeth Perry, 30–61. Cambridge, MA: Harvard University Asia Center, 2011.

Pi Yan et al. "Guonei gaoxiao yichuanxue jiaocai fazhan yanjiu" [Research on the development of Chinese high-school genetics textbooks]. *Yichuan* 31.1 (Jan 2009): 102–12.

Pingyang xian nonglin ju. "Pingyang xian 1964 nian kaizhan wandao tianyang ping kexue shiyan yundong de jidian tihui" [A few observations from Pingyang county's 1964 scientific experiment movement experiences with raising duckweed in rice paddies]. *Zhejiang nongye kexue* 1965.7: 361–63.

Pu Chih-lung [Pu Zhelong]. "The Biological Control of Insect Pests in Agriculture and Forestry in China." In *Biological Insect Control in China and Sweden*, edited by The Royal Swedish Academy of Sciences, 3. Stockholm: The Royal Swedish Academy of Sciences, 1979.

Pu Chih Lung [Pu Zhelong] and Tsui-Ying Lee. Letter to Dr. and Mrs. Richards. 2 February 1951. Box 56. Folder "Pu, Chih Lung." Collection 938, Entomology and Economic Zoology, University of Minnesota Archives.

Pu Zhelong. Draft letter from Pu Zhelong to Grishin (with comments by party officials). 14 June 1961. Sun Yat-sen University Archives.

———. Draft letter from Pu Zhelong to Grishin. 1 November 1961. Sun Yat-sen University Archives.

———, ed. *Haichong shengwu fangzhi de yuanli he fangfa* [Principles and methods of biological control of insect pests]. Beijing: Kexue chubanshe, 1978.

———. "Kexue yanjiu yao wei wuchan jieji zhuanzheng fuwu" [Scientific research must serve the dictatorship of the proletariat]. *Dongwu xuebao* 21.3 (September 1975): 213–15.

Pu Zhelong and Liu Zhifeng. "Chiyanfeng daliang fanzhi jiqi duiyu ganzhe mingchong de datian fangzhi xiaoguo" [Mass breeding of *Trichogramma* and its results in controlling sugarcane stem borers in large field studies]. *Kunchong xuebao* 11.4 (1962): 409–14.

———. "Woguo zhequ qunzhong liyong chiyanfeng fangzi ganzhe mingchong qingkuang" [Use of *Trichogramma* by the masses to control sugarcane stem borers in sugarcane-producing areas of China]. *Kunchong zhishi* 1959.9: 299–300.

Pu Zhelong, Mai Xiuhui, and Huang Mingdu. "Liyong pingfu xiaofeng fangzhi lizhichun shiyan chubao" [Preliminary report of experiments in the use of *Anastatus* to control lychee stinkbug]. *Zhiwu baohu xuebao* 1.3 (1962): 301–6.

Pu Zhelong, Zhu Jinliang, and Wu Weiji. "Ganshu xiaoxiang bichong (*Cylas formicarius* Fabr.) tianjian huaxue fangzhi shiyan" [Field experiments in the chemical control of sweetpotato weevil]. *Zhongshan daxue xuebao* 1961.2: 79–80.

Qi Shuying and Wei Xiaowen. *Yuan Longping zhuan* [Biography of Yuan Longping]. Taiyuan: Shanxi renmin chubanshe, 2002.

Qi Zhaosheng, Zhang Zepu, and Wang Dexiu. "Mian hongzhizhu de zonghe fangzhi fa" [Integrated control of red spider mites in cotton]. *Nongye kexue tongxun*, 1952.5: 20–22.

Qin, Liyan. "The sublime and the profane: a comparative analysis of two fictional narratives about sent-down youth." In *The Chinese Cultural Revolution as History*, edited by Joseph Esherick, Paul Pickowicz, and Andrew Walder, 240–66. Stanford, CA: Stanford University Press, 2006.

Qu Chunlin, Xiang Biao, and Wang Jue. "Zajiao shuidao wenhua" [Hybrid rice culture]. *Hunan shehui kexue* 2006.5: 159–62.

Reardon-Anderson, James. *The Study of Change: Chemistry in China, 1840–1949*. Cambridge: Cambridge University Press, 1991.

Rofel, Lisa. *Other Modernities: Gendered Yearnings in China after Socialism*. Berkeley: University of California Press, 1999.

Rogaski, Ruth. "Addicted to Science." *Historical Studies in the Natural Sciences* 42.5 (2012): 581–89.

———. "Nature, Annihilation, and Modernity: China's Korean War Germ-Warfare Experience Reconsidered." *Journal of Asian Studies* 61.2 (May 2002): 381–415.

Rosenberg, Gabriel. *The 4-H Harvest: Sexuality and the State in Rural America*. Philadelphia: University of Pennsylvania Press, 2016.

Rothwell, Matthew. "Transpacific Solidarities: A Mexican Case Study on the Diffusion of Maoism in Latin America." In *The Cold War in Asia: The Battle for Hearts and Minds*, edited by Zheng Yangwen, Hong Liu, and Michael Szonyi, 185–216. Leiden: Brill, 2010.

Rowe, William T. "Political, Social and Economic Factors Affecting the Transmission of Technical Knowledge in Early Modern China." In *Cultures of Knowledge: Technology in Chinese History*, edited by Dagmar Schäfer, 25–44. Leiden: Brill, 2012.

Saha, Madhumita. "State Policy, Agricultural Research and Transformation of Indian Agriculture with Reference to Basic Food-Crops, 1947–75." PhD diss., Iowa State University, 2012.

Saha, Madhumita, and Sigrid Schmalzer. "Science and Agrarian Modernization in China and India: Technocracy, Revolution, and Social Transformation." Paper prepared for the collaborative project "ReFocus: New Perspectives on Science and Technology in 20th-century India and China."

Saich, Tony, and Biliang Hu. *Chinese Village, Global Market*. New York: Palgrave Macmillan, 2012.

Schmalzer, Sigrid. "Breeding a Better China: Pigs, Practices, and Place in a Chinese County." *Geographical Review* 92.1 (January 2002): 1–22.

———. "Insect Control in Socialist China and the Corporate Unites States: The Act of Comparison, the Tendency to Forget, and the Construction of Difference in 1970s U.S.-Chinese Scientific Exchange." *Isis* 104 (2013): 303–29.

———. *The People's Peking Man: Popular Science and Human Identity in Twentieth-Century China*. Chicago: University of Chicago Press, 2008.

———. "Self-Reliant Science: The Impact of the Cold War on Science in Socialist China." In *Science and Technology in the Global Cold War*, edited by Naomi Oreskes and John Krige, 75–106. MIT Press, 2014.

———. "Speaking about China, Learning from China: Amateur China Experts in 1970s America." *Journal of American-East Asian Relations* 16.4 (2009): 313–52.

Schneider, Laurence. *Biology and Revolution in Twentieth-Century China.* Lanham, MD: Rowman & Littlefield, 2003.

———. "The Rockefeller Foundation, the China Foundation, and the Development of Modern Science in China." *Social Science and Medicine* 16 (1982): 1217–21.

Schram, Stuart. *The Thought of Mao Tse-tung.* Cambridge: Cambridge University Press, 1989.

Schurmann, Franz. *Ideology and Organization in Communist China.* Berkeley: University of California Press, 1966.

Science for the People. *China: Science Walks on Two Legs.* New York: Avon, 1974.

Secord, James. "Knowledge in Transit." *Isis* 95.4 (2004): 654–72.

Selden, Mark. *The Yenan Way in Revolutionary China.* Cambridge, MA: Harvard University Press. 1971.

Seybolt, Peter, ed. *The Rustication of Urban Youth in China: A Social Experiment.* New York: M. E. Sharpe, 1975.

Shan Ren. "Xiangcun de weilai" [The future countryside]. *Shaonian kexue* 1979.1: 38–44.

Shanxi sheng Xinxian diqu geweihui, Nonglin shuili ju keji xiaozu, eds. *Xinxian diqu nongye kexue shiyan* [Xin County region agricultural scientific experiment]. N.p., 1971.

Shanxi Yanbeiqu zhuanyuan gongshu. "Yanbei zhuanshu zhaokai kuihua zengchan laonong zuotanhui" [Yanbei District Commissioner's Office convenes old peasant conference on increasing production in sunflowers]. *Nongye kexue tongxun* 1954.6: 300–301.

Shaoyang di gewei nongyeju, ed. *Wei geming zhong hao zajiao shuidao* [Plant hybrid rice for the revolution]. N.p., 1978.

Shapin, Steven. *A Social History of Truth: Civility and Science in Seventeenth-Century England.* Chicago: University of Chicago Press, 1994.

Shapiro, Judith. *Mao's War against Nature: Politics and the Environment in Revolutionary China.* Cambridge: Cambridge University Press, 2001.

She Shiguang, ed. *Dangdai Zhongguo de qingnian he gongqingtuan* [Youth and the Communist Youth League in contemporary China]. Beijing: Dangdai Zhongguo chubanshe, 1998.

Shen, Grace Yen. *Unearthing the Nation: Modern Geology and Nationalism in Republican China.* Chicago: University of Chicago Press, 2014.

Shen Dianzhong. *Sixiang chenfu lu* [Record of the ebb and flow of my thoughts]. Shenyang: Liaoning renmin chubanshe, 1998.

Sheng nongkezhan weiyuanhui. "Guanyu shenqing qunzhongxing nongye kexue shiyan jingfei de baogao" [Report on the application for funding for mass agricultural scientific experiments]. 16 May 1970. Guangdong Provincial Archives, 2681-0185-146~147.

Shepard, H. H. Letter from H. H. Shepard to the China Foundation for the Promotion of Education and Culture. 22 December 1939. Box 10. Folder "China—Miscellaneous, 1937–1945." Collection 938, Entomology Papers, Division of Entomology and Economic Zoology, University of Minnesota Archives.

Shi Weimin, ed. *Zhiqing riji xuanbian* [Selected diaries of educated youth]. Beijing: Zhongguo shehui kexue chubanshe, 1996.

Shiva, Vandana. "Seeds of Suicide and Slavery versus Seeds of Life and Freedom." *Aljazeera.* 30 March 2013. http://www.aljazeera.com.

———. *The Violence of Green Revolution: Third World Agriculture, Ecology, and Politics.* London: Zed Books, 1991.

Shuidao xiongxing buyu yanjiu xiaozu. "1972 nian shuidao xiongxing buyu shiyan xiaojie" [Summary of 1972 experiments on male sterility in hybrid rice]. *Nongye keji ziliao* 1972.2: 1–8.

Siddiqi, Arif. *The Red Rockets' Glare: Spaceflight and the American Imagination.* Cambridge: Cambridge University Press, 2010.

Sihui xian Dasha gongshe geweihui et al. "Sihui xian Dasha gongshe 1975 nian zaozao shuidao bingchonghai zonghe fangfa" [Integrated methods for (controlling) disease and insect pests in early rice in Big Sand Commune, Sihui County]. 1976. Sihui Municipal Archives.

Sinnott, Edmund Ware, Leslie Clarence Dunn, and Theodosius Grigorievich Dobzhansky. *Principles of Genetics.* New York: McGraw Hill, 1950.

Smith, Aminda. *Thought Reform and China's Dangerous Classes: Reeducation, Resistance, and the People.* Lanham, MD: Rowman & Littlefield, 2013.

Smith, Jenny Leigh. *Works in Progress: Plans and Realities on Soviet Farms, 1930–1963.* New Haven, CT: Yale University Press, 2014.

Smith, Steve A. "Local Cadres Confront the Supernatural: The Politics of Holy Water (Shenshui) in the PRC, 1949–1966." *China Quarterly* 188 (2006): 999–1022.

———. "Talking Toads and Chinless Ghosts: The Politics of 'Superstitious' Rumors in the People's Republic of China, 1961–1965." *American Historical Review* 111.2 (April 2006): 405–27.

Song Yiching and Ronnie Vernooy, eds. *Seeds and Synergies: Innovating Rural Development in China.* Warwickshire, UK: Practical Action Publishing, 2010.

Song Yiqing. "Nongmin liuzhong keyi tigao zuowu pinzhong duoyangxing" [Peasant seeds can increase the diversity of crop varieties]. 1 January 2014. Accessed 4 June 2015. http://www.shiwuzq.com/plus/view.php?aid=141.

Stavis, Benedict. *Making Green Revolution: The Politics of Agricultural Development in China.* Ithaca, NY: Rural Development Committee, Cornell University, 1974.

———. *The Politics of Agricultural Mechanization in China.* Ithaca, NY: Cornell University Press, 1978.

Stern, Vernon F., Ray F. Smith, Robert van den Bosch, and Kenneth S. Hagen. "The Integrated Control Concept." *Hilgardia* 29.2 (October 1959): 81–101.

Stiffler, Douglas. " 'Three Blows of the Shoulder Pole': Soviet Experts at Chinese People's University, 1950–1957." In *China Learns from the Soviet Union, 1949-Present,* edited by Thomas P. Bernstein and Hua-yu Li, 303–25. Lanham, MD: Lexington Books, 2010.

Stone, Glenn Davis. "Agricultural Deskilling and the Spread of Genetically Modified Cotton in Warangal." *Current Anthropology* 48.1 (February 2007): 67–103.

Stross, Randall. *The Stubborn Earth: American Agriculturalists on Chinese Soil, 1898–1937.* Berkeley: University of California Press, 1986.

Su Shixin. "Gaoshan yangzhi deye changcun: Mianhuai jiechu de kunchong xuejia Pu Zhelong yuanshi" [A character like a lofty mountain whose noble deeds will live forever: Remembering the outstanding entomologist academician Pu Zhelong]. *Kexue Zhongguoren* 2001.2: 1–3.

Sun Changzhou. "Guantian shaobing Chen Changyu" [Chen Changyu, weather sentry]. *Renmin ribao,* 4 April 1981, 4.

Sun Xiuchun. "Yushi xiang qunzhong qingjiao, suishi he laonong shangliang" [Learning from the masses, always consulting with old peasants]. *Renmin ribao,* 4 August 1960, 4.

Sun Zhongchen. "Gao hao kexue shiyan, huiji youqingpai an feng" [Do scientific experiment well, fight back against the wind of the friendship faction]. *Nongcun kexue shiyan* 1976.3: 4–5.

Suttmeier, Richard. *Research and Revolution: Science Policy and Societal Change in China.* Lexington, MA: Lexington Books, 1974.

Sze, Julie. *Fantasy Islands: Chinese Dreams and Ecological Fears in an Age of Climate Crisis.* Berkeley: University of California Press, 2015.

Taiji Mao [pseudonym meaning Extreme Mao]. "Yuan Longping shi Zhongguo zajiao shuidao zhi fu ma?" [Is Yuan Longping the father of Chinese hybrid rice?]. Viewed 5 February 2014. http://www.wengewang.org/read.php?tid=30019.

Tan Shouzhang. *Mao Zedong yu Zhongguo nongye xiandaihua* [Mao Zedong and the modernization of agriculture in China]. Changsha: Hunan daxue chubanshe, 2009.

Tang Ruifu. "Chen Yongkang zhong dao gaochan jishu" [Chen Yongkang's high yield rice-planting technology]. *Nongye jishu tongxun* 1985.1: 2–5.

Teiwes, Frederick C., and Warren Sun. "China's New Economic Policy under Hua Guofeng: Party Consensus and Party Myths." *China Journal* 66 (July 2011): 1–23.

Thomson, James Claude. *While China Faced West: American Reformers in Nationalist China, 1928–1937.* Cambridge, MA: Harvard University Press, 1969.

Tuan shengwei Yangchun xianwei gongzuozu. "Guangdong Yangchun Sanjie dui qingnian kaizhan kexue shiyan huodong de jingyan" [The experience of youth in Sanjie Brigade, Yangchun, Guangdong carrying out scientific experiment activities]. 1965. Guangdong Provincial Archives, 232-1-0084-106~108.

van den Bosch, Robert. *The Pesticide Conspiracy.* Berkeley: University of California Press, 1978. Reprint, 1989.

Vandeman, Ann M. "Management in a Bottle: Pesticides and the Deskilling of Agriculture." *Review of Radical Political Economics* 27.3 (September 1995): 49–55.

Wagner, Rudolf G. "Agriculture and Environmental Protection in China." In *Learning from China? Development and Environment in Third-World Countries,* edited by Bernhard Glaeser, 127–43. New York: Routledge, 1987.

Walder, Andrew G. "Organized Dependency and Cultures of Authority in Chinese Industry." *Journal of Asian Studies* 43.1 (1983): 51–76.

Wang, Y. Z., and Y. T. Chen. "The Eco-Unit Settlement Adapted to the Vernacular Culture: A Case Study of Dwelling Design in the Chaoshan Area of Guangdong Province, China." In *The Sustainable World,* edited by C. A. Brebbia, 265–74. Southampton, UK: WIT Press, 2011.

Wang, Zuoyue. "The Cold War and the Reshaping of Transnational Science in China." In *Science and Technology in the Global Cold War,* edited by Naomi Oreskes and John Krige, 343–69. Cambridge, MA: MIT Press, 2014.

———. "Transnational Science during the Cold War: The Case of Chinese/American Scientists." *Isis* 101.2 (2010): 367–77.

Wang Chunling. "Jianjue zou yu gongnong xiang jiehe de daolu" [Resolutely traveling the road of uniting with workers and peasants]. *Kexue shiyan* 1974.3: 3.

Wang Jinling et al. "Hunhe geti xuanzefa zai dazadoujiao yuzhong gongzuo zhong de yingyong" [The use of mixing individual selection methods in breeding hybrid soybeans]. *Dongbei nongxueyuan xuebao* 1960.1: 1–7.

Wang Yuying et al. "Lunzuo, jianzuo, tanzhong, hunzuo zai daban nongye zhong de zuoyong" [The use of crop rotation, intercropping, interplanting, and mixed planting in large-scale agriculture]. *Renmin ribao,* 29 April 1961, 7.

Williams, James H. "Fang Lizhi's Big Bang: Science and Politics in Mao's China." PhD diss., University of California, Berkeley, 1994.

Wu, Chenfu F. Letter to William A. Riley. 15 February 1941. Box 10. Folder "China—

Miscellaneous, 1937–1945." Collection 938, Entomology Papers, Division of Entomology and Economic Zoology, University of Minnesota Archives.

Wu Jun. "Yu shuidao tu zhong linfei de jingji shiyong fangfa wenti" [Questions regarding the economical methods of applying phosphate fertilizer to soil in rice paddies]. *Turang* 1961.9: 58–59.

Wu Tingjie. "'Liuliuliu' bu shi 'wanlingyao'" [BHC is not a "cure-all"]. *Renmin ribao*, 3 September 1955, 6.

Wu Xianzhi. "'Si zi yi fu' de youlai" [The origins of "four selfs and one supplement"]. *Zhongzi shijie* [Seed world] 1984.8: 32.

Xia Yunfeng. "Nongmin weishenme meiyou pubian shiyong 'liuliuliu'" [Why peasants do not yet widely use BHC]. *Renmin ribao*, 10 June 1954, 2.

Xiangpu xian. "Jiaqiang lingdao kaizhan zajiao shuidao shizhong, shifan, tuiguang" [Strengthen leadership and develop experimental planting, modeling, and extension of hybrid rice]. *Guangxi nongye kexue* 1976.2: 21–23.

Xiangzhou xian shuidao zayou liyong tuiguang lingdao xiaozu. "Jiaqiang lingdao kaizhan zajiao shuidao shizhong, shifan, tuiguang" [Strengthen leadership in developing the testing, modeling, and extension of hybrid rice]. *Guangxi nongye kexue* 1976.2: 21–23.

Xiangzhou xian Xiangzhou gongshe Shalan yi dui nongkezu. "Kexue zhongtian duo gaochan zajiao shuidao chao qian jin" [Scientific farming seizes high yield of hybrid rice surpassing 1,000 jin]. *Guangxi nongye kexue* 1976.2: 36–38.

Xie Changjiang. *Yuan Longping*. Guiyang: Guizhou renmin chubanshe, 2004.

———. *Yuan Longping yu zajiao shuidao* [Yuan Longping and hybrid rice]. Changsha: Hunan kexue jishu chubanshe, 2000.

———. *Zajiao shuidao zhi fu: Yuan Longping zhuan* [The father of hybrid rice: A biography of Yuan Longping]. Nanning: Guangxi kexue jishu chubanshe, 1990.

Xie Jieyin. "Woyang xian Zhaowo xiang laonong zuotan yanjiu xiaomai wanzhong baoshou de banfa" [Old peasant conference in Zhaowo Village, Woyang County, researches methods of protecting late-planted wheat]. *Nongye kexue tongxun* 1954.11: 577.

Xinhua she. "Fazhan nongye shengchan de zhongyao cuoshi" [Important measures for developing agricultural production]. *Renmin ribao*, 21 May 1965, 5.

———. "Mei diguo zhuyi xijunzhan zuixing diaocha tuan" [Investigative team on US imperialist crimes of germ warfare]. *Renmin ribao*, 26 March 1952, 1.

———. "Nongcun tiandi guangluo: Qingnian dayou kewei" [The countryside is a big world where much can be accomplished]. *Renmin ribao*, 25 November 1960, 4.

———. "Quanguo nongcun qingnian kexue shiyan huiyi zongjie jingyan tichu renwu" [Summary of experiences and proposed responsibilities at the National Conference on Rural Youth in Scientific Experiment]. *Renmin ribao*, 30 October 30 1965, 3.

———. "Qunzhong huanying de 'guantian shaobing'" [The weather sentry welcomed by the masses]. *Renmin ribao*, 2 April 1976, 3.

———. "Rang kunchongxue wei nongye xiandaihua fuwu" [Make entomology serve agricultural modernization]. *Renmin ribao*, 22 November 1977, 3.

———. "Zai renmin jiaoshi de gangwei shang" [On the job as a teacher of the people]. *Renmin ribao*, 13 October 1972, 3.

———. "Zajiiao shuidao shi zenyang peiyu chenggong de" [How Hybrid rice was successfully cultivated]. *Renmin ribao*, 17 December 1976, 5.

———. "Zhongshan daxue pin Li Shimei wei jiaoshou" [Sun Yat-sen University hires Li Shimei as professor]. *Renmin ribao*, 22 June 1958, 1.

Xinjiang Weiwuer zizhi qu nongken zongju shuidao zayou xiezuozu. "Yebai gengxing 'sanxi' xuanyu he liyong" ["Three line" selection and use of wild abortive non-glutinous rice]. *Beifang shuidao* 1976.Z2: 28–32.

Xiong Weimin and Wang Kedi. *Hecheng yi ge danbaizhi: Jiejing niuyi daosu de rengong quan hecheng* [Synthesize a protein: The story of total synthesis of crystalline insulin project in China]. Jinan: Shandong jiaoyu chubanshe, 2005.

Xu, Zhun. "The Political Economy of Agrarian Change in the People's Republic of China." PhD diss., University of Massachusetts, Amherst, 2012.

Xu Guanren and Xiang Wenmei. "Liyong xiongxing buyu xi xuanyu zazhong gaoliang" [Use of male-sterile lines in breeding hybrid gaoliang]. *Zhongguo nongye kexue* 1962.2: 15–20.

Xu Jiatun. "Shixian nongye kexue jishu gongzuo geminghua de jige wenti: Zai Jiangsu sheng nongye kexue jishu gongzuo huiyi de jianghua" [A few questions in revolutionizing agricultural science and technology work: Speech at the Jiangsu Provincial Agricultural Science and Technology Work Conference]. *Zhongguo nongye kexue* [Chinese agricultural science] 1965.10: 1–8, 6.

Xue Pangao. "Dui tu zhuanjia jin kexueyuan dang yanjiuyuan de fansi" [Reflections on native experts entering the academy as researchers]. *Beida kexueshi yu kexue zhexue*, 12 March 2008. Viewed 7 January 2009. http://hps.phil.pku.edu.cn/2008/03/2729/.

Yan Hairong and Chen Yiyuan. "Cong dadou weiji kan shiwu zhuquan" [What the soybean crisis tells us about food sovereignty]. *Wuyou zhi xiang wangkan.* 10 September 2013. Accessed 4 June 2015. http://www.wyzxwk.com/Article/shehui/2013/09/306001.html.

Yang, Bin. "'We Want to Go Home!' The Great Petition of the Zhiqing, Xishuangbanna, Yunnan, 1978–1979." *China Quarterly* 198 (June 2009): 401–21.

Yang, Hsin-pao. "Promoting Cooperative Agricultural Extension Service in China." In *Farmers of the World: The Development of Agricultural Extension*, edited by Edmund deS. Brunner, Irwin T. Sanders, and Douglas Ensminger, 46–60. New York: Columbia University Press, 1945.

Yang Jisheng. *Tombstone: The Great Chinese Famine, 1958–1962.* Translated by Stacy Mosher and Guo Jian. New York: Farrar, Strauss, and Giroux, 2012. [Original published in 2008.]

Yang Shouren et al. "Xianjingdao zajiao yuzhong yanjiu (di'er bao)" [Research on hybridization of Xianjing rice (second report)]. *Zuowu xuebao* 1962.2: 13–18.

Yang Yuefu. "Kexue zhi chun: Quanguo kexue dahui suxie" [Springtime for science: Sketches of the National Conference on Science]. *Renmin ribao*, 4 April 1978, 4.

Yao, Shuping. "Chinese Intellectuals and Science: A History of the Chinese Academy of Sciences (CAS)." *Science in Context* 3 (1989): 447–73.

Yao Kunlun. *Zoujin Yuan Longping* [Approaching Yuan Longping]. Shanghai: Shanghai kexue jishu chubanshe, 2002.

Ye Weili with Ma Xiaodong. *Growing Up in the People's Republic of China: Conversations between Two Daughters of China's Revolution.* New York: Palgrave Macmillan, 2005.

Yuan Longping. "Cun cao yang chunhui, quan kao dang lingdao" [Little shoots growing under spring sunshine, everything depends on party leadership]. *Guangming ribao*, 6 July 2001. http://www.sina.com.cn.

———. "Shuidao de yongxing buyunxing" [Male sterility in rice]. *Kexue tongbao* 1966.4: 185–88.

———. "Zajiao shuidao peiyu de shijian yu lilun" [Practice and theory in hybrid rice cultivation]. *Zhongguo nongye kexue* 1977.1: 27–31.

Yuan Longping, Li Bihu, and Yin Huaqi. "Tantan zajiao shuidao: Dui shuidao sanxi de renshi" [On hybrid rice: Knowledge of three-line rice]. *Shengming shijie* 1977.1: 41–42.

Yuan Longping and Xin Yeyun. *Yuan Longping koushu zizhuan.* Changsha: Hunan jiaoyu chubanshe, 2010.

Yue, Gang. *The Mouth that Begs: Hunger, Cannibalism, and the Politics of Eating in Modern China*. Durham, NC: Duke University Press, 1999.

Yunnan shengchan jianshe bu dui mou bu sanjiehe keyan xiaozu. "Jinjina shumiao shi zenyang peizhi chenggong de?" [How are cinchona saplings successfully cultivated?]. *Kexue shiyan* 1974.3: 6–7.

Zhang Dianqi. "Shehui zhuyi shi an pinxia zhongnong de kaoshan" [Socialism is the rock that supports us poor and lower-middle peasants]. *Renmin ribao*, 20 October 1969, 5.

Zhang Renpeng. "Houlu duizhang Yang Liguo kexue zhongtian chuang gaochan" [Houlu Brigade leader Yang Liguo achieves high yields through scientific farming]. *Xin nongye* 1974.14, 26.

Zhang Yang. *"Di'erci woshou" wenziyu* [The literary inquisition of *The Second Handshake*]. Beijing: Zhongguo shehui chubanshe, 1999.

Zhao'an xian liangzhong chang. "Shuidao nantaigeng 'sanxi' de xuanyong tihui" [Experiences in "three-line" selection of nantai non-glutinous rice]. *Fujian nongye keji* 1976.6: 9–12.

Zhejiang sheng Huangyan xian Haimen qu baodao zu. "Pinxia zhongnong de 'guantianbing': zhishi qingnian Su Fuxing" [The poor and lower-middle peasants' "soldier who manages the heavens": The educated youth Su Fuxing]. *Kexue shiyan* 1974.3: 8–9.

Zhejiang sheng nonglinju liangshi shengchanchu, ed. *Jiji shizhong tuiguang zajiao shuidao* [Actively test and extend hybrid rice]. N.p.: May 1976.

Zheng, Xiaowei. "Images, Memories, and Lives of Sent-down Youth in Yunnan." In *Visualizing Modern China: Image, History, and Memory, 1750-Present*, edited by James Cook et al., 241–58. Lanham, MD: Lexington Books, 2014.

Zhong, Xueping, Wang Zheng, and Bai Di, eds. *Some of Us: Chinese Women Growing Up in the Mao Era*. New Brunswick, NJ: Rutgers University Press, 2001.

Zhonggong Guangdong shengwei Tongzhanbu, Qiaowu zu [Overseas Affairs Group of the United Front Department of the Guangdong Provincial Chinese Communist Party Committee]. Letter to the Zhongshan daxue geweihui Zhenggongzu [Sun Yat-sen University Revolutionary Committee Political Work Group]. 3 March 1973. Sun Yat-sen University Archives.

Zhonggong Hunan sheng Huarong xian weiyuanhui. "Women shi zenyang ban nongcun kexue shiyan wang de" [How we created a rural scientific experiment network]. *Kexue shiyan* 1974.12: 1–3.

Zhonggong Nantong gongshe weiyuanhui. "Nongmin zijue zhangwo nongye kexue de shidai kaishi le" [The era of peasants consciously grasping agricultural science has begun]. *Renmin ribao*, 18 May 1966, 4.

Zhonggong Nanweizi Gongshe Weiyuanhui. "Yi jieji douzheng wei wang banhao sanji nongkewang" [With class struggle as the key link, create the three-level agricultural science network]. *Nongcun kexue shiyan* 1976.8: 6–9.

Zhongguo fupin jijinhui. "Meng kaishi de cunzhuang: Huarun Baise Xiwang Xiaozhen jianjie" [A village to dream of: Introducing Huarun Baise Hope Town]. Sina. 28 September 2010. http://gongyi.sina.com.cn/gyzx/2010–09–28/171420499.html.

Zhongguo kexueyuan caizheng bu. Zengbo qunzhongxing kexue shiyan huodong buzhu jingfei de han [Letter regarding increasing allocations for supplemental funding for mass scientific experiment activities]. 31 October 1975. Guangdong Provincial Archives, 306-A0.05-22-91.

Zhongguo kexueyuan dongwu yanjiusuo. *Zhongguo zhuyao haichong zonghe fangzhi* [Integrated control of major insect pests in China]. Beijing: Kexue chubanshe, 1979.

Zhongguo kexueyuan yichuan yanjiusuo. *Zenyang zhong zajiao gaoliang* [How to plant hybrid sorghum]. Beijing: Kexue chubanshe, 1968.

Zhongshan daxue shengwuxi jiaoyu geming shijian dui. "Mianxiang qunzhong, mianxiang shiji, gaizao shengwuxi" [Turning toward the masses, turning toward practice, transforming the biology department]. *Renmin ribao*, 7 August 1970, 2.

Zhou Peiyuan. "Kexue de weilai jituo zai nimen de shenshang" [The future of science rests on your shoulders]. *Shaonian kexue* 1979.10: 1.

Zhou Yun. "Cong genben shang zhuanbian zuofeng" [Radically transform work styles]. *Renmin ribao*, 18 February 1958, 3.

Zhu Dehui and Zhu Xianli. *Nongyan li de kexue daoli* [Scientific rationality in agricultural maxims]. Beijing: Zhongguo qingnian chubanshe, 1965.

Zhu Ruzuo and Hu Cui. "Nongzuowu haichong huaxue fangzhi yu shengwu fangzhi de jiehe wenti" [Problems in the integration of chemical and biological control in agricultural pests]. *Zhongguo nongye kexue*, 1962.4: 1–8.

Zhu Ruzuo and Hu Yongxi. "Chiyanfeng shenghuo zhi yanjiu" [Research on the life cycle of *Trichogramma*]. *Zhejiangsheng kunchongju niankan* 1935: 164–77.

Zhu Xianli, Chen Jisheng, and Ren Huiru. *Nongyan zhujie* [Annotated agricultural maxims]. Beijing: Tongsu duwu chubanshe, 1957.

Zhun Xu. "The Chinese Agriculture Miracle Revisited." *Economic & Political Weekly* 47.14 (7 April 2012): 51–58.

Zweig, David. *Agrarian Radicalism in China, 1968–1981*. Cambridge, MA: Harvard University Press, 1989.

Index

Italicized page numbers indicate figures.

poor and blank (一穷二白), China as, 37

poor and lower-middle peasants (贫下中农): as experts, 102, 105; and hybrid rice, 81, 83–84, 90; ideological significance of, 14, 100; and Pu Zhelong, 62; and schools, 112; and scientific experiment movement, 142, 164–65; and youth, 164–65, *166*, 169, 186

postcolonial subjectivity, 98

postcolonial theory, 37–38

Prakash, Gyan, 239n33

prevention, in insect control, 56, 243n47

Principles of Genetics, 77, 89, 247n51

propaganda, 14, 235n38; and experience, 18, 137, 148, 164, 176; and historical sources, 15–16, 18, 120, 126, 148, 160; influence of, 15, 32, 172–73, 191; socialist, vs. postsocialist "jargon," 222

propaganda posters, *7–10, 17, 30–31, 101, 110, 137, 162, 166, 168, 169, 215*; for agricultural extension, 6–7, 142; Chinese, in West, 222; depicting ideal official, 136; as historical sources, 16–18, 183, 236n44; postsocialist, 215. *See also* propaganda

Pu Zhelong (蒲蛰龙), 22; character of, 52, 63; and chemical insecticides, 54, 56–57; choice to return to China, 52; environmentalism of, 53, 61, 69, 211, 221; as exception, 48; and Gu Dexiang, 57, 63, 241n17; hosting foreign delegations, 65–67, 244n89; and Mai Baoxiang, 56, 63, 132–33; and Maoist politics, 67–68; and peasants, 59–64, 126, 151; political evaluation of, 62; and political troubles, 52; postsocialist biographies of, 68–72; in Republican era, 50; and Science for the People, 47–48, 241n5; scientific network of, 130; and *The Second Handshake*, 172; shift to applied research, 59; and Soviet advisers, 55; in Sweden, 67; as taxonomist, 58–59, 62; and transnational science, 58, 68, 74, 98; and *tu/ yang* binary, 48, 52–53, 60–72, *71*; and University of Minnesota, 50, *51*, 64–66; violin playing, *69*, 72; and youth, *188*, 204

pumpkin, 21

qi (气), 150

Qianyang (黔阳), 52, 70, 242n19

Qianyang Agricultural School (黔阳农校), 73, 75, 78, 90, 245n1. *See also* Anjiang Agricultural School

Qin Yunfeng, 132

Qinghua University (清华大学), 50

Quotations from Chairman Mao Zedong (毛主席语录), 157

radicalism, Mao-era: and campaign against Confucius, 105; contrasted with decolonialism, 127; decline of, 205; discomfort with, 221; discrediting of, 18, 45, 48, 201–2; and Great Leap Forward, 59; and hybrid rice, 86; influence abroad, 5, 227; legacy of, 24, 218, 222; and Li Shimei, 60, 64, 70,

111, 113; and modernization, 27–28, 119, 237–38n2; and patriarchy, 195; and Pu Zhelong, 60, *71*; and resistance, 34; and science, 3–5, 22, 25, 128; and scientific experiment movement, 40, 44, 112, 114, 142; and scientists, 23, 34; and two-line analysis, 237; and Zhou Enlai, 161. *See also* Maoism

rapeseed, 102, 190

Rat King (老鼠王), 111, 132

rats, 108, 111, 113, 131–32, 248n12

"real people" (as opposed to "poster children" or political labels), 17–18, 24, 42

rebel faction (造反派), 175

Rectification Campaign (整风运动), 36

red (红) and expert (专), 34, 37, 62, 192

Red Guards (红卫兵), 88, 156, 191

reforms, post-Mao economic, 70, 80, 88, 208, 210, 212

religion, 29, 32, 96

"Report on an Investigation of the Peasant Movement in Hunan" (湖南农民运动考察报告), 29

Republican era: agricultural reform efforts, 203, 238n16, 259n61; antisuperstition campaigns, 254n70; attitudes toward science, 239n33; biological control, 55, 242n40; genetics, 247n54

resistance, human: of cadres and technicians, 130, 145–47; to colonialism, 224; and experiment, 152–53; and gender, 120; to ideology or propaganda, 15, 18; of local people, 23, 29, 125, 133; narratives of, 148–52; in nineteenth century, 29; to outside models, 43; of peasants, 15–16, 129, 147–52; and self-reliance, 144–47; and *tu* science, 236n54; to Western slave mentality, 67; of youth, 176, 177

resistance, insect, 12–13, 55–56, 60, 67, 210

resistance, plant: to disease and pests, 114, 223; to lodging, 118

reskilling, 125, 251n94

Resolution on Chinese Communist Party History (关于建国以来党的若干历史问题的决议), 45

rest, 119, 165

returned educated youth (回乡知青). *See* rural educated youth

revisionism (修正主义), 4, 82, 84, 226

revolution in education (教育革命), 60–62, 65

revolutionary spirit, 18, 59, 89, 174, 176

rice, *101*; cultivation practices, 109, 113, 118; cultural significance of, 11; double-cropping of, 29, 145; dwarf varieties, 11, 100, 106, 111, 148, 151; Hu Select, 223; Mexican, 147; new varieties, 140, 145, 149, 167–69, 181, 196–97; pest control in, 57, 62, 66, 131, 159; as self-pollinating, 75; and self-sufficiency, 123; supposedly crossed with corn, 114; traditional varieties, 123–24, 255n86; and water conditions, 131. *See also* hybrid rice; International Rice Research Institute